Handbook for Electrical Safety in the Workplace

FOURTH EDITION

With the complete text of the 2015 edition of *NFPA 70E*, *Standard for Electrical Safety in the Workplace*

Edited by

Michael D. Fontaine, PE, CESCP

Christopher D. Coache

Gil Moniz

Art editor: Jean Blanc

With the complete text of the 2015 edition of *NFPA 70E*, *Standard for Electrical Safety in the Workplace*

National Fire Protection Association®, Quincy, Massachusetts

Product Management: Debra Rose
Development: Jennifer Harvey
Production: Irene Herlihy
Copyediting: Kenneth Ritchie
Permissions: Josiane Domenici
Art Direction: Cheryl Langway

Art Editor: Jean Blanc
Cover Design: Greenwood Associates
Interior Design: M. Palmer Design
Composition: Cenveo Publisher Services
Manufacturing: Ellen Glisker
Printing/Binding: R.R. Donnelley/Willard

Copyright © 2014
National Fire Protection Association®
One Batterymarch Park
Quincy, Massachusetts 02169-7471

Important Notices and Disclaimers: Publication of this handbook is for the purpose of circulating information and opinion among those concerned for fire and electrical safety and related subjects. While every effort has been made to achieve a work of high quality, neither the NFPA® nor the contributors to this handbook guarantee or warrantee the accuracy or completeness of or assume any liability in connection with the information and opinions contained in this handbook. The NFPA and the contributors shall in no event be liable for any personal injury, property, or other damages of any nature whatsoever, whether special, indirect, consequential, or compensatory, directly or indirectly resulting from the publication, use of, or reliance upon this handbook.

This handbook is published with the understanding that the NFPA and the contributors to this handbook are supplying information and opinion but are not attempting to render engineering or other professional services. If such services are required, the assistance of an appropriate professional should be sought.

NFPA 70E®, Standard for Electrical Safety in the Workplace® *("NFPA 70E")*, is, like all NFPA codes, standards, recommended practices, and guides ("NFPA Standards"), made available for use subject to Important Notices and Legal Disclaimers, which appear at the end of this handbook and can also be viewed at *www.nfpa.org/disclaimers.*

Notice Concerning Code Interpretations: This fourth edition of the *Handbook for Electrical Safety in the Workplace* is based on the 2015 edition of *NFPA 70E*. All NFPA codes, standards, recommended practices, and guides ("NFPA Standards") are developed in accordance with the published procedures of the NFPA by technical committees comprised of volunteers drawn from a broad array of relevant interests. The handbook contains the complete text of *NFPA 70E* and any applicable Formal Interpretations issued by the NFPA at the time of publication. This NFPA Standard is accompanied by explanatory commentary and other supplementary materials.

The commentary and supplementary materials in this handbook are not a part of the NFPA Standard and do not constitute Formal Interpretations of the NFPA (which can be obtained only through requests processed by the responsible technical committees in accordance with the published procedures of the NFPA). The commentary and supplementary materials, therefore, solely reflect the personal opinions of the editor or other contributors and do not necessarily represent the official position of the NFPA or its technical committees.

REMINDER: UPDATING OF NFPA STANDARDS

NFPA 70E, Standard for Electrical Safety in the Workplace, like all NFPA codes, standards, recommended practices, and guides ("NFPA Standards"), may be amended from time to time through the issuance of Tentative Interim Amendments or corrected by Errata. An official NFPA Standard at any point in time consists of the current edition of the document together with any Tentative Interim Amendment and any Errata then in effect. In order to determine whether an NFPA Standard has been amended through the issuance of Tentative Interim Amendments or corrected by Errata, visit the Document Information Pages on NFPA's website. The Document Information Pages provide up-to-date, document specific information including any issued Tentative Interim Amendments and Errata. To access the Document Information Page for a specific NFPA Standard go to http://www.nfpa.org/document for a list of NFPA Standards, and click on the appropriate Standard number (e.g., *NFPA 70E*). In addition to posting all existing Tentative Interim Amendments and Errata, the Document Information Page also includes the option to sign-up for an "Alert" feature to receive an email notification when new updates and other information are posted regarding the document.

The following are registered trademarks of the National Fire Protection Association:

National Fire Protection Association®
NFPA®
National Electrical Code®, NFPA 70®, and NEC®
Standard for Electrical Safety in the Workplace® and NFPA 70E®

NFPA No.: 70EHB15
ISBN (book): 978-1-455-90818-9
ISBN (PDF): 978-1-455-90819-6
ISBN (e-book): 978-1-455-91132-5
Library of Congress Control No.: 2014943179

Printed in the United States of America

15   16   17   18   5   4   3   2

Dedication

This edition of the *Handbook for Electrical Safety in the Workplace* is dedicated to William H. Gatenby, William R. Kruesi, Kent P. Stiner, Richard E. Stevens, and Howard P. Michener for their outstanding contributions to the development of *NFPA 70E*.

In 1974, the National Electrical Manufacturers Association (NEMA) petitioned the Occupational Safety and Health Administration (OSHA) to adopt the 1975 edition of the *National Electrical Code*® as a replacement for the 1971 edition of the *NEC*® adopted under section 6(a) rules of the OSHA Act. OSHA responded that the *NEC*® is primarily directed at an initial installation and that determination of compliance is best judged during the design and construction phases of a project. Instead, OSHA needed regulations that were tailored for elements in electrical safety requirements observable during an OSHA inspection.

A NEMA-sponsored subcommittee composed of William H. Gatenby (Harvey Hubbel, Inc.), William R. Kruesi (General Electric Company), Kent P. Stiner (ITE Imperial Corporation), Richard E. Stevens (National Fire Protection Association), and Howard P. Michener (NEMA Technical Director) met with OSHA on May 8, 1975, and submitted a proposal to OSHA for the development of electrical safety requirements. It was conceived that a document would be put together by a competent group, representing all interests, that would extract suitable portions from the *NEC* and from other documents applicable to electrical safety, such as those prepared by a group operating under a consensus process compatible with the American National Standard Institution's rules and regulations — a technical committee (TC) sponsored by the NFPA and operating under NFPA rules and regulations. With positive encouragement from OSHA, the *NEC* Correlating Committee examined the feasibility of developing such a document to be used as a basis for evaluating electrical safety in the workplace, and *NFPA 70E* was born. Recognition is due to the members of the NEMA subcommittee as the "founding fathers" of *NFPA 70E* and is hereby given.

Contents

Preface   vii

About the Editors   ix

PART 1

NFPA 70E®, Standard for Electrical Safety in the Workplace®,
with Commentary   3

ARTICLE 90

Introduction   5

CHAPTER 1

Safety-Related Work Practices   19

100   Definitions   20
105   Application of Safety-Related Work Practices   50
110   General Requirements for Electrical Safety-Related Work Practices   53
120   Establishing an Electrically Safe Work Condition   79
130   Work Involving Electrical Hazards   101

CHAPTER 2

Safety-Related Maintenance Requirements   167

200   Introduction   168
205   General Maintenance Requirements   169
210   Substations, Switchgear Assemblies, Switchboards, Panelboards, Motor Control Centers, and Disconnect Switches   176
215   Premises Wiring   178
220   Controller Equipment   179
225   Fuses and Circuit Breakers   179
230   Rotating Equipment   181
235   Hazardous (Classified) Locations   182
240   Batteries and Battery Rooms   183
245   Portable Electric Tools and Equipment   184
250   Personal Safety and Protective Equipment   185

CHAPTER 3

Safety Requirements for Special Equipment   189

300   Introduction   189
310   Safety-Related Work Practices for Electrolytic Cells   190
320   Safety Requirements Related to Batteries and Battery Rooms   197
330   Safety-Related Work Practices for Use of Lasers   206
340   Safety-Related Work Practices: Power Electronic Equipment   209
350   Safety-Related Work Requirements: Research and Development Laboratories   215

Informative Annexes

A   Referenced Publications   219
B   Informational References   223
C   Limits of Approach   227
D   Incident Energy and Arc Flash Boundary Calculation Methods   231
E   Electrical Safety Program   249
F   Risk Assessment Procedure   251
G   Sample Lockout/Tagout Procedure   273
H   Guidance on Selection of Protective Clothing and Other Personal Protective Equipment (PPE)   279
I   Job Briefing and Planning Checklist   283
J   Energized Electrical Work Permit   285
K   General Categories of Electrical Hazards   289
L   Typical Application of Safeguards in the Cell Line Working Zone   293
M   Layering of Protective Clothing and Total System Arc Rating   295
N   Example Industrial Procedures and Policies for Working Near Overhead Electrical Lines and Equipment   299
O   Safety-Related Design Requirements   305
P   Aligning Implementation of This Standard with Occupational Health and Safety Management Standards   309

PART 2

Supplements   313

1   *National Electrical Code* Requirements Associated with Safety-Related Work Practices   315

2   Electrical Preventive Maintenance Programs   321

3   Typical Safety Procedure (Procedure for Selection, Inspection, and Care of Rubber Insulating Gloves and Leather Protectors)   329

4   Steve and Dela Lenz: One Family's Experience with an Arc-Flash Incident   341

Index   347

Important Notices and Legal Disclaimers   355

Preface

The 2015 edition of the *Handbook for Electrical Safety in the Workplace* contains the latest information on electrical safety. More than 118 years have passed since March 18, 1896, when a group of 23 persons representing a wide range of organizations met at the headquarters of the American Society of Mechanical Engineers in New York City. Their purpose was to develop a national code of rules for electrical construction and operation. This was the first national effort to develop electrical installation rules for the United States. This successful effort resulted in the *National Electrical Code*® (*NEC*®), the installation code used throughout the United States and in many countries around the world.

With the implementation of the Occupational Safety and Health Act, it became apparent that a separate standard would be necessary to provide requirements for safe work practices for people who might be exposed to electrical hazards. On January 7, 1976, the Standards Council of the National Fire Protection Association appointed the Committee on Electrical Safety Requirements for Employee Workplaces. The Standards Council recognized the importance of the creation of a document that could be used in conjunction with the *National Electrical Code*. To keep these documents well coordinated, the Standards Council decided that the new committee should report to the association through the National Electrical Code Technical Correlating Committee. Although the committee recognized the importance of compliance with all of the requirements of the *NEC*, the first edition of *NFPA 70E*®, *Standard for Electrical Safety in the Workplace*®, dealt primarily with those electrical installation requirements from the *NEC* that were most directly tied to worker safety. In subsequent editions, the document expanded to include safety-related work practices, safety-related maintenance requirements, and safety requirements for special equipment. For the 2009 edition, the installation requirements were removed because OSHA no longer believed that they were necessary because the *NEC* is now widely adopted and used.

For the first few editions, *NFPA 70E* was a four-part document that was essentially four books bound together. Beginning with the 2004 edition, *NFPA 70E* adopted the *NEC Style Manual*, which provided a simple means to integrate the parts of the document into a comprehensive and cohesive standard. Since the *NEC* requirements were deleted from the standard, there are now three chapters. However, the handbook includes a list of pertinent *NEC* sections to assist the user in understanding how the installation requirements of the *NEC* can make a safer work environment.

Until the 2000 edition of *NFPA 70E*, most believed that the only electrical hazard was electric shock. The 2000 edition brought attention to the hazards of arc flash phenomena. The use of the standard has grown tremendously as workers and their employers try to provide protection from this dangerous hazard. The 2005 edition added the requirement for arc flash hazard equipment labels. The 2012 edition included specific requirements and information on shock and arc-flash protection for persons working on direct-current systems. The 2015 edition continues the evolution of the only ANSI-accredited standard on workplace safety and safety-related work practices by harmonizing the document with other national and international standards on risk assessment by replacing the hazard/risk category tables with new arc flash PPE categories tables, updating hazard and risk terminology, and methodology.

Acknowledgments

Electricity can be very dangerous occupational hazard. Almost all members of the workforce are exposed to electrical energy as they perform their duties every day. Since the creation of *NFPA 70E*, electrical workers have become more aware of the hazards of electricity and of how to protect themselves. The 2000 and 2004 editions of the standard increased the awareness of arc flash phenomena. The increased use of this standard resulted in a significant increase in public participation in the form of public inputs and comments. This edition would not be possible without the tireless work of the dedicated professionals who serve on the Committee on Electrical Safety in the Workplace. Their work will save countless lives.

The editors have learned much from the deliberation and discussions of the technical committee. We hope our work in this book accurately reflects the wisdom we accumulated from the committee.

Handbooks are a team effort, and the editors of this book have been supported by an outstanding team of professionals. We wish to acknowledge with thanks the wonderful work of Kim Fontes, Division Manager, Product Development; Debra Rose, Product Manager; Irene Herlihy, Associate Project Manager; Jennifer Harvey, Associate Project Manager; and Josiane Domenici, Project Coordinator, who supported us every step of the way. We acknowledge their patience and professionalism. Mark W. Earley, PE, NFPA's Chief Electrical Engineer, participated with technical development, and his *NEC* expertise was invaluable.

The editors also wish to acknowledge the contributions of former NFPA staff who served as staff liaisons to the committee: Dick Murray, Ken Mastrullo, Joe Sheehan, E. William Buss, and Jeffery Sargent. They were responsible for educating their successors about the importance of electrical safety.

This is the fourth edition of the *Handbook for Electrical Safety in the Workplace*. The editors gratefully acknowledge the work of the editors who assembled the first, second, and third editions. The collective electrical safety expertise of former editors Kenneth G. Mastrullo, Ray A. Jones, Jane G. Jones, E. William Buss, Mark W. Earley, Jeffery Sargent, and Michael D. Fontaine provided the solid foundation on which the 2015 edition of this handbook was built. All of the professionals we have worked with on this project have made a difference in making the workplace safer.

Dennis Rossbach, PhD, CLSO (Boeing), Stephen W. McCluer (Schneider Electric IT Corporation), Rodney West, PE, CFEI (Schneider Electric USA), and Randell Bouton Hirschmann (Oberon Company) assisted in the development of commentary for this edition of the handbook. Lloyd Gordon, PhD (Los Alamos National Laboratory), Bobby J. Grey (Hoydar/Buck, Inc.), David Dini, PE (Underwriters Laboratories, Inc.), Mark McNellis (Sandia National Laboratory), and Daniel Roberts, CSP (Schneider Electric CAN) contributed material or ideas for this edition of the handbook.

About the Editors

Michael D. Fontaine, PE, CESCP

Michael was formerly a Senior Electrical Engineer with NFPA. He has a BSEE, BSBA in Accounting, MSEE, MST, and JD. He is a Registered Professional Engineer (PA & MA), Licensed Attorney (MA & NE), Licensed Real Estate Broker (MA & NE), and an NFPA-certified electrical safety compliance professional. Prior to joining NFPA in 2010, he held various positions in the electrical industry, including electrical engineer, senior electrical engineer, electrical engineering manager, chief electrical engineer, instructor, training material developer, and president of an MEP consulting engineering firm. Michael has developed and presented seminars on *NFPA 70E* and OSHA electrical safety requirements. He has written and presented several webinars on *NFPA 70* and *NFPA 70E*, has developed the certificate examination for the NFPA 70E 2012 training program, and participated in the development of NFPA's *70E* certification program. He has been president or vice-president of several local sections or regional engineering societies. Michael is a technical committee member of CSA Z462, *Workplace Electrical Safety*, serves on several of Underwriters Laboratories (UL) Standard Technical Panels including UL 489, *Molded Case Circuit Breaker, Molded Case Switches and Circuit Breaker Enclosures*, and serves on several ASTM committees. Michael has been a registered professional electrical engineer since 1979 and a licensed attorney since 1995.

Christopher D. Coache

Chris is Senior Electrical Engineer at NFPA, specializing in hazardous locations. Prior to joining NFPA, he was employed for more than 25 years as an electrical engineer and as a compliance engineer in the information technology industry. He has participated in the International Electrotechnical Commission (IEC), Underwriters Laboratories (UL), and the Instrument Society of America (ISA) standards development. He serves as the staff liaison for NFPA 73, *Standard for Electrical Inspections for Existing Dwellings*, NFPA 110, *Standard for Emergency and Standby Power Systems*, and NFPA 111, *Standard on Stored Electrical Energy Emergency and Standby Systems*. He is an editor of the *National Electrical Code Handbook*. Chris is a member of International Association of Electrical Inspectors (IAEI) and IEEE.

Gil Moniz

Gil is Senior Electrical Specialist at NFPA. Prior to joining NFPA in 2013, he served as the Northeast Field Representative for the National Electrical Manufacturers Association, an electrical inspector for the City of New Bedford, Massachusetts, a licensed master electrician in Massachusetts, and licensed journeyman in Massachusetts and Rhode Island. Gil served as Chairman of Code Making Panel 1 for the 2011 and 2014 *NEC* and as a principal member of Code Making Panel 20 for the 2008 *NEC*. He served on the 2008 and 2012 New York State Residential Code Technical Subcommittees, Massachusetts Electrical Code Advisory Committee, New York City Electrical Advisory Board, New York City Electrical Code Revisions and Interpretations Committee, and as an advisor to the Rhode Island Electrical Code Subcommittee.

Jean Blanc

Jean Blanc is an Associate Electrical Engineer at NFPA. Prior to joining NFPA in 2010, he was a design engineer and holds a B. S. in electrical engineering from Northeastern University. As an Associate Electrical Engineer, he provides advisory services and represents NFPA on several of Underwriters Laboratories (UL) Standard Technical Panels. Jean is a member of International Association of Electrical Inspectors (IAEI) and serves on the IEEE Boston Section's Executive Committee as the Membership Development Co-chair. He holds a Construction Supervisor's License from the state of Massachusetts.

PART | 1

NFPA 70E®
Standard for Electrical Safety
in the Workplace®
with Commentary

Part One of this handbook includes the complete text and figures of the 2015 edition of *NFPA 70E, Standard for Electrical Safety in the Workplace* The text, tables, and figures from the standard are printed in black and are the official requirements of *NFPA 70E*. Illustrations from the standard are labeled as "Figures."

In addition to standard text and informative annexes, Part One includes explanatory commentary that provides historical perspective and other background information for specific paragraphs in the standard. This insightful commentary takes the reader behind the scenes, into the reasons underlying the requirements, to provide a clear understanding of how those requirements are to be properly applied.

Commentary text, captions, and tables are printed in color to clarify identification of commentary material. So that the reader can easily distinguish between the illustrations of the standard and those of the commentary, line drawings, graphs, and photographs in the commentary are labeled as "Exhibits."

This edition of the handbook includes a summary of changes for each article of the 2015 edition of *NFPA 70E,* where applicable.

Introduction

Summary of Global Changes

- Replaced *work shoes* with *footwear* throughout the standard.
- ASTM F1506 title change made throughout the standard.
- *Arc flash risk assessment* replaces *arc flash hazard analysis* throughout the standard.
- *Shock risk assessment* replaces *shock hazard analysis* throughout the standard.
- *Electrical hazard risk assessment* replaces *electrical hazard analysis* throughout the standard.
- *Risk assessment* replaces *hazard identification and risk assessment* throughout the standard.
- *Arc flash PPE category* replaces *hazard/risk category* throughout the standard.
- *Arc flash PPE categories* replaces *hazard/risk categories* throughout the standard.
- All references to *HRC* are deleted throughout the standard.
- Formatting changes to all reference standards titles are made throughout the standard.

Summary of Article 90 Changes

90.2(A): Added *safety related maintenance requirements and other administrative controls* to the Scope.

90.2(B)(2): Deleted old Item (2) *Installations underground in mines and self-propelled mobile surface mining machinery and its attendant electrical trailing cable* and renumbered remainder of section.

90.3: Reorganized section.

90.4: Changed Item (9) from *Hazard Analysis, Risk Estimation, and Risk Evaluation* to *Risk Assessment Procedure*.

NFPA 70E®, Standard for Electrical Safety in the Workplace®, provides enforceable obligations for employers and employees for protecting against electrical hazards to which an employee might be exposed. It is an approved national standard — internationally accepted — that defines electrical safety–related work practices. The requirements at the heart of the document are suitable for adoption and implementation by agencies and employers charged with the responsibility of electrical safety plan development, implementation, and maintenance. Article 90, Introduction, identifies the purpose, scope, and arrangement of the document and where procedures regarding formal interpretation may be found.

Following and fostering basic installation safety, maintenance, and prudent work procedure rules are essential to employee safety. This is accomplished by installing the electrical system in accordance with *NFPA 70®, National Electrical Code® (NEC®)*; by maintaining the electrical system in accordance with NFPA 70B, *Recommended Practice for Electrical Equipment Maintenance,* in the absence of the specific manufacturer's instructions; and by following the safety policies (principles), procedures (including practices), and process controls (metrics) identified in this document, while utilizing the hierarchy of safety controls (methods) identified in *NFPA 70E*. The technical committees for the *NEC, NFPA 70E,* and NFPA 70B all report to the NFPA through the Correlating Committee of the *NEC;* this promotes consistency between safety-related work procedures (including practices), installation requirements, and recommended maintenance procedures (including practices) contained in the latest issued versions of the documents.

NFPA 70E doesn't include design requirements, but system design considerations — and the installations made in accordance with the design — can have a considerable impact on worker safety during installation, inspection, operation, maintenance, and even demolition of electrical systems and components. The type and location of disconnect switches, for example, can play a role in the need for use of personal protective equipment (PPE) when creating an electrically safe work condition. It is easier and less costly to implement safety features during the design phase than during (or after) the installation phase. This indicates the need for early discussions between engineers (including designers), installers, maintainers, owners, and users of electrical systems.

NFPA 70E has a long history, beginning in 1976 with the formation of the *NFPA 70E* Technical Committee (TC) on Electrical Safety Requirements for Workplaces. This committee was formed to assist OSHA in the preparation of electrical safety standards. Prior to the issuance of *NFPA 70E*, OSHA had been using the installation requirements of the *NEC* where safety-related work practices were needed. OSHA's Electrical Safety-Related Work Practices became effective in 1990. These were based mainly on Part II, Safety Related Work Practices, of the 1983 edition of *NFPA 70E* and were written mainly in performance-based language.

OSHA looks to the prescriptive-based requirements of *NFPA 70E* to fulfill the performance-based requirements included in its standards, especially since *NFPA 70E* is the American National Standard on the subject and sets the bar for safe work practices. This symbiotic relationship between *NFPA 70E* and OSHA electrical safety standards increases safety in the workplace.

90.1 Purpose.

The purpose of this standard is to provide a practical safe working area for employees relative to the hazards arising from the use of electricity.

The purpose of *NFPA 70E* is to provide a practical, safe working area for employees, safe from unacceptable risk associated with the use of electricity in the workplace. By the use of an appropriate mix of safety controls from the hierarchy of safety controls, as explained in Informative Annex F, the risks associated with the use of electricity in the workplace can be reduce to an acceptable level. (See Informative Annex P commentary for more information on the hierarchy of safety controls.)

Some of the controls included in the hierarchy of safety controls are the following:

- Engineering controls (enclosures)
- Awareness (warnings, such as arc flash hazard equipment label)
- Administrative controls (safe work policies/procedures including practices, and process controls/metrics)
- PPE

By reducing to an acceptable level the risks associated with the use of electricity, injuries — including fatalities — from the use of electricity in the workplace should be able to be managed to the extent that they are practically eliminated.

Merriam-Webster's Collegiate Dictionary essentially defines the word *practical* as "capable of being put to into use or action." The safety controls being discussed here are not impractical or unrealistic; they are sound, viable, workable applications of safety procedures and policies that will be implemented by the employer and employee.

When electrical equipment is used in accordance with its listings or the manufacturer's instructions — providing it has been properly installed and maintained — the risk of injury from the use of electricity should be minimal, especially under normal operating conditions. However, as indicated by use of the word *practical*, the safety can only be what is reasonably actionable, and all risk associated with the use of electricity in the workplace may not be eliminated. The risk, however, must be reduced to an acceptable level.

Where electricity is used in the workplace, prudent decision making is necessary in order for employees to keep clear of situations where the risk of injury is unacceptable. Where the risk of injury is unacceptable, proper training and supervision will educate employees on how to avoid the potentially dangerous situations. Employees need to be qualified to perform required tasks involving interaction with equipment utilizing electricity. The safe work procedures identified in *NFPA 70E* are meant to be used by qualified persons, not by inadequately or untrained workers.

NFPA 70E, like many safety standards, defines requirements that are based on the safety controls identified in the hierarchy of safety controls starting with the most effective safety control and ending with what is considered the least effective safety control (see Informative Annex P for more on the hierarchy of safety controls). This document utilizes or identifies all of the safety controls identified in this hierarchy, but one of the main strategies is that of creating an electrically safe work condition. Only under limited circumstances is energized electrical work allowed to be performed without creating an electrically safe work condition.

Electrical equipment is to be considered energized until an electrically safe work condition is established by taking the steps necessary to verify that de-energization has taken place and will remain in effect. The process of creating an electrically safe work condition involves risks from which the employee is to be protected by following appropriate safe work procedures and by the use of appropriate PPE. Creating an electrically safe work condition is the use of elimination, which is the first of the safety controls covered in this document

Conditions Providing Acceptable Risk The risk associated with normal operation of electrical equipment is minimal, given the following conditions:

- The equipment is properly installed, in accordance with applicable industry codes and standards including the manufacturer's instructions.

- The equipment is properly maintained, in accordance with the manufacturer's instructions and applicable industry codes and standards.

- The equipment doors are closed and secured.

- All the equipment's covers are in place and secured.

- There is no evidence of impending failure, such as arcing, overheating, loose or bound equipment parts, visible damage, or deterioration.

- The equipment is used in accordance with the manufacturer's instructions and its listing.

Under such conditions the risks of an employee being shocked, electrocuted, or injured by an arc flash or blast or from fire caused from the use of electricity is minimal and considered acceptable.

(see Informative Annex P and the inside cover of this handbook for more on the hierarchy of safety controls).

The use of PPE is the last and least effective safety control that can be used before an event happens, but it is still an important safety control for avoiding injury from electrical hazards. This standard covers in detail the selection, use, and care of PPE. *NFPA 70E* establishes safety processes that use policies (principles), procedures (including practices), and program controls (monitoring) to reduce the risk associated with the use of electricity in the workplace to an acceptable level. It is all about practical, accomplishable electrical safety and about the worker going home safe at the end of the day to his or her family.

90.2 Scope.

It is important to understand that the *National Electrical Code* (*NEC*) applies to *installations* (premises wiring systems), and *NFPA 70E* applies to *employee workplaces* that are located on premises that have premises wiring systems. Installations and employee workplaces are located on various types of *premises*. A *premises* is a location that may consist of buildings, other structures, and grounds, or any combination of the three. *NFPA 70E* is not intended to be applicable to workplaces or work environments found in all types of premises. When locations become employee workplaces, 90.2 makes clear which employee workplaces and the premises thereon are intended to be covered by the work policies (practices), procedures (including practices), and process controls (metrics) found in this document.

The point of connection between the utility power system and the owner's electrical system is defined as the service point. The *NEC*, *NFPA 70E*, and NFPA 70B are applicable after the load side of the service point, which is where the owner of the facility or the owner's representative is responsible for the installation, operation, and maintenance during its full life cycle, from initial installation through operation and maintenance, up to and including its decommissioning and legal disposal.

On the utility side of the service point — the part owned, operated, and maintained by the utility company — utility-type safety rules are often followed, such as those included in ANSI C2, *National Electrical Safety Code*. Although not required by *NFPA 70E*, an employee on the utility side of the service point that is not qualified to follow the utility-type safety rules is required by OSHA to follow the safety procedures identified in its general industry safety standards and as defined by *NFPA 70E*.

(A) Covered. This standard addresses electrical safety-related work practices, safety-related maintenance requirements, and other administrative controls for employee workplaces that are necessary for the practical safeguarding of employees relative to the hazards associated with electrical energy during activities such as the installation, inspection, operation, maintenance, and demolition of electric conductors, electric equipment, signaling and communications conductors and equipment, and raceways. This standard also includes safe work practices for employees performing other work activities that can expose them to electrical hazards as well as safe work practices for the following:

(1) Installation of conductors and equipment that connect to the supply of electricity

(2) Installations used by the electric utility, such as office buildings, warehouses, garages, machine shops, and recreational buildings that are not an integral part of a generating plant, substation, or control center

Informational Note: This standard addresses safety of workers whose job responsibilities entail interaction with electrical equipment and systems with potential exposure to energized electrical equipment and circuit parts. Concepts in this standard are often adapted to other workers whose exposure to electrical hazards is unintentional or not recognized as part of their job responsibilities. The highest risk for injury from electrical hazards for other workers involve unintentional contact with overhead power lines and electric shock from machines, tools, and appliances.

Section 90.2(A) defines employee activities that require realistic safety-related work procedures. These activities include planned interaction with the electrical distribution system and connected equipment (such as installation, inspection, operation, maintenance, and demolition), and can also include unanticipated interaction (such as when employees are painting, tree trimming, or using lifts, bucket trucks, ladders, or long tools). This standard is also intended to cover employees when they are interacting with electrically powered machines, tools, and appliances.

Large complexes — such as college campuses, industrial, military, or multi-unit type facilities — and sometimes even smaller facilities frequently include utility-type generating facilities, utility-type substations, or utility-type overhead or underground electrical distribution systems. But often these installations are not electrically self-sufficient and are connected to utility power systems. The point of connection between the utility power system and the owner's electrical system is defined as the service point. Therefore, maintenance to the facilities on the line side of the service point is the responsibility of the utility company, and maintenance on the load side of the service point is the owner's responsibility or that of the owner's agent.

It is the intent of *NFPA 70E* that utilization-type distribution systems in a power plant — including building services such as heating, ventilation, and air conditioning (HVAC), and lighting — be included within its scope. These systems are typically not wired as an integral part of the electric power generation installation. OSHA, with respect to 29 CFR 1910, Subpart S, considers the covered installation to begin where it becomes electrically independent of conductors and equipment used for the generation of electric power. It is also the intent of *NFPA 70E* that the utility support facility (administration building, maintenance shop, etc.) and the customer-owned substation, such as those shown in Exhibit 90.1, are included within the scope of *NFPA 70E*. See 90.2(B) and its associated commentary regarding workplaces that are outside of the scope of the standard.

Dwelling units (residences) that become employee workplaces are not excluded from the scope of this standard (and OSHA's electrical safety-related work procedure

EXHIBIT 90.1

Electric utility support facility and a customer-owned substation for an industrial complex, both covered within the scope of NFPA 70E.

Scenario and Causation.

Joe, a journeyman electrician with over 25 years of experience, and Al, an electrical apprentice, were removing a set of circuit conductors from a circuit breaker installed in a 480Y/277-volt, 3-phase, 4-wire, 800-amp main breaker service panelboard in the electrical room of a public building. The panelboard supplied power to the entire building. Due to the cost associated with having the utility company disconnect the power to the building, it was decided to proceed with this task by simply placing the main circuit breaker into the off position to de-energize the panelboard busbars. The service conductors into the line side terminals of the main circuit breaker were still energized and exposed.

Joe began removing the conductors from a branch circuit breaker in the upper left-hand side of the enclosure. Al was standing on a ladder 4 or 5 feet away and was manually supporting the conduit that contained the circuit conductors. The bare equipment grounding conductor being removed was still connected to the equipment grounding terminal in the enclosure. No precautions were taken to insulate or cover the equipment grounding conductor or the bare ends of the circuit conductors to prevent inadvertent contact with energized parts within the enclosure, nor was any barrier or shielding material used to protect the exposed energized terminals on the line side of the main circuit breaker.

As the conductors were being removed, the equipment ground conductor accidently contacted an energized terminal on the main circuit breaker and an arc flash event ensued.

Results.

Al saw a large flash and heard a loud noise, followed rapidly by two more explosive noises. Then, Al noticed that Joe's synthetic clothing was in flames. Al patted out the flames on his partner with his hands and burned himself in the process. He then led Joe out of the smoke filled room and called 911 for assistance.

Joe sustained second- and third-degree burns over nearly 50 percent of his body. He required surgery for removal of destroyed skin and restorative skin grafts. He needed physical therapy for a year. Joe returned to work part time after several months, and after a year he returned to work full time. Al was treated for second degree burns and released.

Analysis.

If an electrically safe work condition had been established prior to the removal of the circuit conductors, this accident would not have occurred.

If it could have been demonstrated that working within the equipment was acceptable per 130.2(A)(1) or (A)(2), the following steps needed to be implemented in order to ensure that the employees were protected against electrical hazards.

In order to work near energized conductors and circuit parts, justification and an energized work permit is required; otherwise, energized electrical conductor and circuit parts must be put into an electrically safe condition before an employee works within the limited approach boundary or the arc flash boundary of the conductors or circuit parts. Work on energized electrical conductors or circuit parts is to be performed by written permit only. The permit must be approved by the owner, responsible management, or safety officer.

If the work had to be done with the service conductor energized, the exposed parts of the service conductors and main circuit breaker could have been shielded by an appropriate insulating barrier. The exposed parts of the circuit conductors, including the equipment grounding conductor, could have been insulated. Appropriate personal protective equipment (PPE) could have been worn, including PPE for shock protection and PPE for arc flash protection.

Arc-rated clothing should have been worn. Non-arc-rated cotton, polyester-cotton blends, nylon, nylon-cotton blends, silk, rayon, and wool fabrics are flammable and can ignite and continue to burn, resulting in serious injuries. Furthermore, clothing made from flammable synthetic materials that melt at temperatures below 315°C (600°F), such as acetate, acrylic, nylon, polyester, polyethylene, polypropylene, and spandex, either alone or in blends, can melt from arc flash exposure and melt into the skin, aggravating the burn injury. These types of fabrics should not be worn within the arc flash boundary.

Relevant *NFPA 70E* Requirements.

If Joe, Al, and the company they work for had followed all of the requirements of *NFPA 70E*, Joe and Al might not have been injured. Some of the steps they might have taken include the following:

- Create an Electrical Safety Program, per 110.1.
- Perform hazard identification and risk assessment analysis of this task per 110.1(G).
- Train employees to be qualified to recognize and avoid the electrical hazard per 110.2.
- Create an Energized Electrical Work Permit (EWP) per 130.2(B).
- Perform shock and arc flash hazard analysis per 130.4 and 130.5.
- Use proper PPE for protection against shock and arc flash hazards per 130.7.
- Install protective shields or barriers, rubber insulating equipment, or voltage-rated plastic guard equipment per 130.7(D)(1)(f), (g), or (h).

rules). Even though permanent existing wiring of a dwelling unit is not covered by the OSHA electrical safety standards, the construction employer is required to follow the OSHA safety-related work practices in order to protect employees from recognized hazards presented by the electrical installation. In addition to new permanent wiring installed during construction, equipment — such as portable tools and extension cord sets connected to the permanent wiring — is covered by the OSHA electrical safety standards.

(B) Not Covered. This standard does not cover safety-related work practices for the following:

Since MSHA by an agreement with OSHA has endorsed the application of *NFPA 70E* as "Arc Flash Accident Prevention Best Practices," the exemption in previous editions of *NFPA 70E* for "installations underground in mines and self-propelled mobile surface mining machinery and its attendant electrical trailing cable" has been removed. It is now clear that *NFPA 70E* applies to the mining industry workplace. (See Exhibit 90.2.)

NFPA 70E and the Mining Industry Workplace Electrical safety for the coal mine, and for the metal and nonmetal surface mining industry, is covered by the Code of Federal Regulations (CFR) Title 30, either by Part 56, Subpart K (Electricity) or by Part 57, Subpart K (Electricity). The Mine Safety and Health Administration (MSHA) in most instances is responsible for enforcing electrical safety in mining operations. As spelled out in the Mine Act, in some cases MSHA looks to OSHA to fill in the details, and thereby (due to the relationship between *NFPA 70E* and OSHA) *NFPA 70E* comes into play.

(1) Installations in ships, watercraft other than floating buildings, railway rolling stock, aircraft, or automotive vehicles other than mobile homes and recreational vehicles

All of the workplaces indicated in this exemption have the common feature of movement and transportation of people or material. Because such facilities may have special features and requirements, the electrical distribution systems in them do not have to be installed in accordance with the *NEC,* and they are not required to follow the safe work procedures included in this document. However, if the employer determines that the required level of safety can be achieved by following the rules in this document, the employer is not precluded from following them.

•

(2) Installations of railways for generation, transformation, transmission, or distribution of power used exclusively for operation of rolling stock or installations used exclusively for signaling and communications purposes

This exemption deals with installations of generation, transformation, transmission, or distribution of power or the associate installations of signaling and communication systems used solely for rolling stock. Note that this exemption is only applicable to utility-type systems and is not for premises wiring systems. For example, general yard lighting for a railroad yard with rolling stock is covered by this standard. Where the employer in a railroad yard determines that the desired level of safety can be achieved by following the rules in this document, the employer is not precluded from following them.

(3) Installations of communications equipment under the exclusive control of communications utilities located outdoors or in building spaces used exclusively for such installations

Communication equipment installations under the control of private entities that are not utilities and premises wiring systems in building spaces used solely for communications equipment by a utility are not exempt from the safe work practices of this standard.

(4) Installations under the exclusive control of an electric utility where such installations:

    a. Consist of service drops or service laterals, and associated metering, or

    b. Are located in legally established easements or rights-of-way designated by or recognized by public service commissions, utility commissions, or other regulatory agencies having jurisdiction for such installations, or

    c. Are on property owned or leased by the electric utility for the purpose of communications, metering, generation, control, transformation, transmission, or distribution of electric energy, or

    d. Are located by other written agreements either designated by or recognized by public service commissions, utility commissions, or other regulatory agencies having jurisdiction for such installations. These written agreements shall be limited to installations for the purpose of communications, metering, generation, control, transformation, transmission, or distribution of electric energy where legally established easements or rights-of-way cannot be obtained. These installations shall be limited to federal lands, Native American reservations through the U.S. Department of the Interior Bureau of Indian Affairs, military bases, lands controlled by port authorities and state agencies and departments, and lands owned by railroads.

Wherever the apparatus indicated in Exhibit 90.3 is owned by and under a utility company's sole control, the utility exemption given above is in effect, and the utility company is not required to follow the safe work procedures in this standard for qualified persons. However, for persons who are not qualified to work under the work procedures specifically applicable to a utility, OSHA's general industry safety-related work procedure rules apply, and therefore the safe work rules in this standard would apply.

90.3 Standard Arrangement.

This standard is divided into the introduction and three chapters, as shown in Figure 90.3. Chapter 1 applies generally for safety-related work practices; Chapter 2 applies to safety-related maintenance requirements for electrical equipment and installations in workplaces; and Chapter 3 supplements or modifies Chapter 1 with safety requirements for special equipment.

Industrial Installations as Utility Companies In accordance with OSHA's *Note* to 29 CFR 1910.269(a)(1)(i)(A), OSHA considers equivalent generation, transmission, and distribution installations of industrial establishments that follow OSHA's rules regarding utility-type systems in 29 CFR 1910.269 effectively to be utility companies.

Informative annexes are not part of the requirements of this standard but are included for informational purposes only.

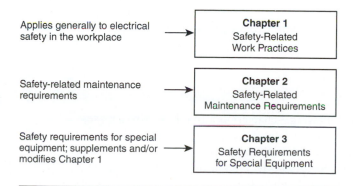

Applies generally to electrical safety in the workplace →

Chapter 1
Safety-Related
Work Practices

Safety-related maintenance requirements →

Chapter 2
Safety-Related
Maintenance Requirements

Safety requirements for special equipment; supplements and/or modifies Chapter 1 →

Chapter 3
Safety Requirements
for Special Equipment

FIGURE 90.3 *Standard Arrangement.*

EXHIBIT 90.3

Typical apparatus falling under the utility scope exclusion in 90.2(B)(4)(a) through (d).

Article 90 defines the purpose and the conditions and situations that *NFPA 70E* is intended to cover. The requirements contained in Chapter 1 and in Chapter 2 generally apply to electrical safety in all workplaces. Chapter 1 is supplemented or modified by the procedures contained in Chapter 3. Chapter 3 applies to special equipment, systems containing special equipment, or areas containing special equipment. Special equipment is often found in specific types of workplaces, such as research and development laboratories, or as part of a specialized system, such as electrolytic cells.

Where the rules identified in Chapter 1 might result in unsafe conditions due to unique or unusual situations, new rules or modification to the rules are provided in Chapter 3. Where construction of specialized equipment or methods of general process operations are required to be different because of unique equipment characteristics and situations, Chapter 3 changes or modifies the work procedures so as to achieve the same goal, the provision of safe work practices for employees while accommodating unique or unusual circumstances.

90.4 Organization.

This standard is divided into the following 3 chapters and 16 informative annexes:
(1)   Chapter 1, Safety-Related Work Practices
(2)   Chapter 2, Safety-Related Maintenance Requirements
(3)   Chapter 3, Safety Requirements for Special Equipment
(4)   Informative Annex A, Referenced Publications
(5)   Informative Annex B, Informational References

(6)   Informative Annex C, Limits of Approach
(7)   Informative Annex D, Incident Energy and Arc Flash Boundary Calculation Methods
(8)   Informative Annex E, Electrical Safety Program
(9)   Informative Annex F, Risk Assessment Procedure
(10) Informative Annex G, Sample Lockout/Tagout Procedure
(11) Informative Annex H, Guidance on Selection of Protective Clothing and Other Personal Protective Equipment
(12) Informative Annex I, Job Briefing and Planning Checklist
(13) Informative Annex J, Energized Electrical Work Permit
(14) Informative Annex K, General Categories of Electrical Hazards
(15) Informative Annex L, Typical Application of Safeguards in the Cell Line Working Zone
(16) Informative Annex M, Layering of Protective Clothing and Total System Arc Rating
(17) Informative Annex N, Example Industrial Procedures and Policies for Working Near Overhead Electrical Lines and Equipment
(18) Informative Annex O, Safety-Related Design Requirements
(19) Informative Annex P, Aligning Implementation of This Standard with Occupational Health and Safety Management Standards

The arrangement of this standard is such that the purpose and scope is given up front, then common rules with support material and then the rules for more defined subject areas. At the end of the standard are the informative annexes — explanatory material covering specific subjects in detail.

Chapter 1, Safety-Related Work Practices, is the heart of the standard. It provides general definitions, rules, and guidelines for electrically safe work practice. Chapter 2 deals with the type and amount of maintenance necessary to ensure electrical safety from potential exposure to electrical hazards. Chapter 3 covers unique and distinctive types of electrical equipment and systems in the workplace, and unique and distinctive workplace locations that call for specialized procedures.

The informative annexes at end of the standard provide additional explanatory material, and each covers a separate and distinctive subject area. This explanatory material identifies references, clarifies or provides details on necessary processes such as performing a risk assessment, or provides useful forms such as a sample energized electrical work permit. Additionally the informative annexes explain how approach limits were derived, show sample procedures on subjects such as lockout/tagout, and provide additional guidance on calculation methods, selection of PPE, safety-related design, and the like.

90.5 Mandatory Rules, Permissive Rules, and Explanatory Material.

(A) Mandatory Rules. Mandatory rules of this standard are those that identify actions that are specifically required or prohibited and are characterized by the use of the terms *shall* or *shall not*.

Mandatory rules are obligatory requirements either compelling or barring an action, either by the employee or by the employer. Mandatory rules in *NFPA 70E* are necessary to keep employees safe in the work environment. They are particularly important where the risk of an incident from electrical hazard is determined to be unacceptable. These rules were developed based on the collective knowledge and experience of all those who contributed to and participated in the development of this standard.

(B) Permissive Rules. Permissive rules of this standard are those that identify actions that are allowed but not required, are normally used to describe options or alternative methods, and are characterized by the use of the terms *shall be permitted* or *shall not be required*.

Permissive rules typically point out options that may be allowed or realistic alternative steps that can be taken to meet the objective.

It should be noted that permissive rules use mandatory language and as such constitute rules (they use the terms *shall be permitted* or *shall not be required*). When evaluating choices among options that are permitted, it is always advisable for the employer to make sure that the option selected does not increase the risk of injury from a potential electrical hazard above what is considered acceptable. The term *may* is used where discretionary judgment on the part of an authority having jurisdiction is required.

(C) Explanatory Material. Explanatory material, such as references to other standards, references to related sections of this standard, or information related to a rule in this standard, is included in this standard in the form of informational notes. Such notes are informational only and are not enforceable as requirements of this standard.

Brackets containing section references to another NFPA document are for informational purposes only and are provided as a guide to indicate the source of the extracted text. These bracketed references immediately follow the extracted text.

> Informational Note: The format and language used in this standard follow guidelines established by NFPA and published in the *National Electrical Code Style Manual*. Copies of this manual can be obtained from NFPA.

Explanatory material is presented in extracts, informational notes, and informative annexes.

The only place where extracted material appears in *NFPA 70E* is in Article 100, Definitions. The material comes from the 2014 edition of *the NEC*. Brackets containing document number and section references identify material extracted from another NFPA document. The bracketed references directly follow the extracted text. The extracted material included in the body of this document is part of the rules of this document and is enforceable.

The content of informational notes does not include rules, mandatory or permissive, but rather provides information intended to help the user understand the rule to which the informational note applies, or to point to the location where other notable material can be found. An informational note can include references to other standards, references to related sections of this standard, or information related to a rule in this standard.

A footnote is not an informational note. When a footnote is included below a table in the body of the document, it is part of the requirements of this standard and is enforceable. The limitations put in footnotes are as important as the requirements included in the table — in fact, the contents of the table are only applicable within the restrictions included in the footnotes.

In addition, there are 16 informative annexes in this document. Each informative annex deals with a different subject, and all include information to help the user of the document understand and implement the requirements contained in the body of the document so as to improve electrical safety.

Definitions Extracted from the *NEC* The only material in *NFPA 70E* currently extracted from another NFPA document consists of the definitions extracted from the *NEC*. These extracted definitions are kept intact, except where items such as informational notes or references are clearly inapplicable. Since *NFPA 70E* and the *NEC* are closely related, it is imperative that common definitions are consistent except where there is substantial reason for some deviation or to make the material consistent with *NFPA 70E*. Under this process, the context of the original material is not violated or compromised and the extract is reviewed by the *NEC* Correlating Committee.

90.6 Formal Interpretations.

To promote uniformity of interpretation and application of the provisions of this standard, formal interpretation procedures have been established and are found in the NFPA Regulations Governing Committee Projects.

The procedures for formal interpretations (FIs) of the requirements of *NFPA 70E* are outlined in Section 6 of the *NFPA Regulations Governing the Development of NFPA Standards*. These regulations are included in the *NFPA Standards Directory*, which is published annually and can be downloaded from the NFPA website at nfpa.org. Two general forms of FIs are recognized: (1) those that are interpretations of the literal text, and (2) those that are interpretations of the intent of the technical committee at the time the particular text was issued.

Because most of the interpretations requested do not qualify for processing as an FI in accordance with the regulations named above, many interpretations are rendered as the personal opinions of NFPA electrical staff. Such opinions are rendered in writing only in response to written request. In lieu of submitting an FI, most requests for an informal interpretation come in through the organization's advisory services call center or through questions entered online via the Technical Question Tab of the *NFPA 70E* Doc Info Pages (http://www.nfpa.org/codes-and-standards/document-information-pages?mode=code&code=70e&tab=questions).

Informal interpretation by NFPA technical staff of the requirements included in an NFPA issued document is one of the benefits of association membership. Frequently, the staff member answering the technical question is the staff liaison to the technical committee of the document in question. However, correspondences with NFPA staff are not to be construed as *formal* interpretations. An FI can be requested through the process named here. The opinion given by staff is not meant to be provided as a professional consultation or service and is not to be relied upon as such. If a professional opinion is required, the requester should retain the services of a professional in the jurisdiction where the information is to be used. Also, NFPA technical staff do not issue opinions on the technical documents of other organizations even if the document is referenced in an NFPA document.

FIs of *NFPA 70E* rules, when issued, are published in several venues — *NFPA News*, NFPACodesOnline.org (the *National Fire Code®* subscription service), and various trade publications, as well as on nfpa.org.

Safety-Related Work Practices

In the previous edition of *NFPA 70E*, *Standard for Electrical Safety in the Workplace*, hazard was separated from risk. This edition continues this process by redefining hazard and risk terminology to be consistent with other national and international standards and by replacing the hazard/risk categories tables with the new arc flash PPE categories tables.

Where electricity is used in the workplace, *NFPA 70E* provides the safety-related work rules necessary to eliminate or reduce the risk associated with potential exposure to electrical hazards to an acceptable level.

Electrical safety is like a three-legged stool. First, it depends on a safe installation. *NFPA 70*, *National Electrical Code* (*NEC*), stipulates the installation rules necessary to assure that an electrical installation is safe to operate when it is installed. The *NEC* ensconces safe installation rules for premises wiring systems for residential, commercial, institutional, industrial, and certain utility facilities.

Second, these facility installations are kept safe by following the preventive maintenance recommendations included in NFPA 70B, *Recommended Practice for Electrical Equipment Maintenance,* or those required by the manufacturer. The preventive maintenance program for the facility should include an inspection and testing program that is performed by competent in-house or outside personnel, or by competent independent contractors. *NFPA 70B* identifies the recommended procedures necessary to maintain electrical equipment in a safe use condition.

The third leg of the stool is *NFPA 70E*; it provides the employer with the direction required to create an electrical safety program (ESP) for the facility. With the employer's ESP following the rules laid out in *NFPA 70E*, the facility owner's or operator's employees ought to be able to select and execute the safety-related work procedures required when the electrical equipment they are interacting with is in other than a normal operating condition.

Workers inappropriately interacting with energized electrical conductors and circuit parts can be exposed to an unacceptable risk of exposure to electrical hazards. Exposure to an unacceptable risk of electrical hazards can result in an electrical shock injury or electrocution, arc flash burn or thermal burn (from hot equipment or fire), injury from the molten metal or shrapnel or pressure blast associated with an arc flash, or a secondary injury as the result of a fall or from trying to avoid the exposure.

Injuries from electrical exposure may be categorized as catastrophic, critical, medium, minor, or slight and can be defined as follows:

- A *catastrophic* injury is one that results in death or permanent total disability.
- A *critical* injury is one where there is a permanent partial disability or a temporary total disability of 3 months or more.
- A *medium* injury is one that results in loss of work — a loss work injury requiring medical treatment.

- A *minor* injury is one where minor medical treatment is possible.
- A *slight* injury is one where first aid or minor treatment is involved.

Chapter 1 consists of five articles, which include the necessary information and essential safety-related work procedures required for employees to work safely where there is an unacceptable risk of injury from an exposure to electrical hazards in the workplace. The information and safety-related work procedures included in *NFPA 70E* should form the basis of the employer's electrical safety program. The following is a brief description of each article:

- Article 100 offers the terms and meanings necessary to understand this standard. Everyday terminology found in a standard English language dictionary is not delineated in Article 100.
- Article 105 establishes the overall scope, purpose, and organization of Chapter 1 and distinguishes the responsibility of employers and employees.
- Article 110 covers the overall requirements for formulating and implementing an ESP, including policies, rules, measurement and recording of performance, risk assessment, emergency procedures, employee training, use of equipment, job briefing, program and field auditing, documenting, and the relationships between various independent contractors and their employees or tradesmen. Other parts essential to an ESP are found in Article 120 and Article 130.
- Article 120 provides the information necessary for creating an electrically safe work condition (ESWC). It provides the requirements for establishing a lockout/tagout program — using lack of voltage verification — for a facility where electrical hazards are involved. Verified de-energization through the creation of an ESWC is the only assured way of eradicating the potential of electrical hazards in the workplace. An electrical lockout/tagout program where there are potential electrical hazards using verified de-energization is part of the overall ESP and must be included as part of its documentation.
- Article 130 defines the situations under which an electrically safe work condition must be established and describes the situations under which energized electrical work can be justified and the necessary requirements for working safely with energized electrical equipment. Policies, procedures, and program controls covering these areas are also part of the ESP and need to be documented.

Article 100 Definitions

Summary of Article 100 Changes

Accessible, Readily (Readily Accessible): Revised to correlate with the definition in the 2014 *NEC*.

Arc Rating, Informational Note No. 1: Revised informational note to explain that all arc-rated clothing is also flame resistant.

Barricade: Deleted the phrase *to a hazardous area* from the end of the definition.

Boundary, Restricted Approach: Replaced the term *risk* with *likelihood* and added the term *electric* to clarify the type of shock.

Device: Revised to correlate with the definition in the 2014 *NEC*.

Enclosure: Replaced the phrase *energized parts* with *energized electrical conductors or circuit parts* for improved usability.

Equipment: Deleted the term *material* from list.

Grounding Conductor, Equipment (EGC): Revised to correlate with the definition in the 2014 *NEC*.

Hazard: Added to provide clarity and consistent use throughout the standard.

Hazardous: Added to provide clarity and consistent use throughout the standard.

Incident Energy: Revised to include the term *thermal*.

Incident Energy Analysis: Replaced the phrase *arc flash hazard analysis* with *arc flash risk assessment* to provide clarity.

Luminaire: Revised to correlate with the definition in the 2014 *NEC*.

Premises Wiring (System): Added informational note to correlate with the definition in the 2014 *NEC*.

Qualified Person: Revised for consistency with the OSHA definition.

Raceway: Revised to correlate with the definition in the 2014 *NEC*.

Risk: Added to provide clarity and consistent use throughout the standard.

Risk Assessment: Added definition and informational note to provide clarity and consistent use throughout the standard.

Service Point: Added informational note to correlate with the definition in the 2014 *NEC*.

Switchboard: Revised to correlate with the definition in the 2014 *NEC*.

Article 100 contains definitions of technical terms considered fundamental to the proper understanding of key requirements. Although these terms may be used or defined differently in another standard, these definitions are the ones that apply within the scope of *NFPA 70E*. Understanding of the vocabulary used in this standard is crucial to comprehending its requirements, and to the flow of information between parties, and therefore can increase electrical safety in the workplace. Many of the defined terms included in *NFPA 70E* come from the *NEC*, and many of definitions in Article 100 are taken directly from the *NEC*. A reference in brackets indicates an extraction from the *NEC*. The definition of each of these extracted terms is identified by a bracketed reference of "[**70**:100]" indicating it comes from Article 100 of that document. The remainder of the defined terms have been developed by the Technical Committee on Electrical Safety in the Workplace.

Scope.

This article contains only those definitions essential to the proper application of this standard. It is not intended to include commonly defined general terms or commonly defined technical terms from related codes and standards. In general, only those terms that are used in two or more articles are defined in Article 100. Other definitions are included in the article in which they are used but may be referenced in Article 100. The definitions in this article shall apply wherever the terms are used throughout this standard.

Official NFPA definitions are under the purview of the NFPA Standards Council, not the committee that is responsible for this document. Official definitions are generally the same in all NFPA documents. Many NFPA documents rely on or cross-reference each other, and having official definitions facilitates the consistency across documents.

Commonly defined general terms, including those in general English language dictionaries, are not defined in *NFPA 70E* unless they are used in a unique or restricted manner. Commonly defined technical terms, such as *volt (V)* and *ampere (A)*, are found in ANSI/IEEE 100, *IEEE Standard Dictionary of Electrical and Electronic Terms*, Seventh Edition (2000).

Not all definitions are found in Article 100. Additional definitions can be found in Chapters 2 and 3, in the article they are used in. Those articles follow the format of the *NEC Style Manual*, in which the definition section is listed as "XXX.2, Definition(s)." For example, the definitions applicable to lasers can be found in 330.2, Definitions. Although requirements are not included in definitions, an understanding of the definitions is paramount to understanding the rules in this standard.

Accessible (as applied to equipment). Admitting close approach; not guarded by locked doors, elevation, or other effective means. [**70:**100]

Access to equipment is necessary for operation, maintenance, and service; but this definition is also used to identify equipment that is accessible to unqualified persons. Exhibit 100.1 illustrates examples of equipment considered to be accessible. The main rule for access to a disconnecting means, including circuit breakers, according to 240.24(A) and 404.8(A) of the *NEC*, is shown as part (a). In (b), the busway installation is in accordance to 368.17(c). Three exceptions to the 6 ft. 7 in. height rule in *NEC* 404.8(A) are illustrated in part (c), where the installation of the busway switches are located according to Exception No. 1; in part (d), where a switch is installed adjacent to a motor according Exception No. 2; and in part (e), where a hookstick-operated isolating switch is installed according to Exception No. 3.

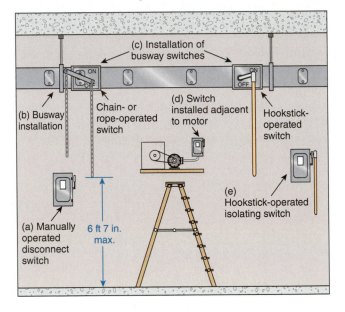

EXHIBIT 100.1

Some illustrations of "accessible (as applied to equipment)." Item (a) is a disconnecting means that is compliant with the basic requirement, while (b), (c), (d), and (e) are busways and switches considered accessible even though they are located above the 6 ft 7 in height rule.

Accessible (as applied to wiring methods). Capable of being removed or exposed without damaging the building structure or finish or not permanently closed in by the structure or finish of the building. [**70:**100]

Wiring methods located behind a panel designed to allow removal by tool or by hand are not considered permanently enclosed. They are also considered to be an exposed wiring method. Exhibit 100.2 illustrates examples of wiring methods and equipment that are considered to be accessible despite being located above a suspended ceiling.

Accessible, Readily (Readily Accessible). Capable of being reached quickly for operation, renewal, or inspections without requiring those to whom ready access is requisite to actions

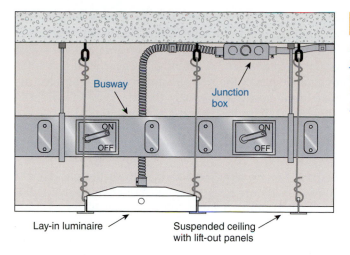

EXHIBIT 100.2

Examples of busways and junction boxes considered accessible even if located behind hung ceilings having lift-out panels.

such as to use tools, to climb over or remove obstacles, or to resort to portable ladders, and so forth. [**70:**100]

The definition of *readily accessible* does not preclude using locks on service equipment doors or doors of rooms containing service equipment, provided a key or lock combination is available to those for whom ready access is necessary. If a tool is necessary to gain access, the equipment is not readily accessible.

Approved. Acceptable to the authority having jurisdiction.

See the definition of the term *authority having jurisdiction (AHJ)* for a better understanding of the approval process. An understanding of terms such as *listed* and *labeled* can help the user to understand the approval process. Typically, approval of listed equipment is more readily given by an authority having jurisdiction where the authority accepts a laboratory's listing mark. The importance of the role of the authority having jurisdiction in the North American safety system cannot be overstated. The authority having jurisdiction verifies that an installation or in the case of *NFPA 70E* that the established work procedure complies with the standard. The definition of the term *authority having jurisdiction* and the accompanying informational note are helpful to understand code enforcement and the inspection process.

Arc Flash Hazard. A dangerous condition associated with the possible release of energy caused by an electric arc.

Informational Note No. 1: An arc flash hazard may exist when energized electrical conductors or circuit parts are exposed or when they are within equipment in a guarded or enclosed condition, provided a person is interacting with the equipment in such a manner that could cause an electric arc. Under normal operating conditions, enclosed energized equipment that has been properly installed and maintained is not likely to pose an arc flash hazard.

Informational Note No. 2: See Table 130.7(C)(15)(A)(a) for examples of activities that could pose an arc flash hazard.

An arc flash hazard exists if a person is or might be exposed to a significant thermal hazard. A significant thermal hazard is one with an incident (thermal) energy of 1.2 calories per square centimeter (cal/cm²) or more. Personal protective equipment (PPE) with a rating that exceeds the potential thermal hazard must be worn. Use of PPE for exposures with less than 1.2 cal/cm² incident energy might be deemed appropriate by the employer and employee.

An arcing fault contained within equipment could generate a pressure wave sufficient to destroy the integrity of the enclosure under certain conditions. The phrase *interacting with the equipment* could mean opening or closing a disconnecting means, pushing a reset button, or latching the enclosure door of a door-and-door trim. However, if equipment is installed in accordance with the requirements of the *NEC* and the manufacturer's instructions, and it is maintained adequately and operated normally, the chance of one of these actions initiating an arcing fault is remote. An arc flash hazard is more likely to exist where the trim of functioning electrical equipment is fully or partially removed, exposing energized electrical conductors and circuit parts, or where a circuit breaker is being racked in or out of switchgear.

Exhibit 100.3 shows an arc flash incident from the initial fault to the extinguishment of the arc. An arc flash incident can result in intense heat, shrapnel, molten metal, light, poisonous oxides, and pressure waves, as illustrated in Exhibit 100.4.

EXHIBIT 100.3

Photos of an arc flash. (Courtesy of Westex)

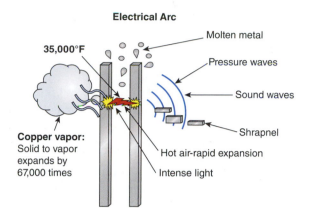

Electrical Arc

35,000°F

Molten metal

Pressure waves

Sound waves

Shrapnel

Copper vapor:
Solid to vapor
expands by
67,000 times

Hot air-rapid expansion

Intense light

EXHIBIT 100.4

Physical results of an arc flash incident. (Adapted from Cooper Bussmann, a Division of Cooper Industries, PLC)

Arc Flash Suit. A complete arc-rated clothing and equipment system that covers the entire body, except for the hands and feet.

> Informational Note: An arc flash suit may include pants or overalls, a jacket or a coverall, and a beekeeper-type hood fitted with a face shield.

An increased awareness of the hazards associated with electrical equipment in the workplace has led to the rapid development of protective schemes and equipment for employees by the PPE industry. Historically, the term *arc flash suit* has not been used consistently throughout the industry. This definition and its associated informational note clarify the specific arc-rated components that comprise an arc flash suit and that a face shield alone or in combination with a balaclava cannot be used when a flash suit is required.

An arc flash suit does not provide protection for an employee's hands and feet, and therefore all body parts, including the hands and feet, must be covered by adequate protective equipment.

Arc Rating. The value attributed to materials that describes their performance to exposure to an electrical arc discharge. The arc rating is expressed in cal/cm² and is derived from the determined value of the arc thermal performance value (ATPV) or energy of breakopen threshold (E_{BT}) (should a material system exhibit a breakopen response below the ATPV value). Arc rating is reported as either ATPV or E_{BT}, whichever is the lower value.

> Informational Note No. 1:   Arc-rated clothing or equipment indicates that it has been tested for exposure to an electric arc. Flame resistant clothing without an arc rating has not been tested for exposure to an electric arc. All arc-rated clothing is also flame-resistant.

> Informational Note No. 2:   *Breakopen* is a material response evidenced by the formation of one or more holes in the innermost layer of arc-rated material that would allow flame to pass through the material.

> Informational Note No. 3:   ATPV is defined in ASTM F1959/F1959M, *Standard Test Method for Determining the Arc Rating of Materials for Clothing,* as the incident energy (cal/cm²) on a material or a multilayer system of materials that results in a 50 percent probability that sufficient heat transfer through the tested specimen is predicted to cause the onset of a second degree skin burn injury based on the Stoll curve.

> Informational Note No. 4: E_{BT} is defined in ASTM F1959/F1959M, *Standard Test Method for Determining the Arc Rating of Materials for Clothing,* as the incident energy (cal/cm²) on a material or a material system that results in a 50 percent probability of breakopen. Breakopen

is defined as a hole with an area of 1.6 cm² (0.5 in²) or an opening of 2.5 cm (1.0 in.) in any dimension.

The definition of arc rating correlates with the definitions in ASTM F1506, Standard Performance Specification for Flame Resistant Textile Materials for Wearing Apparel for Use by Electrical Workers Exposed to Momentary Electric Arc and Related Thermal Hazards, for protective apparel and ASTM F1891-12, Standard Specification for Arc and Flame-Resistant Rainwear, for protective raingear. In the 2012 edition of NFPA 70E, the term flame-resistant (FR) was replaced by the term arc rated (AR). This definition is consistent with the terminology used in these ASTM standards and provides consistency in the selection of protective apparel. For electrical hazards, arc-rated PPE is necessary rather than flame-resistant (FR) apparel, because it has been specifically tested for protection against the thermal effects of an arc flash event. See Informational Note No. 1.

Manufacturers determine the arc rating for protective equipment. Labels on arc-rated clothing should include the rating, as shown in Exhibit 100.5. The arc flash suit arc rating should be visibly marked to ensure that correctly rated gear is used. An arc flash PPE category marking alone (such as 1 or 3) is not an arc rating (arc ratings are given in cal/cm²) and is only appropriate where Tables 130.7(C)(15)(A)(a) and 130.7(C)(15)(A)(b), or Tables 130.7(C)(15)(B) and 130.7(C)(16), are used to select PPE.

EXHIBIT 100.5

An arc flash suit with the arc rating clearly marked on the suit. (Courtesy of Salisbury by Honeywell)

An important difference between arc rated (AR) and flame resistant (FR) garments is clarified in Informational Note 1. Although all AR clothing is also FR, the inverse is not always true.

Informational Notes No. 3 and No. 4 provide guidance on how arc thermal performance value (ATPV) and energy of breakopen threshold (E_{BT}) relate to the testing that is performed on PPE in accordance with ASTM F1506. The lower of these two values, as determined by testing, is the one used to label the PPE. Where flammable garments are used under AR garments, the E_{BT} rating should not be exceeded. These informational notes point out that the testing is based on a 50 percent probability of success. Where arc-rated equipment is used at less than its rating, there can be a substantial increase in the probability of success. See Exhibit M.3 associated with the commentary to Informative Annex M.

Where energized work is performed, it must be performed by qualified persons using appropriate safe work procedures and appropriate PPE for both shock and arc flash protection. Many fabrics, including non-arc-rated cotton, polyester-cotton blends, nylon, silk, and wool fabrics, are flammable and can ignite and continue to burn, resulting in serious injuries. Furthermore, synthetic materials that melt at a temperature below 600°F (315°C), such as nylon, polyester, polypropylene, and spandex, can melt from arc flash exposure and melt into the skin, aggravating the burn injury. These types of fabrics are not permitted to be worn within the arc flash boundary, except in fiber blends that are arc rated.

Attachment Plug (Plug Cap) (Plug). A device that, by insertion in a receptacle, establishes a connection between the conductors of the attached flexible cord and the conductors connected permanently to the receptacle. [**70:**100]

The contact blades of general purpose attachment plugs have specific shapes, sizes, and configurations so that the plug cannot be inserted into a receptacle or cord connector of a different voltage or current rating. Locking and non-locking attachment plugs are available with options such as integral switches, fuses, or ground-fault circuit-interrupter protection.

Authority Having Jurisdiction (AHJ). An organization, office, or individual responsible for enforcing the requirements of a code or standard, or for approving equipment, materials, an installation, or a procedure.

> Informational Note: The phrase "authority having jurisdiction," or its acronym AHJ, is used in NFPA documents in a broad manner, since jurisdictions and approval agencies vary, as do their responsibilities. Where public safety is primary, the authority having jurisdiction may be a federal, state, local, or other regional department or individual such as a fire chief; fire marshal; chief of a fire prevention bureau, labor department, or health department; building official; electrical inspector; or others having statutory authority. For insurance purposes, an insurance inspection department, rating bureau, or other insurance company representative may be the authority having jurisdiction. In many circumstances, the property owner or his or her designated agent assumes the role of the authority having jurisdiction; at government installations, the commanding officer or departmental official may be the authority having jurisdiction.

The authority having jurisdiction where electrical safety of employees is involved is the party (or parties) responsible for enforcing electrical safety requirements such as *NFPA 70E*. This could be the head of a federal or state governmental agency or designee; the commanding officer of a military base or another governmental departmental official; a party designated by the owner or operator of a facility, or a party designated by an employer; a party from an insurance company inspection department; or a combination of several of these.

Automatic. Performing a function without the necessity of human intervention.

An automatic function or operation could be initiated by a mechanical operation. For instance, a high-pressure cutoff switch requires no human interaction to open the circuit upon detection of high pressure.

Balaclava (Sock Hood). An arc-rated hood that protects the neck and head except for the facial area of the eyes and nose.

•

An arc-rated balaclava, as shown in Exhibit 100.6, is knitted from yarn. It fits tightly against the wearer's head and neck, and few air pockets exist between the balaclava and the wearer's skin. Only balaclavas having a tested arc rating in calories per square centimeter (cal/cm²) can be used to meet the requirements of *NFPA 70E*. The balaclava shown in Exhibit 100.6 combined with a hard hat and arc-rated face shield (rated not less than 12 cal/cm²) could be used in lieu of an arc-rated arc flash suit hood where the incident energy exposure does not exceed 12 cal/cm² or the tested arc rating of the combined items of PPE. See Exhibit 130.11 part (a). Balaclavas intended only for warmth must not be worn as arc flash protection.

EXHIBIT 100.6

A balaclava sock hood with an arc rating of 12 cal/cm². (Courtesy of Salisbury by Honeywell)

Barricade. A physical obstruction such as tapes, cones, or A-frame-type wood or metal structures intended to provide a warning and to limit access.

The purpose of a barricade is to provide warning to approaching individuals that a hazardous condition exists. It is intended to limit approach to an unsafe condition (see barrier). A barricade might consist of yellow warning tape as shown in Exhibit 100.7.

Barrier. A physical obstruction that is intended to prevent contact with equipment or energized electrical conductors and circuit parts or to prevent unauthorized access to a work area.

A barrier must be of sufficient integrity to eliminate the chance of unsafe contact. A barrier constructed from voltage-rated materials could be in physical contact with an energized conductor. Barriers constructed of other materials — such as wood, metal, or fiberglass — must be installed with a safe distance between an energized conductor or circuit part and unauthorized personnel.

Bonded (Bonding). Connected to establish electrical continuity and conductivity. [**70:**100]

Bonding is establishing an electrical connection between conductive elements of an electrical installation. Bonding does not necessarily rely on the presence of a ground connection.

EXHIBIT 100.7

Yellow tape used to warn unqualified people about a work area. (Courtesy of Salisbury by Honeywell)

Bonding Conductor or Jumper. A reliable conductor to ensure the required electrical conductivity between metal parts required to be electrically connected. [**70:**100]

A bonding jumper is an electrical conductor that is installed to join discontinuous or potentially discontinuous portions of conductive elements. For instance, both concentric- and eccentric-type knockouts can impair the electrical conductivity between the metal parts. Installing a bonding jumper between metal raceways and metal parts ensures continuous electrical conductivity. Although a bonding jumper is often interpreted to mean a short conductor, bonding jumpers may be several feet in length.

Boundary, Arc Flash. When an arc flash hazard exists, an approach limit at a distance from a prospective arc source within which a person could receive a second degree burn if an electrical arc flash were to occur.

> Informational Note: A second degree burn is possible by an exposure of unprotected skin to an electric arc flash above the incident energy level of 5 J/cm^2 (1.2 cal/cm^2).

The arc flash boundary, determined either through calculation or through the use of Table 130.7(C)(15)(A)(b) or Table 130.7(C)(15)(B), separates an area in which a person is potentially exposed to a second-degree burn injury from an area in which the potential for injury does not include a second-degree burn. Arc flash burns may still occur outside the arc flash boundary, but they should not be second-degree or worse. The arc flash boundary may be thought of as coming into existence when electrical equipment is in other than a normal operating state. See 130.2(4) and its associated informational note.

The arc flash boundary defines the point at which arc-rated PPE is necessary to avoid a second-degree burn. All body parts closer to an arc flash hazard than the arc flash boundary must be protected from the potential thermal effects of the hazard. For example, an employee's hand and arm within the arc flash boundary must be protected from the thermal hazard. If the employee's head is also within the arc flash boundary, the employee's head (including the back of the head) must be protected from the thermal hazard. [See 130.7(C) for information on PPE.]

Boundary, Limited Approach. An approach limit at a distance from an exposed energized electrical conductor or circuit part within which a shock hazard exists.

The limited approach boundary is a shock protection boundary that is not related to arc flash or incident energy. The arc flash boundary may be greater than, less than, or equal to the limited approach boundary. This boundary defines the approach limit for unqualified employees and is intended to eliminate the risk of contact with an exposed energized electrical conductor or circuit part. When an employee is closer than this minimum distance, special considerations are necessary for protection. Except as specified in 130.2(B)(3), working within the limited approach boundary is only acceptable if an energized work permit has been completed and authorized. If an unqualified employee is required to work within the limited approach boundary, the employee must be directly and continuously supervised by a qualified person.

•

Boundary, Restricted Approach. An approach limit at a distance from an exposed energized electrical conductor or circuit part within which there is an increased likelihood of electric shock, due to electrical arc-over combined with inadvertent movement, for personnel working in close proximity to the energized electrical conductor or circuit part.

This shock protection boundary is the approach limit for qualified employees. The arc flash boundary may be greater than, less than, or equal to the restricted approach boundary. A qualified employee required to cross the restricted approach boundary must be protected from unexpected contact with the conductors or circuit parts that are energized and exposed. Qualified employees should have the knowledge and ability to avoid unexpected contact with an exposed energized conductor or circuit part. A complete and authorized energized electrical work permit is required before employees are allowed to work within the restricted approach boundary, except as permitted by 130.2(B)(3).

Branch Circuit. The circuit conductors between the final overcurrent device protecting the circuit and the outlet(s). [**70:**100]

It is important to note the difference between branch circuits and feeders. The conductors that originate at overcurrent protective devices in the panelboard and supply the receptacles are branch-circuit conductors. The conductors from the service equipment and the generator to the panelboards are feeders.

Building. A structure that stands alone or that is cut off from adjoining structures by fire walls with all openings therein protected by approved fire doors. [**70:**100]

Definitions of the terms *fire wall* and *fire door* are the responsibility of building codes. Generically, a fire wall is a wall with a fire resistance rating and structural stability that separates buildings or subdivides a building to prevent the spread of fire. Fire doors (and fire windows) are used to protect openings in walls, floors, and ceilings against the spread of fire and smoke within, into, or out of buildings.

Cabinet. An enclosure that is designed for either surface mounting or flush mounting and is provided with a frame, mat, or trim in which a swinging door or doors are or can be hung. [**70:**100]

Cabinets are enclosures that an electrical assembly, such as a panelboard, is installed into to enclose the energized electrical conductors and circuit parts under normal equipment operating conditions.

Circuit Breaker. A device designed to open and close a circuit by nonautomatic means and to open the circuit automatically on a predetermined overcurrent without damage to itself when properly applied within its rating. [**70:**100]

> Informational Note: The automatic opening means can be integral, direct acting with the circuit breaker, or remote from the circuit breaker.

Conductive. Suitable for carrying electric current.

As used in this standard, the term *conductive* refers to any material that is capable of conducting electrical current. If a material does not have an established voltage rating — such as voltage-rated rubber products — the material should be considered to be conductive.

Conductor, Bare. A conductor having no covering or electrical insulation whatsoever. [**70:**100]

Bare conductors are easy to identify, because without insulation they are copper or aluminum colored.

Conductor, Covered. A conductor encased within material of composition or thickness that is not recognized by this *Code* as electrical insulation. [**70:**100]

Examples of where covered conductors may be used include busbars and overhead service conductors. Covered conductors often resemble insulated conductors but should always be treated as bare conductors because the covering does not have an assigned insulation or voltage rating.

Conductor, Insulated. A conductor encased within material of composition and thickness that is recognized by this *Code* as electrical insulation. [**70:**100]

A conductor with an insulating covering material generally is required to pass the testing specified in a product standard. One such product standard is UL 83, *Standard for Safety Thermoplastic-Insulated Wires and Cables*. Only wires and cables that meet the minimum fire, electrical, and physical properties required by the applicable standards are permitted to be marked with the letter designations found in the *NEC*. See *NEC* 310.104 for the minimum required insulation thickness for each insulation type, voltage level, and application.

Controller. A device or group of devices that serves to govern, in some predetermined manner, the electric power delivered to the apparatus to which it is connected. [**70:**100]

A controller can be a remote-controlled magnetic contactor, switch, circuit breaker, or device that normally is used to start and stop motors and other apparatus. Stop-and-start stations and similar control circuit components that do not open the power conductors to the motor are not considered controllers.

Current-Limiting Overcurrent Protective Device. A device that, when interrupting currents in its current-limiting range, reduces the current flowing in the faulted circuit to a magnitude substantially less than that obtainable in the same circuit if the device were replaced with a solid conductor having comparable impedance.

One important circuit characteristic that affects incident energy is the level of arcing current, which is dependent on the available fault current. Limiting the amount of current (energy) that is permitted through an overcurrent device during a faulted condition reduces the amount of available incident energy during an arcing fault. Installing a current-limiting overcurrent protective device is one method to reduce incident energy.

The *NEC* recognizes two levels of overcurrent protection within branch circuits: branch-circuit overcurrent protection and supplementary overcurrent protection. Supplementary devices are always used in addition to the branch circuit overcurrent protective device.

Cutout. An assembly of a fuse support with either a fuseholder, fuse carrier, or disconnecting blade. The fuseholder or fuse carrier may include a conducting element (fuse link), or may act as the disconnecting blade by the inclusion of a nonfusible member.

A cutout is usually associated with protection of a distribution conductor similar to those commonly used in utility distribution systems.

De-energized. Free from any electrical connection to a source of potential difference and from electrical charge; not having a potential different from that of the earth.

The term *de-energized* describes a condition of electrical equipment and should not be used for other purposes. De-energized does not describe a safe condition.

Device. A unit of an electrical system, other than a conductor, that carries or controls electric energy as its principal function. [**70:**100]

Switches, circuit breakers, fuseholders, receptacles, and lampholders that distribute or control, but do not consume, electricity are considered to be devices. Devices that consume incidental amounts of energy, such as a GFCI receptacle with a pilot light, are also considered devices.

Disconnecting Means. A device, or group of devices, or other means by which the conductors of a circuit can be disconnected from their source of supply. [**70:**100]

Disconnecting means can be one or more switches, circuit breakers, or other rated devices used to disconnect electrical conductors from their source of energy. Only a disconnecting means that is load rated should be used to disconnect an operating load.

Disconnecting (or Isolating) Switch (Disconnector, Isolator). A mechanical switching device used for isolating a circuit or equipment from a source of power.

These devices are intended to be operated after interrupting and removing the load current.

Dwelling Unit. A single unit providing complete and independent living facilities for one or more persons, including permanent provisions for living, sleeping, cooking, and sanitation. [**70:**100]

Electrical Hazard. A dangerous condition such that contact or equipment failure can result in electric shock, arc flash burn, thermal burn, or blast.

Informational Note: Class 2 power supplies, listed low voltage lighting systems, and similar sources are examples of circuits or systems that are not considered an electrical hazard.

The arc flash hazard considers only the thermal aspects of an arcing fault. Other known arcing fault hazards include flying parts, molten metal, intense light, poisonous oxides, and generated pressure waves (blasts). Additional electrical hazards might be associated with an arcing fault.

Power limited circuits are not normally considered to be an electrical hazard, and electrical equipment energized at less than 50 volts is not normally considered to be an arc flash hazard. However, the effects of an arcing fault are related to available incident energy. In some instances, an arcing fault hazard might be significant at this lower voltage. If exposure to an electric arc exists, an electrically safe work condition and PPE in accordance with the requirements of Article 130 may be necessary.

Electrical Safety. Recognizing hazards associated with the use of electrical energy and taking precautions so that hazards do not cause injury or death.

Electrical safety is a condition that can be achieved by doing the following:

- Identifying all of the electrical hazards
- Generating a comprehensive plan to mitigate exposure to the hazards
- Providing protective schemes, including training for both qualified and unqualified persons

Electrically Safe Work Condition. A state in which an electrical conductor or circuit part has been disconnected from energized parts, locked/tagged in accordance with established standards, tested to ensure the absence of voltage, and grounded if determined necessary.

Establishing an electrically safe work condition (ESWC) is the only work procedure that ensures that an electrical injury cannot occur. Establishing an ESWC is an example of elimination — one of the controls in the hierarchy of safety controls (see the commentary in Informational Annex P and the inside cover of this handbook for more on the hierarchy of safety controls). Normal operation of a disconnecting means is not in and of itself considered a hazardous activity. See 130.2(A)(4). However, under certain conditions the operation of a disconnecting means and verifying absence of voltage are considered to be activities where the risk of injury from electrical hazards is unacceptable. Until the ESWC is established, an unacceptable risk of injury from the use of electrical energy exists.

In order to work near energized conductors and circuit parts, justification to perform energized work and an energized work permit are required. Otherwise, an ESWC must exist before an employee works within the limited approach boundary or the arc flash boundary. The written energized work permit must be approved by the owner, responsible management, or safety officer.

Enclosed. Surrounded by a case, housing, fence, or wall(s) that prevents persons from accidentally contacting energized parts. [**70:**100]

Enclosed equipment cannot be directly contacted unintentionally by a person. However, conductors or equipment that are enclosed within a fence structure could be touched with the long handle of a tool or with other conductive equipment.

Enclosure. The case or housing of apparatus — or the fence or walls surrounding an installation to prevent personnel from accidentally contacting energized electrical conductors or circuit parts or to protect the equipment from physical damage.

Enclosures are required to be marked with a number that identifies the environmental condition for which they may be used. Commentary Table 100.1, which is *NEC* Table 110.28, summarizes the intended uses of the various standard types of enclosures rated 600 volts nominal or less for nonhazardous locations. Enclosures that comply with the requirements for more than one type may be marked with multiple designations. Enclosures marked with a type can also be marked with the associated term described in the table informational note. For example, Type 1: "Indoor Use Only" or Type 12: "Driptight."

COMMENTARY TABLE 100.1 *Enclosure Selection*

Provides a Degree of Protection Against the Following Environmental Conditions	For Outdoor Use — Enclosure-Type Number									
	3	**3R**	**3S**	**3X**	**3RX**	**3SX**	**4**	**4X**	**6**	**6P**
Incidental contact with the enclosed equipment	X	X	X	X	X	X	X	X	X	X
Rain, snow, and sleet	X	X	X	X	X	X	X	X	X	X
Sleet*	—	—	X	—	—	X	—	—	—	—
Windblown dust	X	—	X	X	—	X	X	X	X	X
Hosedown	—	—	—	—	—	—	X	X	X	X
Corrosive agents	—	—	—	X	X	X	—	X	—	X
Temporary submersion	—	—	—	—	—	—	—	—	X	X
Prolonged submersion	—	—	—	—	—	—	—	—	—	X

Provides a Degree of Protection Against the Following Environmental Conditions	For Indoor Use — Enclosure-Type Number									
	1	**2**	**4**	**4X**	**5**	**6**	**6P**	**12**	**12K**	**13**
Incidental contact with the enclosed equipment	X	X	X	X	X	X	X	X	X	X
Falling dirt	X	X	X	X	X	X	X	X	X	X
Falling liquids and light splashing	—	X	X	X	X	X	X	X	X	X
Circulating dust, lint, fibers, and flyings	—	—	X	X	—	X	X	X	X	X
Settling airborne dust, lint, fibers, and flyings	—	—	X	X	X	X	X	X	X	X
Hosedown and splashing water	—	—	X	X	—	X	X	—	—	—
Oil and coolant seepage	—	—	—	—	—	—	—	X	X	X
Oil or coolant spraying and splashing	—	—	—	—	—	—	—	—	—	X
Corrosive agents	—	—	—	X	—	—	X	—	—	—
Temporary submersion	—	—	—	—	—	X	X	—	—	—
Prolonged submersion	—	—	—	—	—	—	X	—	—	—

*Mechanism shall be operable when ice covered.

Informational Note No. 1: The term *raintight* is typically used in conjunction with Enclosure Types 3, 3S, 3SX, 3X, 4, 4X, 6, and 6P. The term *rainproof* is typically used in conjunction with Enclosure Types 3R, and 3RX. The term *watertight* is typically used in conjunction with Enclosure Types 4, 4X, 6, 6P. The term *driptight* is typically used in conjunction with Enclosure Types 2, 5, 12, 12K, and 13. The term *dusttight* is typically used in conjunction with Enclosure Types 3, 3S, 3SX, 3X, 5, 12, 12K, and 13.

Informational Note No. 2: Ingress protection (IP) ratings may be found in ANSI/NEMA 60529, *Degrees of Protection Provided by Enclosures*. IP ratings are not a substitute for Enclosure Type ratings.

Source: Table 110.28, *NFPA 70*, *National Electrical Code*, 2014 edition, National Fire Protection Association, Quincy, MA.

Energized. Electrically connected to, or is, a source of voltage. [**70:**100]

The term *energized* is associated with all voltage levels, not just those levels that present a shock hazard. Energized equipment or devices are also not limited to those that are connected to a source of electricity. Batteries, capacitors or circuits with induced voltages, or photovoltaic systems must also be considered energized. Electrolytic processes are also considered energized.

Equipment. A general term, including fittings, devices, appliances, luminaires, apparatus, machinery, and the like, used as a part of, or in connection with, an electrical installation. [**70:**100]

Exposed (as applied to energized electrical conductors or circuit parts). Capable of being inadvertently touched or approached nearer than a safe distance by a person. It is applied to electrical conductors or circuit parts that are not suitably guarded, isolated, or insulated.

Some electrical equipment contains conductors that are uncovered and guarded only by the enclosure. Wires and tools inserted through equipment ventilation openings could contact energized conductors; therefore, the level of exposure is determined based on the equipment, task to be performed, and associated tools.

Exposed (as applied to wiring methods). On or attached to the surface or behind panels designed to allow access. [**70:**100]

Wiring methods located above lift-out ceiling panels are considered to be exposed.

Fitting. An accessory such as a locknut, bushing, or other part of a wiring system that is intended primarily to perform a mechanical rather than an electrical function. [**70:**100]

Examples of fittings include condulets, conduit couplings, EMT (electrical metallic tubing) couplings, and threadless connectors.

Fuse. An overcurrent protective device with a circuit-opening fusible part that is heated and severed by the passage of overcurrent through it.

Informational Note: A fuse comprises all the parts that form a unit capable of performing the prescribed functions. It may or may not be the complete device necessary to connect it into an electrical circuit.

Fuses are components that act directly on the current flowing in the electrical circuit. Fuse action does not depend on generating or receiving a signal from another circuit element. Exhibit 100.8 shows examples of various classes of fuses rated 600 volts or below. Note the plainly marked barrels.

EXHIBIT 100.8

Various classes of fuses for use in systems rated 600 volts nominal and less. (Courtesy of Cooper Bussmann, a division of Cooper Industries, PLC)

Ground. The earth. [**70:**100]

The term *ground* is commonly used in North America. For installations within the scope of the *NEC*, ground literally refers to the earth, which is used as a reference point to establish circuit potential. In some electronic circuits, the term is used to describe a reference point, which may not be ground (earth).

Ground Fault. An unintentional, electrically conducting connection between an ungrounded conductor of an electrical circuit and the normally non–current-carrying conductors, metallic enclosures, metallic raceways, metallic equipment, or earth.

Any fault is unintentional and normally unexpected. A fault results in an electrical current being imposed in an unintended circuit, which could include a person. The primary purpose for grounding an electrical system is to provide a path for fault current that excludes a person.

Two types of faults can occur in an electrical circuit — a bolted fault and an arcing fault — and each type results in different hazardous conditions. An example of a bolted fault is a de-energized circuit with safety grounds installed for maintenance purposes, which is then re-energized with the safety grounds still in place. Most electrical equipment is tested under bolted-fault conditions. The most common type of fault is an arcing fault, which might result from a conductive object falling into the circuit or from a component failure. Unless the equipment is rated as arc resistant, it is not tested under arcing fault conditions.

Bolted faults and arcing faults exhibit different characteristics and present different hazards. The primary hazard associated with a bolted fault is shock or electrocution. If a bolted fault exists too long, the mechanical force or pressure produced by the current can cause conductors to move significantly, resulting in an arcing fault. On the other hand, an arcing fault normally is associated with a thermal and physical hazard. Arcing faults also can present a shock or electrocution hazard.

Grounded (Grounding). Connected (connecting) to ground or to a conductive body that extends the ground connection. [**70:**100]

Grounded, Solidly. Connected to ground without inserting any resistor or impedance device. [**70:**100]

Grounded Conductor. A system or circuit conductor that is intentionally grounded. [**70:**100]

In most instances, one conductor of an electrical circuit is intentionally connected to earth. That conductor is the grounded conductor. A grounded system results in the overcurrent protective device functioning in response to the first occurrence of a ground fault in the circuit. An equipment grounding conductor is not a grounded conductor.

Ground-Fault Circuit Interrupter (GFCI). A device intended for the protection of personnel that functions to de-energize a circuit or portion thereof within an established period of time when a current to ground exceeds the values established for a Class A device. [**70:**100]

> Informational Note: Class A ground-fault circuit-interrupters trip when the current to ground is 6 mA or higher and do not trip when the current to ground is less than 4 mA. For further information, see ANSI/UL 943, *Standard for Ground-Fault Circuit Interrupters.*

GFCI devices operate quickly in the 4–6 mA range to minimize the chance of electrocution or serious shock injury. GFCI protection is required by *NEC* Article 590 for all temporary installations involving construction, remodeling, maintenance, repair or demolition

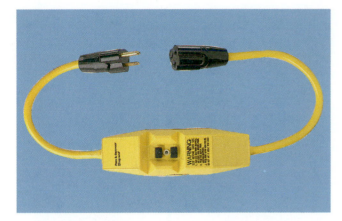

EXHIBIT 100.9

A raintight GFCI with open neutral protection that is designed for use on the line end of a flexible cord. (Courtesy of Pass & Seymour/Legrand®)

EXHIBIT 100.10

A temporary power outlet unit commonly used on construction sites with a variety of configurations, including GFCI protection. (Courtesy of Hubbell Wiring Device–Kellems)

EXHIBIT 100.11

A watertight plug and connector used to prevent tripping of GFCI protective devices in wet or damp weather. (Courtesy of Hubbell Wiring Device–Kellems)

of buildings, structures, or equipment. However, using a GFCI device wherever employees use cord- and-plug-connected equipment is a best practice.

Exhibits 100.9 through 100.12 show examples of ways to implement the ground-fault circuit-interrupter requirements specified in *NEC* 590.6(A) for temporary installations.

Grounding Conductor, Equipment (EGC). The conductive path(s) that provides a ground-fault current path and connects normally non–current-carrying metal parts of equipment together and to the system grounded conductor or to the grounding electrode conductor, or both. [**70:**100]

> Informational Note No. 1: It is recognized that the equipment grounding conductor also performs bonding.

EXHIBIT 100.12

A 15-ampere duplex receptacle with integral GFCI that also protects downstream loads. (Courtesy of Pass & Seymour/ Legrand®)

Informational Note No. 2: See 250.118 of *NFPA 70*, *National Electrical Code*, for a list of acceptable equipment grounding conductors.

Under normal conditions a person typically must be exposed to a potential difference of 50 volts or more for a shock hazard to exist. One of the primary purposes for an equipment grounding conductor (EGC) is to reduce the potential difference (voltage) between surfaces that an employee is likely to touch. An effective ground-fault current path reduces the possibility that a potential difference of 50 volts or more can exist, thereby minimizing the chance of shock injury or electrocution.

In a solidly grounded electrical system, the EGC also provides an effective ground-fault current path that facilitates the operation of circuit overcurrent protective devices when a ground fault occurs. This function of the EGC is extremely important in minimizing the duration of a ground fault in a circuit.

Grounding Electrode. A conducting object through which a direct connection to earth is established. [**70:**100]

Grounding Electrode Conductor. A conductor used to connect the system grounded conductor or the equipment to a grounding electrode or to a point on the grounding electrode system. [**70:**100]

Guarded. Covered, shielded, fenced, enclosed, or otherwise protected by means of suitable covers, casings, barriers, rails, screens, mats, or platforms to remove the likelihood of approach or contact by persons or objects to a point of danger. [**70:**100]

It is unlikely that people or objects will contact equipment, circuits, or conductors that are guarded. Although guarding provides protection from exposure to shock or electrocution, it is still possible for persons to be exposed to arc flash hazards. The fenced-in substation shown in Exhibit 100.13 effectively removes the likelihood of approach or contact by persons.

Hazard. A source of possible injury or damage to health.

EXHIBIT 100.13

An equipment enclosure. This equipment is said to be guarded.

The source of this definition is NFPA 79, *Electrical Standard for Industrial Machinery*. Hazards can refer to all aspects of technology, environmental factors, and activities that create risk. In addition, ANSI/AIHA Z10, *American National Standard for Occupational Health and Safety Management Systems*, defines a hazard as a condition, set of circumstances, or inherent property that can cause injury, illness, or death.

Hazardous. Involving exposure to at least one hazard.

Incident Energy. The amount of thermal energy impressed on a surface, a certain distance from the source, generated during an electrical arc event. Incident energy is typically expressed in calories per square centimeter (cal/cm²).

Incident energy calculations are based on thermal energy only. ASTM standards require PPE ratings to be in calories per square centimeter rather than joules per square centimeter, joules per meter squared, or calories per square inch. Regardless of the unit of measurement, selection of proper PPE dictates that the incident energy and PPE thermal rating be expressed using the same term.

Predicting the amount of available incident energy is crucial in selecting appropriate PPE. Properly rated PPE prevents injury from melting or burning clothing or from direct skin exposure due to the increased temperature during an arcing fault. Using PPE rated above the calculated incident energy value can raise the probability of being protected. PPE ratings are based on a 50 percent probability that a second degree burn will not occur, so arc-rated equipment that is used at less than its rating offers added chance of protection. See the commentary to Informative Annex M and the associated Exhibit M.3.

Incident Energy The word *incident* in "incident energy" is used as an adjective to mean falling on or striking on a surface — here, the surface area of the human body in the direction of the energy flow. Power is the rate at which work is performed. *Irradiance* is the power incident on a surface — also called *radiant flux density*. *Radiant exposure (fluence)* is the energy delivered in a specific time interval per unit area. *Fluence* is one of the fundamental units in radiation dosimetry. Therefore, what we really mean when we talk about *incident energy* is *incident energy fluence* or *incident energy radiant exposure*. However, a more accurate and congruent term that could be acceptable to the general users of *NFPA 70E* might be *incident energy exposure*.

Incident Energy Analysis. A component of an arc flash risk assessment used to predict the incident energy of an arc flash for a specified set of conditions.

The calculated or computed incident energy from the installation-specific assessment is the predicted energy that an employee will be exposed if an arc flash incident occurs. This focused information allows the selection of PPE based on the conditions associated with the task being performed on a specific piece of electrical equipment. See the commentary to Informative Annex M and the associated Exhibit M.3.

Insulated. Separated from other conducting surfaces by a dielectric (including air space) offering a high resistance to the passage of current.

> Informational Note: When an object is said to be insulated, it is understood to be insulated for the conditions to which it is normally subject. Otherwise, it is, within the purpose of these rules, uninsulated.

Interrupter Switch. A switch capable of making, carrying, and interrupting specified currents.

Interrupter switches are a disconnecting means rated to interrupt load current. The manufacturer's product label typically provides the interrupting rating of the switch.

Interrupting Rating. The highest current at rated voltage that a device is identified to interrupt under standard test conditions. [**70:**100]

> Informational Note: Equipment intended to interrupt current at other than fault levels may have its interrupting rating implied in other ratings, such as horsepower or locked rotor current.

The interrupting rating is generally expressed in rms symmetrical amperes and is specified by a fault current magnitude only. One of the standard conditions is the power factor (PF) or reactance to resistance (X/R) ratio of the test circuit. PF is related to X/R ratio by the following equation: PF = cos(tan−1 X/R). Where instantaneous phase trip elements are used, the interrupting capacity is the maximum rating of the device with no intentional delay. Where instantaneous phase trip elements are not used, the interrupting capacity is the maximum rating of the device for the rated time interval. If the X/R ratio of the fault circuit is greater than the test X/R ratio, the interrupting rating has to be adjusted by a process called derating, which is essentially multiplying the interrupting rating by a factor less than one.

Circuit breakers also have a short-time current rating, which is the ability of the breaker to remain closed for a time interval under high fault current conditions. The short-time rating is used to determine the ability of the circuit breaker to protect itself and other devices and to coordinate with other circuit breakers so the system will trip selectively.

Isolated (as applied to location). Not readily accessible to persons unless special means for access are used. [**70:**100]

Isolated conductors, devices, or equipment are protected from accidental contact by an unqualified person because a special means, such as a tool or key, is required to gain access. Standard tools such as Phillips head screwdrivers are not considered to be a special means.

Labeled. Equipment or materials to which has been attached a label, symbol, or other identifying mark of an organization that is acceptable to the authority having jurisdiction and concerned with product evaluation, that maintains periodic inspection of production of labeled equipment or materials, and by whose labeling the manufacturer indicates compliance with appropriate standards or performance in a specified manner.

Labeled equipment, devices, or conductors have an identifying mark to indicate that it meets the requirements defined by the appropriate standards. These marks are most often a third party symbol used to indicate not only that an evaluated sample has complied with the standard but that production units continue to comply. The mark must be of an organization acceptable to the AHJ. Labeled equipment must be approved and used in accordance with instructions supplied as part of the labeling or listing. See Exhibit 100.14.

EXHIBIT 100.14

Circuit breaker showing a label specifying limits of its use.

Listed. Equipment, materials, or services included in a list published by an organization that is acceptable to the authority having jurisdiction and concerned with evaluation of products or services, that maintains periodic inspection of production of listed equipment or materials or periodic evaluation of services, and whose listing states that either the equipment, material, or service meets appropriate designated standards or has been tested and found suitable for a specified purpose.

Informational Note: The means for identifying listed equipment may vary for each organization concerned with product evaluation; some organizations do not recognize equipment as listed unless it is also labeled. The authority having jurisdiction should utilize the system employed by the listing organization to identify a listed product.

Listing is the most common method of third-party evaluation of the safety of a product. These organizations evaluate equipment in accordance with appropriate product standards and maintain a list of products that comply. Listed equipment has a label with the mark of the organization that performed the evaluation. NFPA codes and standards do not use the OSHA term *nationally recognized testing laboratory* (NRTL) because the acceptance of the listing organization is the responsibility of the AHJ. Equipment must be used within the parameters of the listing. Even when not required by this standard, listed equipment is often employed to facilitate approval.

Luminaire. A complete lighting unit consisting of a light source, such as a lamp or lamps, together with the parts designed to position the light source and connect it to the power supply. It may also include parts to protect the light source or the ballast or to distribute the light. A lampholder itself is not a luminaire. [**70:**100]

Motor Control Center. An assembly of one or more enclosed sections having a common power bus and principally containing motor control units. [**70:**100]

A motor control center typically contains combination motor starters, fusible safety-disconnect switches, circuit breakers, power panels, solid-state drives, and similar components. A motor control center is not to be confused with switchgear or a switchboard. It is important to understand the difference when using Table 130.7(C)(15)(A)(b) and Table 130.7(C)(15)(B), because the arc flash hazard PPE categories are based on the type of equipment and the task to be performed.

Outlet. A point on the wiring system at which current is taken to supply utilization equipment. [**70:**100]

This term is frequently misused to refer to receptacles. Although a receptacle is an outlet, not all outlets are receptacles. Other types of outlets include lighting outlets, smoke alarm outlets, and motor outlets.

Overcurrent. Any current in excess of the rated current of equipment or the ampacity of a conductor. It may result from overload, short circuit, or ground fault. [**70:**100]

> Informational Note: A current in excess of rating may be accommodated by certain equipment and conductors for a given set of conditions. Therefore, the rules for overcurrent protection are specific for particular situations.

Overload. Operation of equipment in excess of normal, full-load rating, or of a conductor in excess of rated ampacity that, when it persists for a sufficient length of time, would cause damage or dangerous overheating. A fault, such as a short circuit or ground fault, is not an overload. [**70:**100]

Panelboard. A single panel or group of panel units designed for assembly in the form of a single panel, including buses and automatic overcurrent devices, and equipped with or without switches for the control of light, heat, or power circuits; designed to be placed in a cabinet or cutout box placed in or against a wall, partition, or other support; and accessible only from the front. [**70:**100]

Premises Wiring (System). Interior and exterior wiring, including power, lighting, control, and signal circuit wiring together with all their associated hardware, fittings, and wiring devices, both permanently and temporarily installed. This includes: (a) wiring from the service point or power source to the outlets; or (b) wiring from and including the power source to the outlets where there is no service point.

Such wiring does not include wiring internal to appliances, luminaires, motors, controllers, motor control centers, and similar equipment. [**70:**100]

> Informational Note: Power sources include, but are not limited to, interconnected or stand-alone batteries, solar photovoltaic systems, other distributed generation systems, or generators.

Premises wiring includes all wiring downstream of the service point to the outlets. However, a premises wiring system does not have to be supplied by an electric utility. For example, a generator or photovoltaic system can supply a stand-alone premises wiring system. The premises wiring for these systems includes the power source as well as the wiring to the outlets.

Qualified Person. One who has demonstrated skills and knowledge related to the construction and operation of electrical equipment and installations and has received safety training to identify and avoid the hazards involved.

A qualified employee must understand the construction and operation of the equipment or circuit associated with the planned work task. An employee could be qualified to perform one work task and not qualified to perform a different task. An employee could also be qualified to work on one piece of equipment but not another similar piece of equipment.

Many state and local government licensing programs have training requirements that must be met for a person to be considered qualified. The applicant must be examined initially and then periodically after procuring a license. The license in and of itself does not make a person qualified for all tasks or equipment that may be encountered. Electrical work requires continuing education and demonstration of the necessary skills in order to maintain the requisite skill level to work safely. A qualified person must understand electrical hazards associated with the scheduled work task and must be trained to understand and apply the details of the electrical safety program and procedures provided by the employer. A qualified person should be able to understand an arc flash risk assessment and the proper application and the limitations of PPE.

Part of being a qualified person is recognizing that energized electrical work is permitted only under the conditions specified in 130.2(A). A qualified person must have the ability to recognize all electrical hazards that might be associated with the work task being considered and must be able to react appropriately to all hazards associated with the task. The qualified person must understand the limitations of test equipment, such as voltage testers, how to select appropriate equipment, and how to apply that equipment to the planned work task.

The OSHA definition for qualified person (29 CFR 1910.399) includes the phrase *has demonstrated skills*. This requires that the person actually demonstrate the ability to perform the task. It may be necessary to demonstrate the ability to perform the task while using appropriate PPE to ensure that the restricted lighting and field of view of the flash suit hood or the dexterity limitations of voltage-rated gloves with leather protectors do not hinder the employee.

This definition promotes the consistent use throughout the document of terminology associated with hazard and risk, and aligns with OSHA's definition.

Raceway. An enclosed channel of metal or nonmetallic materials designed expressly for holding wires, cables, or busbars, with additional functions as permitted in this standard. [**70:**100]

Receptacle. A receptacle is a contact device installed at the outlet for the connection of an attachment plug. A single receptacle is a single contact device with no other contact device on the same yoke. A multiple receptacle is two or more contact devices on the same yoke. [**70:**100]

This term is frequently misused. A receptacle is a device installed at an outlet. However, not all outlets are receptacle outlets.

Risk. A combination of the likelihood of occurrence of injury or damage to health and the severity of injury or damage to health that results from a hazard.

Risk is not only the likelihood that an incident could occur, but it also includes the severity (seriousness) of the resulting injury or damage that may result from the hazard. A risk

Case Study

Scenario and Causation.

Mark, an electrician, and John, a helper with no formal electrical training, were installing a new three-phase circuit between an existing, energized 480-volt panelboard and a new piece of machinery. During the process, Mark attempted to install a missing bolt from a circuit breaker mount onto an energized busbar. Mark used protective insulating gloves with leather protectors but did not use any other PPE. While attempting to screw the bolt into place, he lost control of the bolt. Either the bolt or the mount then contacted another busbar phase, resulting in an arc flash. John was standing some distance away when the incident happened. While John was demonstrating to the facility owner what had happened with the arc flash incident involving Mark, a second electrical arc flash occurred at the panelboard. The cause of the second arc flash is unknown. Fortunately the facility owner was standing some distance away and was not hurt.

Results.

Mark temporarily lost his sight and received first- and second-degree burns to his head, face, and forearm. An arc flash event can produce extremely intense bright light and arc flash blindness, resulting from the ultraviolet light emitted by vaporizing metal, which can be temporary or permanent. John was not wearing any PPE at the time of his incident. He received second-degree burns to his head, face, neck, arm, and hand. The nylon jacket he was wearing caught fire and melted to his skin causing third-degree burns. John required surgery for removal of destroyed skin and restorative skin grafts.

Analysis.

Accidents typically occur due to a chain of events or mistakes that lead up to the actual event. If one of the links in the chain of events is eliminated, the accident most likely would not have happened.

There are no facts presented to suggest that it was either infeasible or that a greater hazard would have been created if the power to the panelboard had been turned off to establish an electrically safe work condition for performing this task. The facts do suggest that neither Mark nor John were qualified to perform the tasks to which they had been assigned, and it cannot be ascertained whether either had received any electrical safety training. Therefore, neither Mark nor John should have been within the limited approach boundary or the arc flash boundary of the energized electrical conductors and circuit parts within the panelboard. That both men were unqualified for the task they were assigned and that the equipment being worked on was energized are direct causes of this incident. If an electrically safe work condition had been established prior to the attempted installation of the circuit breaker, this accident would not have occurred.

Relevant *NFPA 70E* Requirements.

If Mark, John, and the company they worked for had followed all of the requirements of *NFPA 70E*, they might not have been injured. Some of the steps they should have taken include the following:

- Create an electrical safety program per 110.1.
- Train employees to be qualified to recognize and avoid the electrical hazard per 110.2.

- Create an electrically safe work condition (ESWC) by disconnecting the power to the panelboard per 130.2.
- Verify that an ESWC has been achieved per 120.1.

If it could have been demonstrated that working within the energized equipment was acceptable per 130.2(A)(1) or (A)(2), the following additional steps needed to be implemented in order to ensure that the employees are protected against electrical hazards:

- Perform a shock risk assessment per 130.4(A) and arc flash risk assessment per 130.5.
- Perform risk assessment procedure of this task per 110.1(G).
- Create an Energized Electrical Work Permit (EWP) per 130.2(B).
- Use proper PPE for protection against shock and arc flash hazards per 130.7.
- Install protective shields or barriers, rubber insulating equipment, or voltage-rated plastic guard equipment per 130.7(D)(1)(6), (7), or (8).

could have a high probability of an incident occurring, but the result could be a minor injury. Conversely, it may have a low likelihood of occurring but present the potential for severe injury. See the commentary on Informative Annex F and Informative Annex P for more information.

Risk Assessment. An overall process that identifies hazards, estimates the potential severity of injury or damage to health, estimates the likelihood of occurrence of injury or damage to health, and determines if protective measures are required.

Informational Note: As used in this standard, *arc flash risk assessment* and *shock risk assessment* are types of risk assessments.

In accordance with the new definition of risk, former 110.1(F) was renumbered as 110.1(G) and retitled from "Hazard Identification and Risk Assessment Procedure" to "Risk Assessment Procedure," because risk assessment now includes hazard identification. Risk assessment is about risk mitigation. Implementing risk control involves application of the safety controls from the hierarchy of safety controls in accordance with the process identified in Informative Annex F and Informative Annex P (see also the inside front cover of this handbook).

Service Drop. The overhead conductors between the utility electric supply system and the service point. [70:100]

The *NEC* contains requirements covering the interface (service point) between the utility conductors (service-drop conductors) and the conductors (overhead service conductors) covered by the *Code*. Conductors on the load side of the service point are overhead service conductors, or overhead service-entrance conductors.

Service Lateral. The underground conductors between the utility electric supply system and the service point. [70:100]

Service Point. The point of connection between the facilities of the serving utility and the premises wiring. [70:100]

Informational Note:　The service point can be described as the point of demarcation between where the serving utility ends and the premises wiring begins. The serving utility generally specifies the location of the service point based on the conditions of service.

The service point is the point of demarcation on the wiring system where the serving utility ends and the premises wiring begins. It is the transition point where the premises wiring system begins and the rules of *NFPA 70E* and the *NEC* apply. Only conductors physically located on the premises wiring side (load side) of the service point are covered by the *NEC*. The location of the service point generally is determined by the serving utility and may vary from utility to utility and among different types of occupancies. The service point might be at the transformer or substation for an industrial installation, or at the weatherhead and drip loop for a small commercial occupancy. Exhibit 100.15 is one example of a service point location.

EXHIBIT 100.15

Service point.

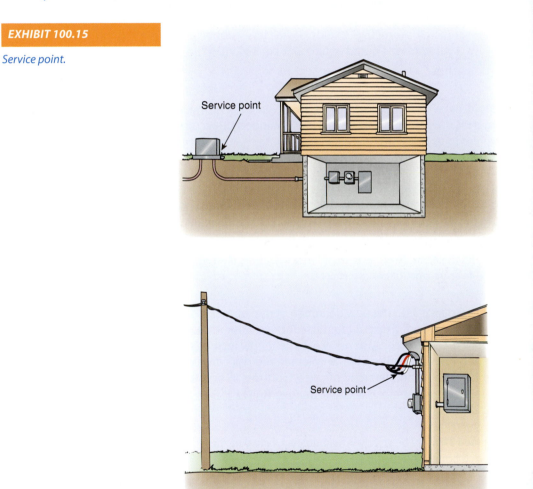

Shock Hazard. A dangerous condition associated with the possible release of energy caused by contact or approach to energized electrical conductors or circuit parts.

A person's tolerance to an electrical current through their body varies depending on the path of the current through the body and on the applied voltage. The tolerated level also varies from person to person. The perception of a 60 Hz alternating electrical current occurring near 1 mA and 5 mA is a typical level considered to cause an involuntary reaction. Pain may be felt at a slightly higher current, and involuntary muscle contractions may occur around 20 mA. Direct current and higher frequencies have an effect on tolerance levels. Although exposure to any energized component may present a shock hazard, a shock hazard is generally considered to exist at a voltage of 50 volts or greater. See also 340.5.

Short-Circuit Current Rating. The prospective symmetrical fault current at a nominal voltage to which an apparatus or system is able to be connected without sustaining damage exceeding defined acceptance criteria. [**70:**100]

The short-circuit current rating is marked on equipment, such as the one shown in Exhibit 100.16. Wire, bus structures, switching, protection and disconnect devices, and distribution equipment will be damaged or destroyed if their short-circuit ratings are exceeded.

EXHIBIT 100.16

Short-circuit rating clearly labeled on equipment.

The basic purpose of overcurrent protection is to open the circuit before equipment and conductors or conductor insulation is damaged when an overcurrent condition occurs as the result of an overload, a ground fault, or a short. But merely providing overcurrent protective devices with sufficient interrupting ratings does not ensure adequate short-circuit protection for the equipment. Overcurrent protective devices should be selected to ensure that the short-circuit current rating of the components are not exceeded should a short circuit or high-level ground fault occur. The overcurrent protective device must limit the let-through energy to within the short-circuit current rating of the electrical components. Adequate short-circuit protection can be provided by fuses, molded-case circuit breakers, and low-voltage power circuit breakers, depending on specific circuit and installation requirements.

The short-circuit current rating (SCCR) is based on the actual symmetrical (rms) fault current not exceeding the SCCR, the period of time the device was tested for, or the X/R ratio of the test circuit. The SCCR is the rms current that can be withstood for a period of time where the X/R of the fault circuit does not exceed that of the test circuit. It could be 3 cycles, 15 cycles, 30 cycles or some other time period depending upon the standard to which the equipment or component was tested.

Utility companies usually determine and provide information on available short-circuit current levels at the service equipment. Literature on how to calculate short-circuit currents

Defined Acceptance Criteria *Defined acceptance criteria* includes the time period that the current can be withstood and the power factor (PF) or reactance to resistance (X/R) ratio of the test circuit. PF is related to X/R ratio by the following equation: PF = cos(tan−1 X/R). If the X/R ratio of the fault circuit is greater than the test X/R ratio, the short-circuit current rating (SCCR) has to be adjusted by a process called derating, which is essentially multiplying the SCCR by a factor less than one. SCCR was formerly known as "withstand rating," and some other standards may still use that terminology.

at each point in any electrical distribution system can be obtained by contacting the manufacturers of overcurrent protective devices or by referring to IEEE 141-1993 (R1999), *IEEE Recommended Practice for Electric Power Distribution for Industrial Plants* (Red Book). The IEEE Red Book is being replaced with the IEEE 3001 *Standards: Power Systems Design.* As of the drafting of this commentary, there is an approved draft of IEEE 3001.5, *IEEE Recommended Practice for Application of Power Distribution Apparatus in Industrial and Commercial Power Systems.*

Single-Line Diagram. A diagram that shows, by means of single lines and graphic symbols, the course of an electric circuit or system of circuits and the component devices or parts used in the circuit or system.

A single-line diagram illustrates a complete system or a portion of a system using graphic symbols. Recorded copies of this drawing should be marked by red-lining or other method to illustrate all changes made to the system. The availability of an up-to-date and accurate single-line diagram that identifies sources of energy and locations of disconnecting means is key to implementing procedures for establishing an electrically safe work condition. Exhibit 205.1 in Chapter 2 shows an example of a single-line diagram. (See Figure D.2.2 in Informative Annex D for another example.)

Special Permission. The written consent of the authority having jurisdiction. [**70:**100]

Step Potential. A ground potential gradient difference that can cause current flow from foot to foot through the body.

Any current path through the body other than from foot to foot is touch potential. Exhibit 100.17 illustrates the difference between step and touch potential. Although the insulating quality of voltage-rated gloves or footwear may be one component of a system to mitigate step and touch potential hazards, these cannot be used as the primary means of employee protection. See 130.7(C)(8) for more information on foot protection.

Structure. That which is built or constructed. [**70:**100]

EXHIBIT 100.17

Step and touch potential.

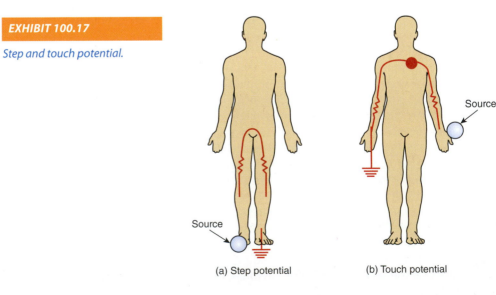

(a) Step potential   (b) Touch potential

Switch, Isolating. A switch intended for isolating an electric circuit from the source of power. It has no interrupting rating, and it is intended to be operated only after the circuit has been opened by some other means. [**70:**100]

Isolating switches, while not intended nor rated to open a circuit under load, can provide a means to install lockout devices after the load current has been removed.

Switchboard. A large single panel, frame, or assembly of panels on which are mounted on the face, back, or both, switches, overcurrent and other protective devices, buses, and usually instruments. These assemblies are generally accessible from the rear as well as from the front and are not intended to be installed in cabinets. [**70:**100]

Switchgear or motor control centers do not meet this definition. However, many electrical employees apply the term *switchgear* to switchboards and motor control centers. It is important to understand the difference when using Table 130.7(C)(15)(A)(b) and Table 130.7(C)(15)(B), because the arc flash hazard PPE categories are based on the type of equipment being serviced and the task to be performed.

Switchgear, Arc-Resistant. Equipment designed to withstand the effects of an internal arcing fault and that directs the internally released energy away from the employee.

Only switchgear specifically identified as being arc resistant provides protection from internal arcing fault when the equipment is closed and operating normally. If doors and covers (including fasteners) are not completely closed, workers are exposed to the hazards associated with an arcing fault just as if no arc-resistant rating existed.

Switchgear, Metal-Clad. A switchgear assembly completely enclosed on all sides and top with sheet metal, having drawout switching and interrupting devices, and all live parts enclosed within grounded metal compartments.

Switchgear, Metal-Enclosed. A switchgear assembly completely enclosed on all sides and top with sheet metal (except for ventilating openings and inspection windows), containing primary power circuit switching, interrupting devices, or both, with buses and connections. This assembly may include control and auxiliary devices. Access to the interior of the enclosure is provided by doors, removable covers, or both. Metal-enclosed switchgear is available in non-arc-resistant or arc-resistant constructions.

Switching Device. A device designed to close, open, or both, one or more electric circuits.

A switching device can be any device that is rated for this use. Splicing devices located in mid-conductor sometimes are rated for use as a disconnecting device and therefore are acceptable as switching devices.

Touch Potential. A ground potential gradient difference that can cause current flow from hand to hand, hand to foot, or another path, other than foot to foot, through the body.

Any current path between two points on a body, other than from foot to foot, is called touch potential. (See Exhibit 100.17 and the commentary following the definition of the term *step potential*.)

Ungrounded. Not connected to ground or to a conductive body that extends the ground connection. [**70:**100]

Unqualified Person. A person who is not a qualified person.

Any person who has not received the specific training to perform a task, to recognize that an electrical hazard exists and how to avoid that hazard, and who has not shown demonstrated ability is an unqualified person. An employee qualified to perform a specific task may be unqualified to perform other tasks. The characteristics of being qualified and unqualified are task dependent.

Utilization Equipment. Equipment that utilizes electric energy for electronic, electromechanical, chemical, heating, lighting, or similar purposes. [**70**:100]

•

Voltage (of a Circuit). The greatest root-mean-square (rms) (effective) difference of potential between any two conductors of the circuit concerned. [**70**:100]

> Informational Note: Some systems, such as three-phase 4-wire, single-phase 3-wire, and 3-wire direct-current, may have various circuits of various voltages.

Voltage, Nominal. A nominal value assigned to a circuit or system for the purpose of conveniently designating its voltage class (e.g., 120/240 volts, 480Y/277 volts, 600 volts). [**70**:100]

> Informational Note No. 1: The actual voltage at which a circuit operates can vary from the nominal within a range that permits satisfactory operation of equipment.

> Informational Note No. 2: See ANSI C84.1, *Electric Power Systems and Equipment — Voltage Ratings (60 Hz).*

Working On (energized electrical conductors or circuit parts). Intentionally coming in contact with energized electrical conductors or circuit parts with the hands, feet, or other body parts, with tools, probes, or with test equipment, regardless of the personal protective equipment (PPE) a person is wearing. There are two categories of "working on": *Diagnostic (testing)* is taking readings or measurements of electrical equipment with approved test equipment that does not require making any physical change to the equipment; *repair* is any physical alteration of electrical equipment (such as making or tightening connections, removing or replacing components, etc.).

A person required to perform a task involving intentional contact with an energized electrical conductor or circuit part is considered to be working on the energized part. The person is subject to all associated requirements, including the selection of the appropriate level of PPE. Regardless of the use of PPE, measuring voltage with probes is an example of a diagnostic task involving working on energized parts that exposes a worker to an electrical hazard.

Two distinctly different types of tasks are included within this definition: diagnostic testing and repair. The two categories suggest that different procedural approaches may be in order depending on the complexity of the task and the employee's exposure to electrical hazards.

Article 105 Application of Safety-Related Work Practices

Article 105 establishes the overall scope, purpose, and organization of Chapter 1. It differentiates between the responsibilities of the employer and those of the employee(s). The employer has the responsibility of providing safety-related work procedures, training employees in the practices, supervising the employees, auditing, and documenting. The employees are responsible for applying the work procedures in accordance with their training and demonstrated ability.

OSHA's electrical safety standards are derived from *NFPA 70E*. Since *NFPA 70E* is the American National Standard for electrical safety in the workplace, it sets the minimum consensus requirements for safe electrical work procedures. *NFPA 70E* establishes the minimum standard of care. OSHA's electrical safety standards are mainly performance based, whereas *NFPA 70E* is mainly prescription based. *NFPA 70E* fleshes out how the performance-based requirements in the OSHA standards should be met by providing and defining the minimum standard industry practices necessary for electrical safety. One could say that OSHA is the law, but *NFPA 70E* outlines how to comply with OSHA's electrical safety requirements.

> **OSHA Standards and the De Minimis Violation** OSHA's electrical safety standards were derived from an early edition of *NFPA 70E*. OSHA attempts to keep its standards' language flexible and simple, to allow for future revisions of referenced standards. Because sometimes the standard referenced is revised on a different timeline than the OSHA standard referencing it, it is OSHA's policy to issue a de minimis violation when "an employer complies with a . . . consensus standard rather than with the [OSHA] standard in effect at the time of the inspection and the employer's action clearly provides equal or greater employee protection or the employer complies with a written interpretation issued by the OSHA Regional or National Office."

105.1 Scope.

Chapter 1 covers electrical safety-related work practices and procedures for employees who are exposed to an electrical hazard in workplaces covered in the scope of this standard.

Chapter 1 outlines the electrical safety procedures necessary to protect employees from the dangers inherent with the use of electricity. The chapter also defines the responsibilities for creating an electrical safety program (ESP) and implementing the procedures included in the program. For covered workplaces, as defined in 90.2(A), an electrical safety procedure should include all types of employees who interact with systems, equipment, components, or parts that use electricity. The applicability of this standard does not depend on the employee's craft; on whether the facility is classified as residential, commercial, institutional, or industrial; on whether the employee is qualified or not; on the business of the employer; or on the employees' level of training. What does depend on one or more of these factors, however, is the applicability of the specific work procedures involved and identified for use in this standard, and the particular ESP developed as required by this standard. The fundamentals identified in this standard are intended to protect all types of employees in workplaces identified. However, the fundamentals identified in this standard are also intended to protect employees in workplaces excluded under 90.2(B), if the employer chooses to adopt the work procedures.

Unacceptable risk of injury is present from the inherent properties associated with the use of electricity when the safety controls employed from the hierarchy of safety controls are insufficient to protect the employee(s) involved with the task at hand in the opinion of the authority having jurisdiction. As OSHA pointed out in 1986 in the preamble to their electrical safety requirements for the construction industry, ". . . electricity has long been recognized as a serious workplace hazard, exposing employees to such dangers as electric shock, electrocution, fires, and explosions." (See 51 FR 25295.) When protections are insufficient, injuries associated with the use of electricity are not limited to unique employees, specific industries, or a particular segment of specific industry. For work locations covered by *NFPA 70E*, no industry segment or work discipline is omitted.

105.2 Purpose.

These practices and procedures are intended to provide for employee safety relative to electrical hazards in the workplace.

Informational Note: For general categories of electrical hazards, see Informative Annex K.

Employees and others who interact with energized electrical conductors and circuit parts in the workplace that are not considered to be under normal operating conditions may be exposed to unacceptable risk from the use of electricity. In fact, when electrical equipment is not properly installed or maintained, even the risk associated with normal operation of electrical equipment may be unacceptable.

The Circle of Safety

As explained in the scope, *NFPA 70E* deals with electrical safety-related work practices and procedures, the *NEC* deals with safe electrical equipment installation practices, and NFPA 70B deals with proper maintenance of electrical equipment. *NFPA 70E* stipulates necessary electrical safety-related practices and procedures, which reduce the risk associated with the use of electricity to an acceptable level for equipment that has been installed in accordance with the *NEC* and maintained in accordance with the manufacturer's recommendations or NFPA 70B. Furthermore, information on maintenance of electrical equipment necessary for electrical safety can be found in Chapter 2. NFPA 70B also provides safety-related requirements for equipment that has not been properly maintained or installed, as well as safety-related work practices for some cases where the electrical equipment has been installed in accordance with an installation code other than the *NEC*.

The requirements of *NFPA 70E* are suitable for use in situations where the potential harm from the use of electricity poses unacceptable risk. This standard specifies the electrical safety-related work procedures required to protect all types, categories, and classes of employees and others — whether or not they are qualified or unqualified to interact with energized electrical equipment — under normal or other than normal operating conditions. *Normal operation* is defined in 130.2(A)(4).

105.3 Responsibility.

The employer shall provide the safety-related work practices and shall train the employee, who shall then implement them.

Employee electrical safety requires a collaborative effort between workers and management. Section 105.3 defines the major division of responsibilities between employer and employee. Employers are required to have an electrical safety program (ESP) for employees that includes the requirements identified in 110.2 and 110.3. The employees are required to put into practice policies and procedures of the ESP, including use of required tools and safety equipment, which includes the training so as to be able to perform the task at hand safely. The best results are achieved by collaboration and cooperation between labor and management, including the use of cross-functional teams and outside experts as determined necessary.

Development and implementation of ESPs and their work procedures requires the knowledge, expertise, and commitment of both management and labor. Management must provide the assets needed. Also, management coordinates planning, scheduling, and best methods of meeting deadlines. The required tasks are normally performed by employees, because employees tend to have familiarity and expertise in the use of the actual systems and equipment. Sometimes the required expertise is lacking and it has to be developed. It is in the organization's interest and the interest of the individuals involved that each and every employee goes home safe and sound. Commitment of both parties is necessary and required.

105.4 Organization.

Chapter 1 of this standard is divided into five articles. Article 100 provides definitions for terms used in one or more of the chapters of this document. Article 105 provides for application of safety-related work practices. Article 110 provides general requirements for electrical safety-related work practices. Article 120 provides requirements for establishing an electrically safe work condition. Article 130 provides requirements for work involving electrical hazards.

The requirements of Article 105 are applicable to all of Chapter 1 and set the tone for the chapter by defining the scope, purpose, and responsibilities of the parties involved. The procedures contained in Article 110 apply to electrical equipment and circuit parts before an electrically safe work condition (ESWC) is created.

The requirements in Article 120 eliminate the hazards associated with the use of electricity by creating an ESWC, through de-energizing, lockout/tagout, and verification that there is no voltage, as well as temporarily grounding equipment as necessary. The practices and procedures in Article 130 generally cover interaction with electrical equipment and circuit parts when they are in an energized state.

It is important that all employers at a specific worksite are on the same page; fatal consequences may occur if they are not. The work procedure needs to be understood by all involved and be consistent with the practices identified in Chapter 1, as applicable. If special equipment is involved, the work procedures in Chapter 1 need to be modified where required by Chapter 3.

General Requirements for Electrical Safety-Related Work Practices   Article   110

Summary of Article 110 Changes

110.1: Relocated former 110.3 to 110.1 and relocated former 110.1 to 110.3 for improved usability.

110.1(A): Informational Notes: Added informational notes to former 110.3(A) to provide valuable information on safety in the workplace and the implementation of an electrical safety program.

110.1(B): Added subsection on the conditions of maintenance.

110.1(F): Revised former 110.3(E) for usability and clarity by replacing *working within the limited approach boundary and for working within the arc flash boundary* with *electrical hazard*.

110.1(G): Revised former 110.3(F) into a list format.

110.1(H): Revised former 110.3(G) to introduce job briefing requirements for more complex tasks and to delete former 110.3(G)(2) and (G)(3).

110.1(I)(1): Revised former 110.3(H)(1) to improve usability.

110.1(I)(2): Revised former 110.3(H)(2) to improve usability.

110.2(A): Replaced the phrase *who face a risk of* with *exposed to* and added the phrase *when the risk associated with that hazard.*

110.2(B): Replaced the phrase *degree of training* with *type and extent of training* to link necessary training to employee risk.

110.2(C): Reorganized paragraph into subsection format for release, first aid, training, and documentation for usability.

110.2(C)(3): Revised manufacturer responsibility from *certify* training to *verify* training.

110.2(D)(1): Removed reference to *50 volts or more* in (D)(1)(b) and replaced *voltage detector* with *test instrument* in (D)(1)(d). Other modifications were made to correct inconsistencies and increase clarity and usability.

110.2(E) Informational Note No. 1: Added informational note to clarify the content of the training required by 110.2(E).

110.3: Relocated former 110.1 to 110.3 as well as 110.3 to 110.1 for improved usability.

110.3: Changed former 110.1 title from *Relationships with Contractors (Outside Service Personnel, and So Forth)* to *Host and Contract Employers' Responsibilities*.

110.3(B)(3)(b): Replaced the phrase *any unanticipated hazards found* with *hazards identified* and revised for clarity.

110.3(C): Revised former 110.1(C) requirement for a documented meeting between the host employer(s) and the contract employer.

110.4: Changed title from *Use of Equipment* to *Use of Electrical Equipment* to clarify the intent of the section.

110.4(A)(2): Replaced the phrase *to which they will be connected* with *where they are utilized* for clarity.

110.4(A)(3): Replaced the term *used* with *utilized* for clarity.

110.4(A)(4): Added the term *repair* to the title and *qualified person* to the text.

110.4(A)(5): Added the phrase *on a known voltage source* to provide guidance and clarification on test instrument verification.

110.4(B)(1): Added the term *storage* to the title and *stored* to the text to enhance safety by addressing the handling and storage of portable equipment and their connected flexible electric cords.

110.4(B)(3): Added the term *repair* to the title. Relocated 110.4(B)(3)(d) to 110.4.

110.4(B)(3)(b): Deleted the phrase *that might expose an employee to injury.*

110.4(B)(4): Relocated former 110.4(B)(3)(d) to 110.4(B)(4).

110.4(B)(6): Added (B)(6) to ensure users read, understand, and comply with the manufacturer's operating and safety instructions.

110.4(C)(2) and 110.4(C)(3): Revised former 110.4(C)(2) into two sections: (C)(2) for construction and maintenance work, and (C)(3) for outdoor work. Clarified that either GFCI protection or an assured equipment grounding conductor program is applicable to equipment supplied by greater than 125 volts.

110.4(E): Replaced the term *that* with *what is* for editorial purposes.

Section 110.1, Electrical Safety Program, is now the first section in Article 110. This promotes the importance to employees and employers of having a documented electrical safety program (ESP). The wording in the article regarding hazard and risk was upgraded throughout *NFPA 70E* to foster consistency and understanding within the document and with other national and international standards that address hazard and risk. An ESP includes many different types of interlinked components, as illustrated in Exhibit 110.1.

In line with this harmonization, the method of assessing risk for the Arc Flash PPE Categories Method, formerly the Hazard/Risk Category Method, was harmonized with the method used to assess risk for the Incident Energy Method. Definitions were added for terms such as *incident, hazard, hazardous, risk*, and *risk assessment*. Hazard identification is now part of risk assessment.

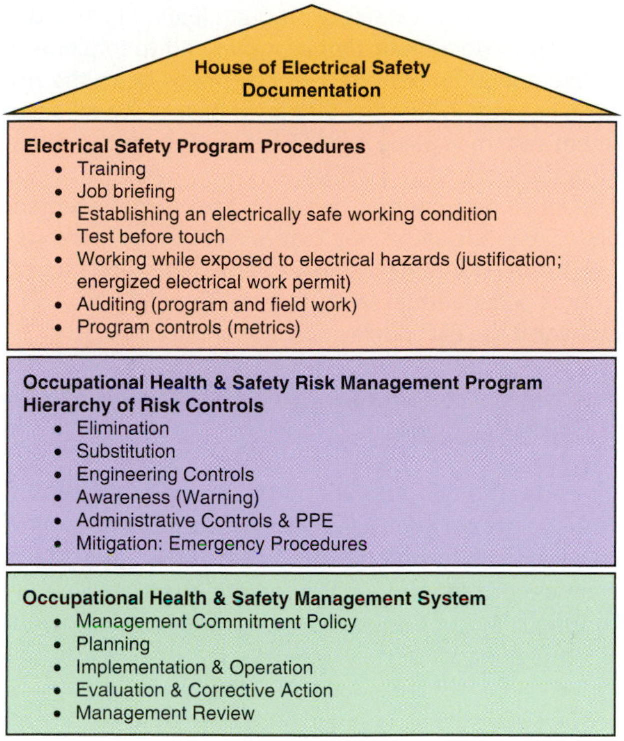

House of Electrical Safety Documentation

Electrical Safety Program Procedures
- Training
- Job briefing
- Establishing an electrically safe working condition
- Test before touch
- Working while exposed to electrical hazards (justification; energized electrical work permit)
- Auditing (program and field work)
- Program controls (metrics)

Occupational Health & Safety Risk Management Program Hierarchy of Risk Controls
- Elimination
- Substitution
- Engineering Controls
- Awareness (Warning)
- Administrative Controls & PPE
- Mitigation: Emergency Procedures

Occupational Health & Safety Management System
- Management Commitment Policy
- Planning
- Implementation & Operation
- Evaluation & Corrective Action
- Management Review

EXHIBIT 110.1

House of electrical safety (electrical safety program).

110.1 Electrical Safety Program.

The revised wording and editorial changes made to 110.1 promote consistency throughout the document with respect to hazard and risk usage. Informational notes were added to provide the user of this standard with valuable information on workplace safety and the implementation of an electrical safety program (ESP), and to emphasize that administrative controls are just one aspect of an overall electrical safety program. The employer's ESP needs to address all situations that can lead to the potential exposure of an employee to an electrical hazard. Where electrical safety is involved, high ethical standards should exist and be enforced.

(A) General. The employer shall implement and document an overall electrical safety program that directs activity appropriate to the risk associated with electrical hazards. The electrical safety program shall be implemented as part of the employer's overall occupational health and safety management system, when one exists.

> Informational Note No. 1: Safety-related work practices such as verification of proper maintenance and installation, alerting techniques, auditing requirements, and training requirements provided in this standard are administrative controls and part of an overall electrical safety program.
>
> Informational Note No. 2: ANSI/AIHA Z10, *American National Standard for Occupational Health and Safety Management Systems*, provides a framework for establishing a comprehensive electrical safety program as a component of an employer's occupational safety and health program.
>
> Informational Note No. 3: IEEE 3007.1, *Recommended Practice for the Operation and Management of Industrial and Commercial Power Systems*, provides additional guidance for the implementation of the electrical safety program.
>
> Informational Note No. 4: IEEE 3007.3, *Recommended Practice for Electrical Safety in Industrial and Commercial Power Systems*, provides additional guidance for electrical safety in the workplace.

It is the employer's responsibility to create an electrical safety program (ESP) that includes policies, procedures, and process controls to document that program and to implement the program through training, auditing, and enforcement that is appropriate to the risk associated with the potential electrical hazards. Where the employer has an overall occupational health and safety management system in place, the ESP is a component of such a program and needs to be integrated in the system.

The employer may have developed their occupational health and safety management system in accordance with ANSI/AIHA Z10 in the same way that *NFPA 70E* provides information on electrical safety-related work procedures, ANSI/AIHA Z10 provides guidance on developing and implementing an overall occupational safety system. The ANSI/AIHA Z10 document provides a framework into which the ESP can fit.

(B) Maintenance. The electrical safety program shall include elements that consider condition of maintenance of electrical equipment and systems.

This section requires that an ESP consider the condition of maintenance of the equipment and its component parts. Without proper maintenance, the equipment cannot be depended on to perform its required safety functions, such as interrupting fault currents within its characteristic time–current curves. Proper maintenance can be achieved by following the manufacturer's instructions or the recommendation included in NFPA 70B.

(C) Awareness and Self-Discipline. The electrical safety program shall be designed to provide an awareness of the potential electrical hazards to employees who work in an environment with the presence of electrical hazards. The program shall be developed to provide the required self-discipline for all employees who must perform work that may involve electrical hazards. The program shall instill safety principles and controls.

Awareness, as used in this section, may have a different meaning when used in the hierarchy of safety controls (in that hierarchy awareness generally means *warnings*). In this section, awareness is attentiveness to self, others, and the situation. According to this section, awareness might mean any of the following:

- Being alert to and understanding when the risk of a potential electrical hazard is unacceptable
- Understanding a possible sequence of events that could lead up to exposure to an electrical hazard
- Being aware of other personnel who could enter the work area
- Being aware that work conditions could change
- Understanding that personal circumstances — such as weariness, exhaustion, boredom, or distraction — can affect performance and result in unsafe conditions

Having an electrical safety program (ESP) is not enough, in itself; employees must follow the policies and effectively implement the procedures. The employer provides the personal and other protective equipment required and ensures that the warning labels required by 130.5(C) are installed. However, the employee has to be able to select, wear and use the equipment appropriate for the specific hazards involved. This is ensured by instilling the safety principles and controls as discussed above.

(D) Electrical Safety Program Principles. The electrical safety program shall identify the principles upon which it is based.

Informational Note: For examples of typical electrical safety program principles, see Informative Annex E.

Electrical safety program principles are the company's safety policies. As such, they are safety guidelines for the employees. The employers' policies might include the following: electrical hazards are to be identified and minimized; every job is to be planned; first time procedures are to be documented; and unexpected events are to be anticipated. Examples of such polices are found in the sample electrical safety program outline found in E.1 of Informative Annex E.

(E) Electrical Safety Program Controls. An electrical safety program shall identify the controls by which it is measured and monitored.

> Informational Note: For examples of typical electrical safety program controls, see Informative Annex E.

Electrical safety program controls are the company's electrical safety metrics for determining if the electrical safety program is effective and efficient. Metrics are measurable points to determine performance. They also can be used to determine if improvements to the safety program are required and, if so, what needs to be changed. Some program controls are found in the sample electrical safety program outline in E.2 redundant?

(F) Electrical Safety Program Procedures. An electrical safety program shall identify the procedures to be utilized before work is started by employees exposed to an electrical hazard.

> Informational Note: For an example of a typical electrical safety program procedure, see Informative Annex E.

The electrical safety program must be documented, whether it is documented in writing or in electronic form. It must include the policies, procedures, and process controls necessary for the employees to work safely.

Electrical safety should be a continuous improvement process, and all procedures should be periodically reviewed. A procedure is the best practice at the current time; some companies use the term *standard operating procedure* (SOP). Over time, new methods might be developed; situations can change. When situations change and the risk of injury exists, the employee must be able to identify and implement the necessary procedures for work where the potential for electrical hazards exists.

(G) Risk Assessment Procedure. An electrical safety program shall include a risk assessment procedure that addresses employee exposure to electrical hazards. The procedure shall identify the process to be used by the employee before work is started to carry out the following:

(1) Identify hazards
(2) Assess risks
(3) Implement risk control according to a hierarchy of methods

> Informational Note No. 1: The hierarchy of risk control methods specified in ANSI/AIHA Z10, *American National Standard for Occupational Health and Safety Management Systems,* is as follows:
>
> (1) Elimination
> (2) Substitution
> (3) Engineering controls
> (4) Awareness
> (5) Administrative controls
> (6) PPE

Informational Note No. 2: The risk assessment procedure may include identifying when a second person could be required and the training and equipment that person should have.

•

Informational Note No. 3: For an example of a risk assessment procedure, see Informative Annex F.

The new definition for the term *risk assessment* in Article 100 states that risk assessment is "an overall process that identifies hazards, estimates the potential severity of injury or damage to health, estimates the likelihood of occurrence of injury or damage to health, and determines if protective measures are required."

A risk assessment procedure is a fundamental principle (guideline) for employees to work safely. This risk assessment procedure must be documented and available to the employees. For this procedure to be effective, it should point out the steps that employees must take where the risk of injury from a potential electrical hazard is unacceptable, and the employees must be trained to perform this procedure correctly each time risk is assessed. Currently, where electrical hazards are involved, there are at least two components to a risk assessment procedure — a shock risk assessment per 130.4 and an arc flash risk assessment per 130.5. Furthermore, justification in accordance with 130.2(A) might also be considered part of the risk assessment procedure. See also the informational note to the definition of the term *risk assessment* in Article 100.

As stated in the definition of the term *risk control* in Article 100, risk assessment involves a determination of whether protective measure are necessary. If they are deemed necessary, these protective measures come from the safety control categories in the hierarchy of safety controls described here. (See also the inside front cover of this handbook for more on the hierarchy of safety controls.)

Where shock hazards are involved, there is the possibility of sudden cardiac arrest (SCA), and it is strongly suggested that an artificial external defibrillator (AED) be available at the scene or close by. The discussion to include Informational Note No. 2 began as a proposal to require a second person for rescue whenever entering the restricted approach boundary.

(H) Job Briefing. Before starting each job, the employee in charge shall conduct a job briefing with the employees involved. The briefing shall cover such subjects as hazards associated with the job, work procedures involved, special precautions, energy source controls, PPE requirements, and the information on the energized electrical work permit, if required. Additional job briefings shall be held if changes that might affect the safety of employees occur during the course of the work.

•

Informational Note: For an example of a job briefing form and planning checklist, see Figure I.1.

Where exposure to potential electrical hazards is involved, the employee in charge should be both qualified and competent for the tasks to be performed (for working at the applicable voltage, for instance). This person might find the job briefing form and checklist located in Informative Annex I helpful. This person also must be able to communicate what is required to the employees involved. (This could require knowledge of a second language.) If the tasks involved require the skills of a licensed electrician, then it might be wise for a licensed electrician to perform the job briefing.

The job briefing needs to be performed before the work tasks are started. However, it should not be performed so far ahead that the employees involved might forget what was covered. If the situation or circumstances change during the course of the work, such that it will affect the work tasks involved, additional job briefings need to be held. Exhibit 110.2 shows a team being briefed just before beginning the job.

EXHIBIT 110.2

Reviewing pertinent drawings as part of the job briefing.

(I) Electrical Safety Auditing.

(1) Electrical Safety Program. The electrical safety program shall be audited to verify that the principles and procedures of the electrical safety program are in compliance with this standard. Audits shall be performed at intervals not to exceed 3 years.

(2) Field Work. Field work shall be audited to verify that the requirements contained in the procedures of the electrical safety program are being followed. When the auditing determines that the principles and procedures of the electrical safety program are not being followed, the appropriate revisions to the training program or revisions to the procedures shall be made. Audits shall be performed at intervals not to exceed 1 year.

(3) Documentation. The audits shall be documented.

Auditing and enforcement is a critical part of any electrical safety program (ESP). It is vital that the ESP be documented — as well as the auditing and enforcement actions — for the benefit of the employees and of the company.

The process control points and actions (the items capable of being measured) need to be determined for there to be effective auditing. For the ESP, the time between audits is not to exceed 3 years. The safety principles (guidelines) and procedures should be compared with the rules in this standard; this not only verifies compliance with the standard, but it is one way to ensure that an ESP is updated to be in accordance with the latest adopted edition of *NFPA 70E*, which is on a 3-year revision cycle.

Field work is the process of going into the field — wherever employees are performing their required tasks and there is the potential of exposure to electrical hazards — to gather information. At intervals not exceeding 1 year, it is important for an employer to confirm employee compliance with its own electrical safety regulations by watching employees perform their electrical safety related task and that they are using PPE appropriate for the task to be performed. The procedures being verified become the process controls (metrics) used to verify the required performance.

When it has been confirmed that the company's ESP principles or procedures are not being followed, corrective action must be taken; this should consist of either applicable modification of the training program or a revision to the procedures, such as increasing the frequency of training or adding follow-up verification of compliance. It is important that a company have an enforcement program or policy of sanctions for violation of the company's safety regulations.

110.2 Training Requirements.

(A) Safety Training. The training requirements contained in this section shall apply to employees exposed to an electrical hazard when the risk associated with that hazard is not reduced to a safe level by the applicable electrical installation requirements. Such employees shall be trained to understand the specific hazards associated with electrical energy. They shall be trained in safety-related work practices and procedural requirements, as necessary, to provide protection from the electrical hazards associated with their respective job or task assignments. Employees shall be trained to identify and understand the relationship between electrical hazards and possible injury.

> Informational Note: For further information concerning installation requirements, see *NFPA 70, National Electrical Code.*

The training requirements in Article 110 become necessary when the risk of an injury from a potential electrical hazard is not reduced to an acceptable level by the pertinent installation requirements when someone is using portable electrical equipment, for example, in contradiction of the manufacturer's instructions or published safety warnings. Normal operation and proper installation go hand in hand, and proper operation involves adequate and proper maintenance. In accordance with 130.2(A)(4), electrical equipment is considered to be operating normally when the following conditions are met:

- It is properly installed.
- It is properly maintained.
- The equipment doors are closed and secured.
- All equipment covers are in place and secured.
- There is no evidence of impending failure.

Proper installation and maintenance entails compliance with applicable industry codes, standards, or recommended practices, and with the manufacturer's instructions. An impending failure can be indicated by arcing, overheating, loose or bound equipment parts, visible damage, unusual noises, or deterioration of condition). When any of these indications exist, the risk from electrical hazards may be unacceptable, and the training requirements of this section come into force.

Each and every employee who may be exposed to unacceptable risk of injury from the use of electricity in the work environment needs to be trained to identify the following:

- Situations that may involve unacceptable risk
- The actual or potential hazards involved with the situation
- The degree or level of the actual or potential hazard
- How the degree or level of the hazard can determine the seriousness of any potential injury
- The means of avoiding or mitigating actual or potential exposure to electrical hazards

The training required depends upon various factors, such as the tasks to be performed, whether the employee is a manager, supervisor, or worker, and whether or not the employee is an unqualified person, or whether the employee is or needs to be a qualified person (see commentary following the definition of *qualified person* in Article 100).

Subjects that various categories of employees could be trained in, depending upon their job functions and the tasks assigned, include the following:

- How to determine when energized electrical work can be justified
- How to determine if an energized electrical work permit (EEWP) is required and how to fill one out
- How to create an electrically safe work condition (ESWC)
- How to perform an electrical hazard analysis, including a shock risk assessment and arc flash hazard assessment, an incident energy analysis, arc flash PPE category method, and determination of the arc flash boundary
- Understanding and using arc flash hazard analysis equipment labels
- Understanding how to select, care for, and use PPE, including the characteristics of different fabrics and the types of arc ratings
- How to select, care for, and use insulating or insulated tools and equipment, including test instruments
- How to select and use alerting techniques, such as safety signs and tags, barricades, and attendants
- Understanding safe work practices where overhead or underground power lines are involved, and where vehicles or lifts are involved
- Understanding job briefings, audit requirements, documentation, and maintenance requirements
- Understanding flexible cord sets and cord- and plug-connected equipment requirements, including GFCI and equipment grounding requirements

Even though 110.2(A) stipulates the essential types of training required, the key concepts in training are (1) the employee obtaining a comprehension of the relevant factors, and (2) the employee demonstrating the ability to work safely. The key to working safely is to always remain vigilant and not let one's guard down even for a second.

(B) Type of Training. The training required by this section shall be classroom, on-the-job, or a combination of the two. The type and extent of the training provided shall be determined by the risk to the employee.

Training can be of the classroom or on-the-job type, or a combination of the two if it imparts the knowledge necessary for employees to perform their duties safely. If an employer needs to communicate work instructions or other workplace information to employees at a certain vocabulary level or in a language other than English, electrical safety training also needs to be provided to those employees in the same manner.

It is incumbent on the employer to verify that the employees have acquired the knowledge and skills necessary to perform their job tasks safely. Retraining needs to be provided when practices

Power System Analysis Training   Where the incident energy (calculation) method is used for determining the degree of the hazard for a potential arc flash incident for complicated electrical systems, very specialized training may be required and, in certain jurisdictions under specific circumstances, licensure or registration may be required. The individual performing the calculations may need to be trained in power system analysis, including short circuit analysis, system coordination, and the use of specialized software programs. This type of training tends to require a higher level of understanding, such as a graduate level of education, rather than an undergraduate level. The employee performing the task does not have to possess this level of understanding, he or she has to be able to read and interpret the required arc flash hazard equipment label in order to define the degree of the hazard involved where a shock hazard or arc flash hazard is present. The person determining the degree of the hazard where complicated calculations are involved needs to be competent, not qualified, since they are not going to be interacting with the equipment or systems. In regard to the above, it may be helpful to understand the following definitions:

- **Licensure.** The granting of licenses, especially to practice a profession or trade.
  Note: Licensure normally requires the passing of an examination with a minimum passing score and that certain qualifications be met. Where licensure is in effect and in force, certain professions or trades may only be performed for covered tasks or structures by licensed individuals.
- **Certification.** The act of making something official.
  Note: A certain set of skills may be certified by an organization, but certification does not equate to licensure. Certification generally requires the passing of an examination with a minimum passing score and that certain qualifications be met. However, certifications are issued by private organizations, not by government entities, such as NFPA's Certified Electrical Safety Compliance Professional certification. The value given to certification may depend upon the reputation of the organization issuing the certification.
- **Registration.** The act or process of entering names on an official governmental list entitling the party to practice a profession, such as a Registered Professional Electrical Engineer.
  Note: Registration is similar to licensure in that it requires the passing of an examination with a minimum passing score and that certain qualifications be met.

When developing and implementing an electrical safety training program, the employer may determine that some, or all, or a combination of some of the following parties — consultants, engineers, manufacturer's representative or agents, safety professionals, teachers (including those with experience in curriculum development), trade persons, contract employees, independent contractors, employees (workers), managers, and supervisors — may be helpful in developing the training material, the training methods, and in presenting the material in a cogent fashion. The instructor needs to be competent in teaching methods, trade practices, and knowledge on the subjects presented.

indicate that the employee has not retained or acquired the requisite understanding or skill. Employees should be tested to demonstrate ability in performing the required tasks safely and in a timely manner. If a reasonable person would conclude that the employer has not conveyed the training to employees in a manner or type they were capable of understanding, then the instruction or training has failed to meet the employer's training obligations and needs to be revised.

The instructor could be an engineer with knowledge of the procedures, a teacher with knowledge of the procedures, a supervisor, or an experienced worker. Job-type knowledge of the work procedures are critical for an instructor. The degree of training should vary based on the tasks that the employee will be performing.

The training should also be appropriate to the task the employee is performing. Where the risk of receiving an electrical injury from energized electrical equipment is minimal (acceptable) — such as when working on equipment rated 600 volts or less that is properly installed, maintained, and enclosed — the electrical safety training required may be minimal.

The training of employees working on equipment rated over 600 volts, however, in most situations requires additional training. There are some situations where the risk of a potential injury is so great — such as where there is substantial arc blast potential energy available — that the work task should not be performed energized or the time in front the exposed energized equipment should be kept to an absolute minimum, even with the use of arc-rated PPE.

Possible Types of Electrical Safety Training

The following list is not meant to be a complete list of all of the types of training that may be available for safety training. It is provided to help the reader understand the variety of training methods and material that may be available and may possibly be used. In actuality, an appropriate training mix is going to include some or all of the following, as well as other methods not mentioned.

- **Apprenticeship**. Training that uses the methods and means of training an apprentice, such as on-the-job training and classroom or other type of training activities usually involving supervision on the job by a senior person (mentor or supervisor).
- **Classroom**. Training performed in a classroom or classroom-type location.
- **E-book or Tablet Based**. Training that uses an e-book or tablet-type computer for training.
- **Laboratory or Shop**. Training performed in a laboratory or shop room away from the job location.
- **On-line**. Training performed while on the Internet.
- **On-the-Job**. Training performed on the job.
- **Self-paced**. Training for which the pace of learning is determined by the student.
- **Trade School**. Training performed in a trade school–type environment.
- **Web-based**. Training performed through a website or web-based applications.

(C) Emergency Response Training.

The emergency response training requirements of *NFPA 70E* first require that employees exposed to shock hazards be trained in methods to safely release victims in contact (hung-up) with energized electrical conductors and circuit parts. Before first aid can be provided, the power has to be turned off or the victim has to be safely moved from contact or the vicinity of the energized electrical conductors and circuit parts. Next, this section deals with training requirements for first aid, including cardiopulmonary resuscitation (CPE) and artificial external defibrillator (AED) use, since contact with an energized conductor or circuit part can result in sudden cardiac arrest (SCA). Finally, the section deals with training verification and emergency response training documentation requirements.

(1) Contact Release. Employees exposed to shock hazards shall be trained in methods of safe release of victims from contact with exposed energized electrical conductors or circuit parts. Refresher training shall occur annually.

(2) First Aid, Emergency Response, and Resuscitation.

(a) Employees responsible for responding to medical emergencies shall be trained in first aid and emergency procedures.

(b) Employees responsible for responding to medical emergencies shall be trained in cardiopulmonary resuscitation (CPR). Refresher training shall occur annually.

(c) Employees responsible for responding to medical emergencies shall be trained in the use of an automated external defibrillator (AED) if an employer's emergency response plan includes the use of this device. Refresher training shall occur annually.

Preparing a Shock Rescue Kit In case of an accidental electrical shock incident, quick response is essential — seconds gained or lost can decide the life or death of a victim. A shock rescue kit designed to have all necessary personal and other required protective equipment readily available to the rescue personnel can save vital time. The shock rescue kit could include items such as an insulated rescue stick or hook, voltage detector, cable cutter with insulated handles, insulating platform or pad, insulated rubber gloves, insulated dielectric overshoes (boots), safety and first aid instructions, roll of adhesive warning tape, and a can of talcum powder. The composition of a kit is dependent on the situation in which it would be used.

No matter how well we prepare, where human beings are involved accidents occur; employers need to have policies, procedures, and trained personnel available to deal with an injury or illness on the job.

OSHA requires that employers ensure availability of medical personnel for advice and consultation on matters of workplace health, and that — in the absence of a doctor, clinic, infirmary, or hospital near the workplace — an adequately trained person with readily accessible first aid supplies be available to render first aid to an injured employee. (See 29 CFR 1910.151 and 29 CFR 1926.50.) Initial medical treatment is usually administered in an emergency room, but where a doctor, clinic, infirmary, or hospital is not in reasonable proximity, the employer's emergency medical response system should include a combination of trained first-aid responder(s) with adequate supplies, emergency medical services (EMS) provider(s) for transport and advance life support services, and finally, full medical services. When in doubt, help should always be called for.

(3) Training Verification. Employers shall verify at least annually that employee training required by this section is current.

First-aid training courses should include instruction in both general and workplace hazard-specific knowledge and skills. CPR training should incorporate AED training if an AED is available at the worksite. First-aid training should be repeated periodically to maintain and update knowledge and skills. First-aid responders may have long intervals between learning and using CPR and AED skills. Instructor-led retraining for life-threatening emergencies should occur at least annually. Retraining for non-life-threatening response should occur periodically. See OSHA Bulletin 3317-06N (2006), *Best Practices Guide: Fundamentals of a Workplace First-Aid Program.*

In accordance with OSHA Instruction CPL-02-00-002, *American Red Cross Agreement*, dated October 22, 1978, persons who have a current training certificate(s) in the American Red Cross Basic, Standard, or Advanced First Aid Course are considered adequately trained to render first aid by OSHA; and the American Red Cross Standard Course sets the minimum level required for first-aid training. Assessment of successful completion of the first-aid training program should include instructor observation of acquired skills and written performance assessments. *NFPA 70E* requires that the employer verify at least annually that training for employees who are first-aid responders is current.

(4) Documentation. The employer shall document that the training required by this section has occurred.

Documentation reinforces the importance of training, refresher training, and training verification for the employer's first-aid responders.

(D) Employee Training.

Training can be defined as the process by which someone is taught the skills that are needed for the job. Training can be seen as teaching (presenting the required knowledge) and learning (absorbing the knowledge and being able to perform the required skill) together. Until the skill is learned, the employee has not been trained. To be qualified, the person has to show that they have retained the required knowledge and have the demonstrated ability to perform the task safely. To be a competent person, it might be said that the student should be able to show proficiency in performing the task, so as to be able to do the task safely and efficiently. However, the prime goal is not efficiency here, it is safety.

(1) Qualified Person. A qualified person shall be trained and knowledgeable in the construction and operation of equipment or a specific work method and be trained to identify and avoid the electrical hazards that might be present with respect to that equipment or work method.

In order to be considered qualified for a particular task or work assignment, an employee should have internalized the requisite knowledge regarding the electrical system involved and of the required procedures. The employee must have received the safety training identified in 110.2(A), 110.2(C), and 110.2(D)(1) and (2), and as determined through use of the decision tree shown in Exhibit 110.3. This decision tree of prerequisites is provided in a bulleted fashion to assist the employer and employees in understanding what training is necessary to be considered a qualified person, depending on the requirements of the specific tasks (e.g., responding to medical emergencies).

(a) Such persons shall also be familiar with the proper use of the special precautionary techniques, applicable electrical policies and procedures, PPE, insulating and shielding materials, and insulated tools and test equipment. A person can be considered qualified with respect to certain equipment and methods but still be unqualified for others.

Electrical safety is a continuous improvement process. Qualified persons have to be alert and always aware of the potential dangers. They should maintain the skills listed in 110.2(D)(1) and should be constantly upgrading their skills to be able to work safely. All of the items listed in this section of *NFPA 70E* should be covered in the policies, procedures, and process controls (metrics) of the electrical safety program.

(b) Such persons permitted to work within the limited approach boundary shall, at a minimum, be additionally trained in all of the following:
   (1) Skills and techniques necessary to distinguish exposed energized electrical conductors and circuit parts from other parts of electrical equipment
   (2) Skills and techniques necessary to determine the nominal voltage of exposed energized electrical conductors and circuit parts
   (3) Approach distances specified in Table 130.4(D)(a) and Table 130.4(D)(b) and the corresponding voltages to which the qualified person will be exposed
   (4) Decision-making process necessary to be able to do the following:
      a. Perform the job safety planning
      b. Identify electrical hazards
      c. Assess the associated risk
      d. Select the appropriate risk control methods from the hierarchy of controls identified in 110.1(G), including personal protective equipment

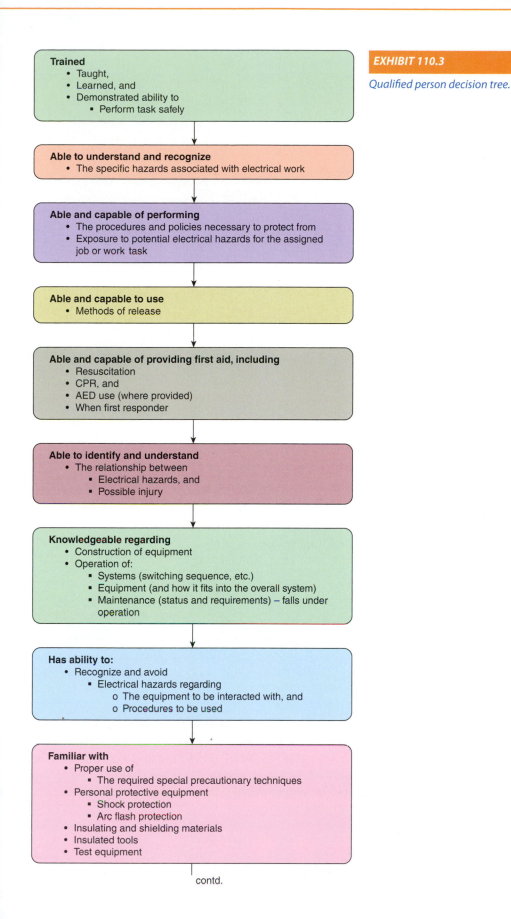

contd.

EXHIBIT 110.3

Qualified person decision tree.

Is permitted to work
- Operating at 50 volts or more
- Within the limited approach boundary of
- Energized conductors and circuit parts (this is what system and equipment is made of)
- At a minimum
  - Has capability (knowledge and techniques)
    - Necessary to distinguish
      - Exposed energized electrical conductors and circuit parts
      - From other parts of electrical equipment (those that are not suitably guarded, isolated or insulated). See definition of exposed (as applied to energized electrical conductors and circuit parts) in Article 100
  - Has capability (knowledge and techniques)
    - Necessary to determine
      - Nominal system voltage of
        - Exposed energized electrical conductors and circuit parts
  - Knows and understands
    - The shock approach boundaries, and
    - How to determine
      - The limits of approach applicable
      - Before approaching exposed energized conductors or circuit parts
    - ac approach limits given in Table 130.4(D)(a)
    - dc approach limits given in Table 130.4(D)(b)
  - Understands
    - The decision making process, and
    - Is able to effective utilize it to determine
    - The degree and extent of
      - The hazards present (particularly electrical hazards), and
      - The personal protective equipment (PPE) required
    - When acting as the person in charge (LOTO)
      - Is capable of accomplishing the job planning necessary to
        - Perform task safely
        - Including the required procedures (SOPs) and
          - Necessary sequence of work
        - Identifying the hazards and severity
        - Determining required personnel
          - Including first responders
        - Obtain and have on hand
          - Required first aid kit including
            - Emergency supplies determine necessary, and
            - Automatic external defibrillator, if determined necessary
        - Determine required PPE
          - Shock protection, as necessary
          - Arc flash protection, as necessary
        - Determine required test equipment
          - Voltage measuring instruments
          - Multi-meter, etc.
        - Determine additional equipment, components, parts, and tools required

EXHIBIT 110.3

Continued

This section deals with the limited approach boundary (LAB), which implies shock risk. Also covered is the protection required for other potential electrical hazards; when there is unacceptable risk of injury as a result of a potential electric shock exposure, there may also be a risk of injury from a potential arc flash exposure, and the employee has to be protected from all potential hazards when performing the task.

Also, the risk assessment procedure used needs to determine if other than electrical hazards will be present while the task is being performed. If there are the employee has to be protected from them also. Where employees are interacting with electrical equipment in a manner that may expose them to an unacceptable risk from potential electrical hazards, it may turn out that the arc flash boundary (AFB) is outside the LAB. In this case, it would be prudent for the employee to consider the arc flash protection required before considering the shock protection.

Foremost, the employee performing the task should be able to identify exposed energized electrical conductors and circuit parts from those that are considered to be effectively guarded, insulated, or isolated. The employee also should be able to determine the nominal system voltage of the exposed energized components and circuit parts concerned. Normally, the nominal system voltage is identified on the nameplate that comes with the equipment. However, in unusual situations it is possible for the equipment to be under-rated or over-rated. Where the nominal voltage cannot be readily determined in this way, checking the one-line diagram for the facility, or checking the nameplates of upstream or downstream equipment, may be helpful. Also, color-coded cable or color-coding, where used, is another method of identifying nominal system voltage.

Once the nominal system voltage level is determined, the employee must be aware of the distances specified in the shock approach boundary tables in 130.4 for the type of voltage applied (ac or dc) and follow the rules applicable at each boundary point — the limited approach boundary (LAB) and the restricted approach boundary (RAB). Because there were no additional rules required at the prohibited approach boundary (PAB) in the 2012 edition of *NFPA 70E*, that shock approach boundary has been eliminated from this edition of the standard to simplify the shock risk assessment process. The arc flash hazard equipment label of 130.5(D) is also a source for determining these shock approach boundaries. Where ac voltage is involved, it is prudent to remember that Table 130.4(D)(a) is based on line-to-line voltages, and where a line-to-neutral voltage is given, it needs to be multiplied by a factor of 1.7321.

As stated in 110.1(G), risk assessment includes identifying the hazards — electrical or otherwise — for the task at hand, assessing the associated risk, and then selecting the appropriate safety control (the risk control method from the categories identified in the hierarchy of safety controls). Information helpful to this process can be found in the information included in Informative Annex F and Informative Annex P (and their associated commentary) and on the inside cover of this handbook. This standard is concerned with electrical hazards, but employees need to be concerned with all potential hazards for the work tasks they are assigned.

(c) An employee who is undergoing on-the-job training for the purpose of obtaining the skills and knowledge necessary to be considered a qualified person, and who in the course of such training demonstrates an ability to perform specific duties safely at his or her level of training, and who is under the direct supervision of a qualified person shall be considered to be a qualified person for the performance of those specific duties.

Part of this training is about learning how to select and use the personal and other protective equipment required to perform the task safely. Depending on the situation, the training may be done in incremental steps. Once the instructor is sure that the student is ready, the student would be allowed to perform the procedure with the instructor close by to help out as necessary. Until the student shows a level of confidence at the task — the ability to perform the task safely with the minimum ability required in the instructor's opinion — the student should not be allowed to perform the procedure without the instructor watching close by. However, once this level of knowledge and ability is reached, the student is then considered a qualified person to perform that particular task.

(d) Tasks that are performed less often than once per year shall require retraining before the performance of the work practices involved.

This most likely means that an organization should keep track of the tasks that its employees are qualified to perform, and it should schedule retraining to make certain their qualifications do not lapse. But when a qualification does lapse, the employer can use on-the-job training to requalify the person, either in house or at another company location, or by the use of outside training or a simulator.

(e) Employees shall be trained to select an appropriate <mark>test instrument</mark> and shall demonstrate how to use a device to verify the absence of voltage, including interpreting indications provided by the device. The training shall include information that enables the employee to understand all limitations of each <mark>test instrument</mark> that might be used.

The employee has to know how to safely use the test instrument to perform the required measurement task. This use requires an understanding and a capability of interpreting all readings and indications given by the test instrument. It also includes knowing the different settings, limitations, and operating modes of the device, as well as the proper terminal into which leads are supposed to be installed. If the device has a duty rating, the employee needs to know how to comply with the rating. It should be assured that the employee is familiar with and understands all of the required information contained in the installation, operating, and maintenance instructions provided with the test instrument.

The ratings of test instruments to be used must be suitable for the parameters of the system and the location where they will be applied — the nominal system voltage, the available fault current, the overvoltage potential (transient rated), and so forth.

Inspection of Test Instruments The employee has to know how to inspect the test instrument and any associated appurtenances, such as leads and probes, to make sure they are suitable for use. Damaged parts — such as fuses, probes, or leads — need to be replaced with appropriately rated replacement parts before the instrument can be used. If broken, the instrument should be taken out of service and returned to an authorized repair shop for repair and recertification. Defective test instruments and parts should promptly be taken out of service and identified as such by an appropriate means, or be legally and properly disposed of in a manner that ensures that there is no chance of reusing them. Part of this inspection is to make sure that the test instrument is listed, labeled, or certified by a nationally recognized testing laboratory (NRTL).

Test instruments within an organization should be standardized. The entity should determine the manufacturer, type, model, and catalog number for the test instruments to be used, and the types of systems and the locals they can be used on. The entity should also determine if it will be providing the test instruments or if it will allow employees to use their own test instruments, which may create some unique problems and situations of its own. Required procedures and policies need to be developed and refined before any test instrument is used. Each test instrument may require its own unique set of policies and procedures.

(f)   The employer shall determine through regular supervision or through inspections conducted on at least an annual basis that each employee is complying with the safety-related work practices required by this standard.

In order to ensure safety, compliance with the requirements of this standard and the employer's electrical safety program is an ongoing and continuous activity. Where there is potential exposure to electrical hazards, each employee should receive consistent observation by a mentor (such as a supervisor or senior person) over short enough periods, or with checkups conducted on at least an annual basis. This may be part of the required field work audit required per 110.1(I)(2).

(2) Unqualified Persons. Unqualified persons shall be trained in, and be familiar with, any electrical safety-related practices necessary for their safety.

Where interaction with energized components and circuit parts is possible, employees need to receive and learn the knowledge and skills necessary for their safety, even if they not considered a qualified person. The following is a list of some of the things that unqualified persons need to be aware of:

- Not to perform housekeeping duties inside the limited approach shock boundary
- Not to leave hinged doors to electrical equipment opened
- Not to remain around electrical equipment where there is evidence of impending failure (and to know what the signs of impeding failure may be, and to warn the appropriate party)

- Not to use flammable material near electrical equipment that can create a spark
- Not to remove plugs from receptacles where the combination is not load-break rated
- Not to use damaged electrical equipment (fixed or portable); damaged cables, cords, or connectors; or damaged receptacles
- After an automatic trip of a circuit breaker, not to reset it unless it tripped because of an overload, not a short circuit
- To use switches that are load rated
- To be aware of the use of alerting techniques, such as safety signs and tags, barricades, warning attendants, and to know to remain outside the shock protection boundaries or arc flash protection boundaries when energized work is being performed
- How to remove a plug from a receptacle, such as by turning the device or circuit off and removing the plug by the plug cap
- The proper distance to keep elevated vehicles away from overhead power lines
- To have a basic understanding of the relationship between exposure to potential electrical hazards and possible bodily injury
- To have a basic understanding and recognition of potential hazards

(3) Retraining. Retraining in safety-related work practices and applicable changes in this standard shall be performed at intervals not to exceed three years. An employee shall receive additional training (or retraining) if any of the following conditions exists:

(1) The supervision or annual inspections indicate that the employee is not complying with the safety-related work practices.

(2) New technology, new types of equipment, or changes in procedures necessitate the use of safety-related work practices that are different from those that the employee would normally use.

(3) The employee must employ safety-related work practices that are not normally used during his or her regular job duties.

Generally, retraining is required when changes in situations, policies, procedures, or equipment require modified or new procedures to be used, such as one of the following:

- When a new edition of this standard is published and, as a result, the work procedures used to perform the task safely change
- When the first line of supervision (or higher) or annual field inspections determine that an employee is not following the required electrical safety policies or procedures
- When new technology, new equipment, or upgrades to old equipment require new or changed electrical safety procedures to be followed by an employee
- When an employee must use a new procedure, or an unfamiliar procedure, because the task is not part of the employee's original job assignment
- When the employee is performing a new task or one that has not been done in the past year [see 110.2(D)(4)]
- When training has not been performed within the last 3-year interval to ensure that the employee still has a full comprehension and ability to perform the required procedures correctly

(E) Training Documentation. The employer shall document that each employee has received the training required by 110.2(D). This documentation shall be made when the employee demonstrates proficiency in the work practices involved and shall be maintained for the duration of the employee's employment. The documentation shall contain the content of the training, each employee's name, and dates of training.

Informational Note No. 1: Content of the training could include one or more of the following: course syllabus, course curriculum, outline, table of contents or training objectives.

Informational Note No. 2: Employment records that indicate that an employee has received the required training are an acceptable means of meeting this requirement.

This section states that the training given to the employee is to be documented, and the documentation must be retained for at least the employee's full time of employment. Other governmental rules may require the safety training documentation to be retained for a longer period of time, at least 7 years after the employee leaves. As employees come and go, it is wise to keep the records for an extended or substantial period of time. As a required minimum, the employee's file or the organization's training records must include the substance or gist of the training that was given, the dates that the training was provided, and the names of the employees that received the training. Where outside training is used, this minimum information must also be added to the entity's training records or the employee's file.

110.3 Host and Contract Employers' Responsibilities.

Generally, the host employer may be seen as the party contracting for the premises where the work is to be performed, or an authorized or apparent agent of the contracting party, even if the agreement is not in writing. There are circumstances where one division of a facility could be seen as the host employer to another division's employees, and another division's employees seen as the contract employer. The concept of *host employer* and *contract employer* have some flexibility, since the terms are not defined within the body of the standard. The host employer may not always be the owner or landlord of the facility. There are also situations where the host employer is the on-site employer (party) acting in the role of a landlord (in this case, not the owner of the facility), and an outside employer (party) is the one that is obligated to perform the work. A service contractor, such as a vending machine operator sending employees to a site, would be considered a contract employer.

Since the terms *host employer* and *contract employer* are currently not defined in this standard, they have some fluidity to them, and it may be possible to pass from the role of contract employer to host employer and, vice versa, from host employer to contract employer, depending on the particular work agreement involved. The final decision regarding the terms is in the hand of the authority having jurisdiction that is enforcing this standard.

(A) Host Employer Responsibilities.

(1) The host employer shall inform contract employers of the following:
  (1) Known hazards that are covered by this standard, that are related to the contract employer's work, and that might not be recognized by the contract employer or its employees
  (2) Information about the employer's installation that the contract employer needs to make the assessments required by Chapter 1
(2) The host employer shall report observed contract employer–related violations of this standard to the contract employer.

The information to be passed back and forth during the documented meeting required by 110.3(C) should be determined ahead of time and made part of the contractual arrangement. During the bidding phase, each party should make clear the information they require and the information they expect to provide, including the exchange of electrical safety programs. Where there is ambiguity, there is the potential for contractual problems, because the various parties may have different expectations. The means and method of reporting violations should either be identified by the parties involved during the bidding phase or be part of the final contractual negotiations. Additionally, the host employer should include the information needed to submit the energized work permit for the signature of the host employer's management where justification of energized work is required by this standard. See 130.2(A).

Where the host employer has the contractor employer follow their electrical safety-related rules in accordance with 110.3(B)(2) below, provided they do not violate the requirements of this standard, it is important to know the rules included in Article 120 and Article 130 as well as the requirements in Chapter 3 where special equipment, systems, or facilities are involved (such as in research or development laboratories). Having staff members or outside help that have a comprehensive understanding of this standard is helpful in meeting its requirements. This handbook should prove helpful in this regard.

(B) Contract Employer Responsibilities.

(1) The contract employer shall ensure that each of his or her employees is instructed in the hazards communicated to the contract employer by the host employer. This instruction shall be in addition to the basic training required by this standard.

The contract employer is often referred to as the independent contractor. While the host employer is generally expected to have greater knowledge of the special conditions, unique situations or features of equipment, and specific hazards other than electrical hazards present in its facility, the contract employer should provide the necessary training on correct work procedures and necessary personal and other protective equipment to perform the job safely. For example, the contract employer is responsible for ensuring that employees implement the host employer's lockout/tagout procedures.

(2) The contract employer shall ensure that each of his or her employees follows the work practices required by this standard and safety-related work rules required by the host employer.

(3) The contract employer shall advise the host employer of the following:

   (1) Any unique hazards presented by the contract employer's work

   (2) Hazards identified during the course of work by the contract employer that were not communicated by the host employer

   (3) The measures the contractor took to correct any violations reported by the host employer under 110.3(A)(2) and to prevent such violation from recurring in the future

Where conditions and situations regarding equipment or procedures need to change as a result of the work being performed by the contract employer, the contract employer has to make the host employer aware of the new conditions — for instance, where a new electrical service has been installed or the utility company has increased the size of the transformer serving the facility. The contract employer has an obligation to inform the host employer of the resulting increase in hazard level and that a new risk assessment should be performed for the facility. Furthermore, where the contract employer notices that such changes might cause the newly installed equipment to be used above its ratings, the contract employer should inform the host employer of the situation; the contract employer may even prefer not to perform the work.

(C) Documentation. Where the host employer has knowledge of hazards covered by this standard that are related to the contract employer's work, there shall be a documented meeting between the host employer and the contract employer.

110.4 Use of Electrical Equipment.

(A) Test Instruments and Equipment.
(1) Testing. Only qualified persons shall perform tasks such as testing, troubleshooting, and voltage measuring within the limited approach boundary of energized electrical conductors or circuit parts operating at 50 volts or more or where an electrical hazard exists.

A limited approach boundary exists only if the ac circuit voltage is 50 volts or greater or the dc circuit voltage is 100 volts or greater. However, careful consideration should be given to circuits operating under 50 volts ac and 100 volts dc where conditions such as a wet environment and high available fault current are encountered. If it is determined that an electrical hazard does exist, only those persons considered as qualified are permitted to perform tasks such as testing, troubleshooting, voltage measuring, or similar diagnostic work. To be qualified to perform testing, the employee must be trained to understand the electrical hazards associated with the work task, select the necessary PPE, and use the specific test instrument and understand its limitations. Exhibit 110.4 shows an example of test instrument used for measuring voltage.

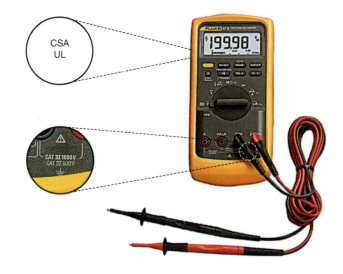

EXHIBIT 110.4

Meter for measuring voltage.
(Courtesy of Fluke Corporation)

(2) Rating. Test instruments, equipment, and their accessories shall be rated for circuits and equipment where they are utilized.

> Informational Note: See ANSI/ISA-61010-1 (82.02.01)/UL 61010-1, *Safety Requirements for Electrical Equipment for Measurement, Control, and Laboratory Use – Part 1: General Requirements*, for rating and design requirements for voltage measurement and test instruments intended for use on electrical systems 1000 volts and below.

(3) Design. Test instruments, equipment, and their accessories shall be designed for the environment to which they will be exposed and for the manner in which they will be utilized.

Test instrument accessories, such as probes and leads, are part of the equipment design and should be treated with the same care as the instrument. Making or modifying accessories is not recommended and should be avoided. There are a variety of special accessories for test instruments available from the test equipment manufacturer.

Test instruments must be selected for the conditions of use. Contacting an exposed energized electrical conductor or circuit part with a test instrument to measure voltage

normally results in a small arc immediately before contact is made or when contact is broken. Therefore, such test instruments must not be used where concentrations of flammable gases, vapors, or dusts occur.

(4) Visual Inspection and Repair. Test instruments and equipment and all associated test leads, cables, power cords, probes, and connectors shall be visually inspected for external defects and damage before each use. If there is a defect or evidence of damage that might expose an employee to injury, the defective or damaged item shall be removed from service. No employee shall use it until a person(s) qualified to perform the repairs and tests that are necessary to render the equipment safe has done so.

If a defect is found during the visual inspection, a tag or label indicating that the instrument is defective and not to be used should be attached to the instrument. The instrument should be removed from service until repairs have been done by a qualified person. A defective lead or probe should not be repaired; it should be destroyed and replaced with a new one.

(5) Operation Verification. When test instruments are used for testing the absence of voltage on conductors or circuit parts operating at 50 volts or more, the operation of the test instrument shall be verified on a known voltage source before and after an absence of voltage test is performed.

When a test instrument is used to test for absence of voltage, a zero reading might mean that no voltage is present during the testing, or it could mean that the instrument has failed. Therefore, proper operation of the test instrument must be verified on a known voltage source before testing for absence of voltage. After the voltage test has been conducted, the proper operation of the test instrument must again be verified on a known voltage source to ensure that a failure did not occur during the testing for absence of voltage.

Although this verification applies to conductors or circuit parts operating at 50 volts or more, under certain conditions (such as wet contact or immersion) even circuits operating under 50 volts can pose a shock hazard. The 50-volt limit does not rely on the frequency of the circuit and applies to 60 Hz as well as high frequency circuits. See Exhibit 120.4 for an illustration of the procedure to test for absence of voltage.

(B) Portable Electric Equipment. This section applies to the use of cord- and plug-connected equipment, including cord sets (extension cords).

For the purposes of this section, portable electrical equipment is electrical equipment that is cord- and plug-connected and is capable of being readily moved from one location to another location with relative ease. Requirements under this section do not include portable cordless equipment powered by battery power packs. However, the cord- and plug-connected battery charger used to recharge the battery power pack for the cordless equipment is included.

(1) Handling and Storage. Portable equipment shall be handled and stored in a manner that will not cause damage. Flexible electric cords connected to equipment shall not be used for raising or lowering the equipment. Flexible cords shall not be fastened with staples or hung in such a fashion as could damage the outer jacket or insulation.

Raising, lowering, or moving portable equipment by the flexible electric cord can damage the cord and equipment, exposing conductors that could leave an employee vulnerable to injury. Such practices should be prohibited as part of the electrical safety plan.

Proper storage of portable equipment and the flexible electric cords connected to the equipment is just as important as proper handling. The cord listing defines indoor storage as protection from sunlight and/or weather. Cords improperly stored (such as those exposed on work trucks) without this necessary protection will potentially suffer UV and weather damage. Flexible cords and cables are required to be protected from damage when being used or handled. In accordance with *NEC* 590.4, flexible cords and cables (extension cords) should be of the hard- or extra-hard-usage type, as identified in *NEC* Table 400.4. In accordance with *NEC* 590.4(J), flexible cords and cables should be supported in place at intervals that ensure that they will be protected from physical damage. Although flexible cords and cables installed as branch circuits or feeders should not be installed on the floor or on the ground, extension cords are permitted to be laid on the floor or ground and to run through doorways, windows, or similar openings, provided they are protected from damage. Where passing through doorways or other pinch points, protection is to be provided to avoid damage; flexible cords might be protected by preventing the doors from closing by means of fasteners or blocking.

(2) Grounding-Type Equipment.

(a) A flexible cord used with grounding-type utilization equipment shall contain an equipment grounding conductor.

An equipment grounding conductor is an integral part of the safety system built into equipment by manufacturers that provides significant protection from shock or electrocution. The flexible cord must have a grounding-type attachment plug, and the receptacle supplying the equipment must be a grounding-type receptacle.

Equipment grounding conductors are not required on tools that are double insulated. Double-insulated tools have two completely separate sets of insulation and must not be connected to the equipment grounding conductor of the supply circuit.

(b) Attachment plugs and receptacles shall not be connected or altered in a manner that would interrupt continuity of the equipment grounding conductor.

Additionally, these devices shall not be altered in order to allow use in a manner that was not intended by the manufacturer.

All components of the circuit must provide the same integrity as the equipment grounding conductor. Plugs and receptacles are constructed, tested, and listed for specific use. Altering these devices in a manner not intended by the manufacturer compromises the integrity of the equipment grounding conductor continuity. Removal of the grounding pin, or twisting the blades or pins of a plug to enable the plug to mate with a receptacle of a different configuration, is prohibited by this requirement.

(c) Adapters that interrupt the continuity of the equipment grounding conductor shall not be used.

Field-fabricated adapters that use a plug of one configuration on one end and a cord cap of a different configuration on the other end that interrupts the continuity of the equipment grounding conductor is prohibited.

(3) Visual Inspection and Repair of Portable Cord- and Plug-Connected Equipment and Flexible Cord Sets.

Damaged portable cord- and plug-connected equipment and flexible cord sets might expose the employee to shock or electrocution. Visual inspections might not identify all possible problems, but a thorough visual inspection does provide significant assurance that the equipment and cord sets will function as intended. The inspection should include a careful observation of each end of the cord or cord set and ensure that all pins and blades are in place and unmodified. During this visual inspection, it is a good practice for the employee to run a hand along the complete surface of the cord. The surface should be smooth with no indentations, cuts, or abrasions. Any anomaly such as an indentation, cut, or abrasion should be inspected further to ensure the integrity of the cord insulation and the conductivity of the equipment grounding conductor. Some examples of cords and receptacles of the grounding type are seen in Exhibits 110.5 and 110.6.

EXHIBIT 110.5

A cord cap. (Courtesy of Pass & Seymour / Legrand ®)

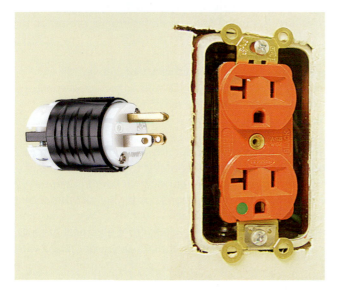

EXHIBIT 110.6

A three-pronged plug (Courtesy of Pass & Seymour / Legrand®) and a three-pronged receptacle.

(a) Frequency of Inspection. Before each use, portable cord- and plug-connected equipment shall be visually inspected for external defects (such as loose parts or deformed and missing pins) and for evidence of possible internal damage (such as a pinched or crushed outer jacket).

Exception: Cord- and plug-connected equipment and flexible cord sets (extension cords) that remain connected once they are put in place and are not exposed to damage shall not be required to be visually inspected until they are relocated.

Although no additional inspection is required until the equipment is moved to another location, periodic inspection of the equipment grounding conductor is recommended.

(b) Defective Equipment. If there is a defect or evidence of damage that might expose an employee to injury, the defective or damaged item shall be removed from service. No employee shall use it until a person(s) qualified to perform the repairs and tests necessary to render the equipment safe has done so.

Defective or damaged portable cord- and plug-connected equipment and flexible cord sets that have been removed from service should be identified as defective by a tag or other method warning personnel that the item should not be used. The warning should remain with the item until repairs have been completed by a person qualified to perform the repairs and tests necessary to render the equipment safe have been completed.

(c) Proper Mating. When an attachment plug is to be connected to a receptacle, the relationship of the plug and receptacle contacts shall first be checked to ensure that they are of mating configurations.

Attachment plugs and receptacles are available in many configurations to ensure proper polarity of connections and prevent the interconnection of different electrical systems, thereby avoiding potentially dangerous conditions. It is also important to check the attachment plug to ensure pins have not been damaged or removed.

(4) Conductive Work Locations. Portable electric equipment used in highly conductive work locations (such as those inundated with water or other conductive liquids) shall be approved for those locations. In job locations where employees are likely to contact or be drenched with water or conductive liquids, ground-fault circuit-interrupter protection for personnel shall also be used.

Informational Note: The risk assessment procedure can also include identifying when the use of portable tools and equipment powered by sources other than 120 volts ac, such as batteries, air, and hydraulics, should be used to minimize the potential for injury from electrical hazards for tasks performed in conductive or wet locations.

If portable electric equipment must be used in highly conductive work locations, it should be suitable for use in those locations. If the environment contains highly conductive material, or the receptacle or plug is wet from water or other conductive liquids, adequately rated gloves are required. In these highly conductive locations it is highly recommended that equipment powered by sources other than 120 volts ac — such as batteries, air, and hydraulics — should be used to minimize the potential for injury from electrical hazards. GFCI protection is required where employees are likely to contact or be drenched with water or conductive liquids.

(5) Connecting Attachment Plugs.

(a) Employees' hands shall not be wet when plugging and unplugging flexible cords and cord- and plug-connected equipment if energized equipment is involved.

An employee unplugging flexible cords and cord- and plug-connected equipment with wet hands is exposed to a shock hazard. Electric tools are not to be used in damp or wet locations unless they are approved for that purpose.

(b) Energized plug and receptacle connections shall be handled only with insulating protective equipment if the condition of the connection could provide a conductive path to the employee's hand (e.g, if a cord connector is wet from being immersed in water).

Any conductive path from an employee's hand to an energized conductor exposes that employee to shock or electrocution. An employee must wear adequately rated insulating gloves when removing the plug from an energized receptacle where the environment contains highly conductive material.

(c)   Locking-type connectors shall be secured after connection.

Locking-type plugs and receptacles, such as the Twist-lock® plug system shown in Exhibit 110.7, and some other portable cord-connecting devices are designed to provide a connection that is secure from accidental withdrawal. These devices must be used per the manufacturer's instructions. For example, the device illustrated in Exhibit 110.7 must be turned to the secure position.

EXHIBIT 110.7

A Twist-Lock® system. (Courtesy of Hubbel Incorporated. Twist Lock® is a registered trademark of Hubbel, Inc.)

(6) Manufacturer's Instructions. Portable equipment shall be used in accordance with the manufacturer's instructions and safety warnings.

Section 110.3(B) of the *NEC* requires listed or labeled equipment to be installed and used in accordance with any instructions included in the listing or labeling.

(C) Ground-Fault Circuit-Interrupter (GFCI) Protection.

(1) General. Employees shall be provided with ground-fault circuit-interrupter (GFCI) protection where required by applicable state, federal, or local codes and standards. Listed cord sets or devices incorporating listed GFCI protection for personnel identified for portable use shall be permitted.

This section permits the use of listed cord sets or devices incorporating listed GFCI protection for personnel identified for portable use. Listed GFCI protection for personnel identified for portable use incorporates open neutral protection. Open neutral protection de-energizes the load if there is a loss in continuity in the neutral conductor leading to the device. A typical Class A GFCI receptacle installed in a box with flexible cord will not provide this additional protection.

(2) Maintenance and Construction. GFCI protection shall be provided where an employee is operating or using cord- and plug-connected tools related to maintenance and construction activity supplied by 125-volt, 15-, 20-, or 30-ampere circuits. Where employees operate or use equipment supplied by greater than 125-volt, 15-, 20-, or 30-ampere circuits, GFCI protection or an assured equipment grounding conductor program shall be implemented.

(3) Outdoors. GFCI protection shall be provided when an employee is outdoors and operating or using cord- and plug-connected equipment supplied by 125-volt, 15-, 20-, or 30-ampere circuits. Where employees working outdoors operate or use equipment supplied by greater than 125-volt, 15-, 20-, or 30-ampere circuits, GFCI protection or an assured equipment grounding conductor program shall be implemented.

The environmental conditions encountered outdoors subject personnel to an elevated exposure to electrical shock hazards, including the possibility of electrocution. In order to mitigate this hazard, GFCI protection is required where employees working outdoors are using cord-and-plug-connected equipment (such as power tools, portable lighting, and other equipment) supplied by a 125-volt, 15-, 20-, or 30-ampere circuit. The GFCI protection required can be a device installed as part of the premises wiring system or a listed portable GFCI device.

Where the cord- and plug-connected equipment is supplied by a circuit rated greater than 125 volts, 15, 20, or 30 amperes, GFCI protection or an assured equipment grounding conductor program must be implemented.

Section 590.6(B)(2) of the *NEC* describes the requirement for an assured equipment grounding conductor program as follows:

> A written assured equipment grounding conductor program continuously enforced at the site by one or more designated persons to ensure that equipment grounding conductors for all cord sets, receptacles that are not a part of the permanent wiring of the building or structure, and equipment connected by cord and plug are installed and maintained in accordance with the applicable requirements of 250.114, 250.138, 406.4(C), and 590.4(D) [of the *NEC*].

> (a) The following tests shall be performed on all cord sets, receptacles that are not part of the permanent wiring of the building or structure, and cord-and-plug-connected equipment required to be connected to an equipment grounding conductor:

> (1) All equipment grounding conductors shall be tested for continuity and shall be electrically continuous.
> (2) Each receptacle and attachment plug shall be tested for correct attachment of the equipment grounding conductor. The equipment grounding conductor shall be connected to its proper terminal.
> (3) All required tests shall be performed as follows:
>
>   a.   Before first use on site
>   b.   When there is evidence of damage
>   c.   Before equipment is returned to service following any repairs
>   d.   At intervals not exceeding 3 months

> (b) The tests required in item (a) shall be recorded and made available to the authority having jurisdiction.

The OSHA-assured equipment grounding conductor program requirements are very similar to the *NEC* requirements for an assured grounding program.

(D) Ground-Fault Circuit-Interrupter Protection Devices. GFCI protection devices shall be tested in accordance with the manufacturer's instructions.

Installation and testing instructions are provided with all listed GFCI devices.

(E) Overcurrent Protection Modification. Overcurrent protection of circuits and conductors shall not be modified, even on a temporary basis, beyond what is permitted by applicable portions of electrical codes and standards dealing with overcurrent protection.

> Informational Note: For further information concerning electrical codes and standards dealing with overcurrent protection, refer to Article 240 of *NFPA 70, National Electrical Code*.

•

All overcurrent devices must comply with the requirements of the *NEC*. The protection provided may be of the overload, short-circuit, or ground-fault type, or a combination of these types, depending on the application.

Overcurrent devices are important components of a safety system to prevent conductors and equipment from exceeding their rating and ability to safely conduct current. Exceeding the conductor or equipment rating can result in overheating, which can lead to ignition of the insulation or nearby flammable material, or result in device and component failure.

The modification prohibited by this section does not include device replacement as a matter of maintenance or upgrade to a newer device. However, the interrupting rating of the replacement device must be at least as great as the original.

Establishing an Electrically Safe Work Condition

Article 120

Summary of Article 120 Changes

120.1: Replaced the phrase *Process of Achieving* with *Verification of* in the title for clarity.

120.1(5): Replaced the phrase *voltage detector* with *test instrument* to clarify the verification requirements of all test instruments.

120.2: Replaced the phrase *Conductors or Circuit Parts* in the title with *Equipment* to clarify that lockout devices are placed on equipment (disconnecting means) not on the conductors or circuit parts.

120.2(B)(2): Revised former (B)(2) into three sections: (B)(2) Training, (B)(3) Retraining, and (B)(4) Training Documentation to more accurately reflect who is to be trained and retrained.

120.2(D)(3)(c): Deleted the term *energy* from the phrase *electrical energy hazards* to promote consistent use of terminology associated with hazard and risk.

120.2(D)(4): Deleted entire section. The training requirements are already covered in 110.2(B)(2).

120.2(E)(2): Deleted the term *energy* from the phrase *electrical energy hazards* to promote consistent use of terminology associated with hazard and risk.

120.2(E)(4)(e): Added a sentence to address the method of accounting for personnel who are working under the protection of the hold card.

120.3(A): Replaced the phrase *hazardous differences in electrical potential* with *shock hazard (hazardous differences in electrical potential)*. Added a sentence to address the application of temporary protective grounding.

120.1 Verification of an Electrically Safe Work Condition.

An electrically safe work condition shall be achieved when performed in accordance with the procedures of 120.2 and verified by the following process:

The process of establishing an electrically safe work condition includes turning off the power, verifying lack of voltage, and ensuring that the equipment cannot be re-energized while work is being performed. An electrically safe work condition does not exist until all of the six steps in 120.1 have been completed. Until then, workers could contact exposed live parts, and they must wear PPE.

If an electrically safe work condition is achieved and verified, no electrical energy is in the immediate vicinity of the work task(s). Danger of injury from an electrical hazard has been reduced to an acceptable level, PPE is not needed, and unqualified persons can perform non-electrical work such as cleaning and painting near electrical equipment. When the equipment is re-energized, the employees must be capable of executing the task(s) in a manner that will not create unacceptable risk from electrical hazards.

(1) Determine all possible sources of electrical supply to the specific equipment. Check applicable up-to-date drawings, diagrams, and identification tags.

It is crucial that all possible sources of information are used to identify and locate all sources of energy. These sources include one-line diagrams, panelboard schedules, elementary or schematic diagrams, electrical plans, identification signs and tags on electrical equipment, consultation with workers who routinely service the equipment, etc. Although most electrical equipment is supplied by a single source, occasionally there are multiple sources such as emergency or standby generators and photovoltaic or fuel cell systems, or a separate control power source.

Frequently, equipment is labeled. However, a sneak circuit (hidden path) could exist that can only be identified by reviewing diagrammatic-type drawings, which identify additional sources. Diagrammatic-type drawings must be maintained and kept up to date, just as required for one-line diagrams to provide accurate information.

Circuits containing transformers must be checked to verify that all links or disconnects are opened to prevent any potential backfeeds. This is especially important when connecting temporary power in situations when equipment is taken out of service for repair or maintenance.

(2) After properly interrupting the load current, open the disconnecting device(s) for each source.

Before operating any disconnecting device, it should be verified whether or not the device is capable of opening the load current. When the rating of a disconnect device is not sufficient to interrupt the load current, the load must be interrupted by another action prior to opening the device. Otherwise, the disconnecting device might be destroyed, initiating a significant failure that could escalate to an arcing fault within the equipment.

Even when the equipment is load-rated, interrupting large full-load currents can reduce the life of the disconnecting device. Motor-driven equipment should be stopped to reduce the amount of current in the circuit before the disconnecting device is operated. See Exhibit 120.1, which shows the opening of the disconnecting means from the on to the off position.

(3) Wherever possible, visually verify that all blades of the disconnecting devices are fully open or that drawout-type circuit breakers are withdrawn to the fully disconnected position.

Disconnecting devices occasionally malfunction and fail to open all phase conductors when the handle is operated. After operating the handle of the disconnecting device, the employee should open the door or cover, or view the switch blades through a viewing window if available, and observe the physical opening in each blade of the device. See Exhibit 120.2. Opening a door or removing a cover could expose an employee to electrical hazards. Therefore, the employee must be protected from those hazards by PPE.

When the physical opening of the contacts cannot be visually verified, the employee should verify that the device is physically opened by testing for the presence of voltage on the load side of the device with the handle in the open position. Testing for the absence of voltage must include measuring voltage to ground from each conductor and between

EXHIBIT 120.1

Opening the disconnect.

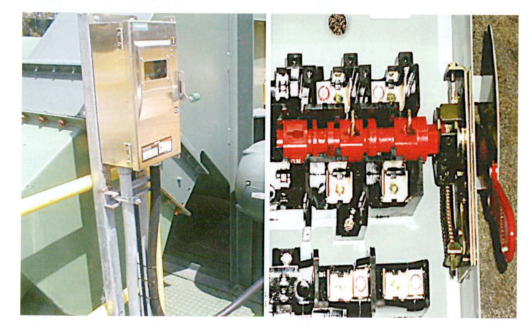

EXHIBIT 120.2

Verifying the opening of contacts.

each conductor and each of the other conductors (phase-to-ground and phase-to-phase), using an appropriate test instrument rated for the circuit voltage. Taking these voltage measurements should be considered working on or near exposed live parts, regardless of the position of the operating handle, and the employee must be protected from those hazards by PPE.

If the disconnecting mechanism does not operate freely, do not try to force it open. De-energize the circuit by means of an upstream device.

(4) Apply lockout/tagout devices in accordance with a documented and established policy.

Only qualified persons should be authorized to apply lockout/tagout devices in accordance with the employer's documented and established policy. The lockout device is typically a specifically identified padlock to keep the disconnecting means open and for the equipment to be worked on isolated from all potentially hazardous energy sources. Tags identify the person(s) responsible for applying and removing the locks. Exhibit 120.3 shows an employee attaching a lock and tag to the open switch.

Applying a lock and tag to a switch.

(5) Use an adequately rated test instrument to test each phase conductor or circuit part to verify it is de-energized. Test each phase conductor or circuit part both phase-to-phase and phase-to-ground. Before and after each test, determine that the test instrument is operating satisfactorily through verification on a known voltage source.

The three-step testing procedure is shown in Exhibit 120.4. First, the test instrument is used to test a known source, to verify that the tester is operating properly, as shown in part (a). Then the test instrument is used to confirm that the equipment to be de-energized has zero voltage, as shown in (b). The tester is used again in (c) to test a known source to confirm that the test instrument has not failed during the testing process.

(a)   (b)   (c)

Test procedure.

> Informational Note: See ANSI/ISA 61010-1, *Safety Requirements for Electrical Equipment for Measurement, Control, and Laboratory Use, Part 1: General Requirements*, for rating and design requirements for voltage measurement and test instruments intended for use on electrical systems 1000 volts and below.

In each instance, the qualified person must determine the absence of voltage on each conductor to which he or she could be exposed. That determination must be made by measuring the voltage to ground and the voltage between all other conductors using a voltage detector rated for the maximum voltage available from any potential source of energy. In an abnormal situation where there is no connection to earth, note that the test instrument might indicate an absence of voltage to ground and thereby indicate that the equipment ground path needs to be re-established.

> The test instrument should be selected on the basis of the service and duty. The rating of the device must be at least as great as the expected voltage. For example, Category III

test instruments can be used on circuits operating at up to 480 volts, and Category II instruments can be used on single-phase 120-volt circuits. The device must have an adequate static discharge rating as discussed in ANSI/ISA 61010-1. See Exhibit 110.4 for an example of what to consider in a voltage-measuring device.

(6)   Where the possibility of induced voltages or stored electrical energy exists, ground the phase conductors or circuit parts before touching them. Where it could be reasonably anticipated that the conductors or circuit parts being de-energized could contact other exposed energized conductors or circuit parts, apply ground connecting devices rated for the available fault duty.

Capacitors used for power factor correction and motor starting are examples of stored energy sources. Energized conductors and equipment in high-voltage installations can induce hazardous voltages in adjacent conductors and equipment that are de-energized. Failure of a conductor support system or conductor can result in the inadvertent re-energizing of a de-energized conductor. For instance, more than one outside overhead line might be installed on a single pole or support structure, and when the higher conductor supported at a higher elevation is damaged, it could fall onto a lower installed conductor and re-energize the lower conductor.

Protective grounding equipment is necessary to protect employees from potentially hazardous voltage, such as that caused by the following accidental instances:

- A long conductor installed in proximity to the de-energized conductor could induce a hazardous voltage onto the otherwise de-energized conductor (magnetic coupling).
- An employee inadvertently connecting a conductor that is not locked out can add a source of energy and re-energize the circuit under repair.
- Other situations that could reintroduce voltage to a conductor under repair could be identified by the risk assessment procedure.

Protective grounding equipment provides protection only if the rating is sufficient to conduct any the available current. Inadequately rated protective grounding equipment can introduce a hazard that otherwise would not exist. The rating must be established by the manufacturer or another generally accepted rating process. See Exhibit 120.5 for an example of a safety-grounding assembly that is identified for a specific level of fault current.

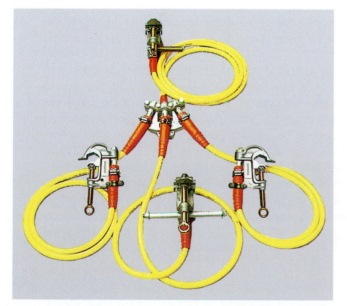

EXHIBIT 120.5

A safety-grounding assembly. (Courtesy of Salisbury by Honeywell)

Case Study

Scenario and Causation.

Nathan was employed by an electrical contractor (contract employer) and working at a wastewater treatment plant for six weeks, installing and testing high-voltage equipment. As he was finishing this phase of the project, Nathan had to complete a few seemingly routine tasks. One of the tasks was to install new relays on the door of a 13,800-volt switchgear cubicle. As he prepared to install the relays on a door of the new equipment, Nathan assembled his tools and opened (disconnected) the main high-voltage switch. The procedure called for him to perform a visual inspection of the switch blades to ensure they had opened properly, but he was interrupted by Dave, the general contractor. Dave was upset that the relays hadn't already been mounted and instructed Nathan to finish this job as soon as possible.

Had Nathan remembered to inspect the switch, he would have discovered that one blade of the three-phase switch had failed to open. Nathan didn't have a functioning tester suitable for this level of voltage. Without it, he had no way to verify if the load side of the switch was energized or de-energized. To mount the relays, Nathan had to drill several holes in the cabinet door. As Nathan stretched into the cabinet to plug in the drill, his right leg got close enough to the 13,800-volt bus for the electricity to jump the air gap and enter his leg. This developed into an arc flash. The extreme temperatures at the arc terminals vaporized the bus material and propagated into a huge fireball and arc blast.

Results.

This arc flash event lasted only 4 milliseconds, yet it ignited Nathan's clothes and he was badly burned. Nathan had third-degree burns over 40 percent of his body. Nathan was wearing a polyester shirt which fused into his skin. He was also wearing a large metal belt buckle, a plastic watch, and his wedding ring. During the incident, the belt buckle burned a hole in his stomach, the watch melted into his wrist, and the wedding ring nearly severed his finger. He was transported to a burn unit, where he began his battle for survival. At the burn center, he underwent several months of excruciatingly painful treatments and skin grafts necessary to recover from his severe burn injuries.

Analysis.

Accidents typically occur due to a chain of events or mistakes that lead up to the actual event. If one of the links in the chain of events is eliminated, the accident most likely would not have happened.

Nathan's injuries could have been prevented if an electrically safe work condition had been established. Additionally, if barricades had been set up and the general contractor (unqualified person) was kept out of the work area, Nathan would not have been distracted from verifying that the disconnecting device was fully opened and all phases had been disconnected. The general contractor was most likely within the limited approach boundary and arc flash boundary of the 13,800-volt disconnecting means, because an electrically safe work condition had not been established. Nathan should have been wearing arc-rated daily wear and not clothing made of nylon, acetate, polyester, or rayon. Although Nathan assumed that the equipment he was working on was de-energized, the fact that he was wearing conductive objects exacerbated his injuries. In accordance with 130.6(D), wearing of jewelry and other conductive articles is prohibited where there is a potential for electrical contact with exposed energized conductors or circuit parts.

Relevant *NFPA 70E* Requirements.

If everyone involved had followed all of the requirements of *NFPA 70E*, Nathan would not have been injured. Some of the steps they should have taken include the following:

- Completing the required steps to establish an electrically safe work condition per 120.1.
- Having and using properly rated test equipment per 110.4(A).
- Maintaining discipline and self-awareness per 110.1(C).
- Following all procedures established by the electrical safety program for employees exposed to an electrical hazard per 110.1(F).
- Preventing unqualified persons from entering the limited approach boundary per 130.4(C)(1).

120.2 De-energized Electrical Equipment That Has Lockout/Tagout Devices Applied.

Each employer shall identify, document, and implement lockout/tagout procedures conforming to Article 120 to safeguard employees from exposure to electrical hazards. The lockout/tagout procedure shall be appropriate for the experience and training of the employees and conditions as they exist in the workplace.

It is the employer's responsibility to establish lockout/tagout procedures that are part of creating an electrically safe work condition that will protect employees from hazardous energy sources during service and maintenance. See 120.1 for verification of achieving an electrically safe work condition.

(A) General. All electrical circuit conductors and circuit parts shall be considered energized until the source(s) of energy is (are) removed, at which time they shall be considered de-energized. All electrical conductors and circuit parts shall not be considered to be in an electrically safe work condition until all of the applicable requirements of Article 120 have been met.

Installing locks and tags does not ensure that electrical hazards have been removed. Until an electrically safe work condition has been established, employees must use work practices that are identical to working on or near exposed live parts. These work practices (including use of PPE) should be appropriate for the voltage and energy level of the circuit as if it were known to be energized.

Generally, an arc flash protection boundary exists until an electrically safe work condition is established. Employees within that boundary must wear arc-rated protective clothing as well as protection for the employee's head, face, and hands as determined by the risk assessment procedure. If it is necessary for an employee to penetrate the restricted approach boundary, the employee must be wearing shock-protective equipment that is rated at least as high as the full-circuit voltage.

Where an electrically safe work condition exists, there is no electrical energy in proximity of the work task(s). Danger of injury from electrical hazards have been removed, and neither protective equipment nor special safety training is required.

Informational Note: See 120.1 for the six-step procedure to verify an electrically safe work condition.

Electrical conductors and circuit parts that have been disconnected, but not under lockout/tagout; tested; and grounded (where appropriate) shall not be considered to be in an electrically safe work condition, and safe work practices appropriate for the circuit voltage and energy level shall be used. Lockout/tagout requirements shall apply to fixed, permanently installed equipment; to temporarily installed equipment; and to portable equipment.

(B) Principles of Lockout/Tagout Execution.

(1) Employee Involvement. Each person who could be exposed directly or indirectly to a source of electrical energy shall be involved in the lockout/tagout process.

Any employee who could be exposed to unacceptable risk of an electrical hazard when executing the work task must be involved in the lockout/tagout process. Temporary and contract employees also must understand how the lockout/tagout procedure that is part of creating an electrically safe work condition influences their exposure to electrical hazards, and they must participate in the process. When multiple employers are involved in the work process, such as when an independent contractor is involved, Section 110.3 requires each employer to share information about hazards and procedures.

> Informational Note: An example of direct exposure is the qualified electrician who works on the motor starter control, the power circuits, or the motor. An example of indirect exposure is the person who works on the coupling between the motor and compressor.

In most cases, electrical energy is converted to other kinds of energy, such as light, heat, and motion. Motors, electrical space heaters, lamps in luminaires, and similar equipment are examples of this converted electrical energy.

(2) Training. All persons who could be exposed or affected by the lockout/tagout shall be trained to understand the established procedure to control the energy and their responsibility in the procedure and its execution. New or reassigned employees shall be trained to understand the lockout/tagout procedure as it relates to their new assignments.

Employers, including independent contractors, must provide training for all employees who might be involved in the process of establishing an electrically safe work condition. Contractors and facility owners must exchange information about creating an electrically safe work condition — along with any information about characteristics specific to the facility — and ensure that their respective employees understand all issues that are important to the other employer.

Each person who is associated with performing the job must be trained to understand their role in the lockout/tagout process and accept individual responsibility for the integrity of the lockout/tagout process. As this section specifies, employees who have been reassigned (either permanently or temporarily) must be trained to understand their new role in the lockout/tagout procedure. All employees must have additional training whenever the established procedure is changed.

(3) Retraining. Retraining shall be performed:

(a) When the established procedure is revised
(b) At intervals not to exceed 3 years

Retraining ensures that employees have an understanding of the established procedure and its implementation. Retraining is required for employees when changes are made to the procedure and shall be performed at intervals not to exceed 3 years. Examples of changes that could initiate retraining are changes in control methods, equipment, job assignment, or when periodic inspection reveals that employees are deviating from the established procedure.

(4) Training Documentation.

(a) The employer shall document that each employee has received the training required by this section.

(b) The documentation shall be made when the employee demonstrates proficiency in the work practices involved.

(c) The documentation shall contain the content of the training, each employee's name, and the dates of the training.

Informational Note: Content of the training could include one or more of the following: course syllabus, course curriculum, outline, table of contents, or training objectives.

Training documentation can be either hard copy or electronic. Whatever format is used, the record must be available for inspection by a third party.

(5) Plan. A plan shall be developed on the basis of the existing electrical equipment and system and shall use up-to-date diagrammatic drawing representation(s).

A plan must be developed for each lockout/tagout that is part of creating an electrically safe work condition. The planner might be the person in charge, the supervisor, or the employee performing the task(s). Qualified persons should be involved in the development of the plan. The plan must identify the location, both physically and electrically, that requires a lockout or tagout device.

Up-to-date accurate single-line diagrams provide essential information that is needed when the employee is creating an electrically safe work condition. If the diagram is inaccurate, however, workers could be injured as a result of one or more sources of energy that are not removed. The purpose of the information is to clearly indicate all sources of electrical energy that are or might be available at any point in the electrical circuit.

(6) Control of Energy. All sources of electrical energy shall be controlled in such a way as to minimize employee exposure to electrical hazards.

All sources of energy must be removed by operating all applicable disconnecting means. Where there is an unacceptable risk of injury from electrical hazards all the steps in 120.1 must be performed. After the disconnecting means is opened, all employees who might be exposed to the electrical hazard should be involved in the process of installing devices that provide adequate assurance that the circuit cannot be re-energized.

(7) Identification. The lockout/tagout device shall be unique and readily identifiable as a lockout/tagout device.

Uniquely and readily identifiable lockout/tagout devices must be durable, substantial, and standardized. They must be capable of withstanding the environment to which they are exposed for the maximum period of time that exposure is expected and substantial enough to prevent removal without the use of excessive force or unusual techniques, such as the use of bolt cutters or other metal cutting tools. Standardization can be by color, shape, or size. The print and format on tagout devices must also be standardized. Employees must be able to recognize a lockout/tagout device, and there cannot be any possibility of confusing lockout/tagout devices with locks or tags used for other purposes, such as information tags and process control locks. Exhibit 120.6 is an example of a lockout station.

EXHIBIT 120.6

Lockout station. [Courtesy of NMC (www.nationalmarker.com)]

(8) Voltage. Voltage shall be removed and absence of voltage verified.

Only qualified persons can perform the test for absence of voltage, and they must be using the necessary personal and other protective equipment. Testing instruments and equipment must be visually inspected for external defects or damage before being used to determine de-energization [29 CFR 1910.334(c)(2)]. The integrity of the test instrument must be verified before and after the voltage test. (See Exhibit 120.4 for an illustration of the correct procedure to test for absence of voltage.)

(9) Coordination. The established electrical lockout/tagout procedure shall be coordinated with all of the employer's procedures associated with lockout/tagout of other energy sources.

OSHA 29 CFR 1910.147 defines *energy sources* as follows: "Any source of electrical, mechanical, hydraulic, pneumatic, chemical, thermal, or other energy." All lockout/tagout procedures shall clearly and specifically outline the scope, purpose, authorization, rules, and techniques to be utilized for the control of hazardous energy. The lockout/tagout procedures should also be coordinated with outside employers — independent contractors who service and/or maintain equipment that require lockout/tagout.

(C) Responsibility.

Section 120.2(C) assigns responsibly for specific tasks to participants in the lockout/tagout procedure that are part of creating an electrically safe work condition. Examples of the topics covered by this section include establishing the lockout/tagout procedures for the organization; providing training and equipment necessary to carry out the procedure, execution and auditing; and designating the person in charge for simple and complex lockout/tagout.

(1) Procedures. The employer shall establish lockout/tagout procedures for the organization, provide training to employees, provide equipment necessary to execute the details of the procedure, audit execution of the procedures to ensure employee understanding/compliance, and audit the procedure for improvement opportunity and completeness.

Not only is the employer responsible for developing, implementing, and enforcing lockout/tagout procedures, the employer must also provide training and initiate audits. The training must ensure that each employee understands the enforcement of an installed lockout or tagout device is part of the program. The audit must ensure that the lockout/tagout procedure is effective and is being properly implemented. The employer is also required to provide the equipment necessary to execute the procedure — locks, multi-lock tree, tags, PPE, and any other tools needed to complete the lockout/tagout procedure.

(2) Form of Control. Two forms of hazardous electrical energy control shall be permitted: simple lockout/tagout and complex lockout/tagout *[see 120.2(D)]*. For the simple lockout/tagout, the qualified person shall be in charge. For the complex lockout/tagout, the person in charge shall have overall responsibility.

> Informational Note: For an example of a lockout/tagout procedure, see Informative Annex G.

The objective in controlling exposure to electrical energy is to ensure that all possible sources of electrical energy are disconnected and cannot be re-energized unexpectedly. The two types of control are the simple lockout/tagout and the complex lockout/tagout, which are administrative controls from the hierarchy of safety controls. See Informative Annex F and Informative Annex P.

In a simple lockout/tagout, a qualified person(s) de-energizes one source of electrical energy and is responsible for their own lockout/tagout. A written lockout/tagout plan is not required.

A complex lockout/tagout typically involves more than one energy source, more than one disconnecting means, and requires a single person with overall responsibility and a written plan prepared in advance for each lockout/tagout application.

(3) Audit Procedures. An audit shall be conducted at least annually by a qualified person and shall cover at least one lockout/tagout in progress and the procedure details. The audit shall be designed to correct deficiencies in the established electrical lockout/tagout procedure or in employee understanding.

The objective of the audit is to make sure that all requirements of the lockout/tagout procedure were properly executed and that employees are familiar with their responsibilities. The audit also must determine whether the requirements contained in the published procedure are sufficient to ensure that the electrical energy is satisfactorily controlled. The audit should also identify and correct any deficiencies in the procedure, employee training, or enforcement of the requirements.

(D) Hazardous Electrical Energy Control Procedure.

(1) Simple Lockout/Tagout Procedure. All lockout/tagout procedures that involve only a qualified person(s) de-energizing one set of conductors or circuit part source for the sole purpose of safeguarding employees from exposure to electrical hazards shall be considered to be a simple lockout/tagout. Simple lockout/tagout plans shall not be required to be written for each application. Each worker shall be responsible for his or her own lockout/tagout.

An example of a simple lockout/tagout procedure is where a single disconnect switch provides energy for equipment, such as a motor, and an employee who intends to service the motor opens the disconnecting means, de-energizing the motor circuit. The only way to guarantee that the disconnecting means remains open and the motor de-energized is to lock the disconnecting means in the open position and affix the appropriate tag. If the disconnecting means is capable of accepting a lockout/tagout device, this would be considered a simple lockout/tagout, and no written lockout/tagout plan is needed. However, a simple lockout/tagout must be a planned activity. Exhibit 120.7 shows a motor disconnect that has been locked open and tagged.

(2) Complex Lockout/Tagout Procedure.

(a) A complex lockout/tagout plan shall be permitted where one or more of the following exist:
   (1) Multiple energy sources
   (2) Multiple crews
   (3) Multiple crafts
   (4) Multiple locations

EXHIBIT 120.7

Motor disconnect that has been locked open and tagged.

(5) Multiple employers

(6) Multiple disconnecting means

(7) Particular sequences

(8) Job or task that continues for more than one work period

The conditions listed in 120.2(D)(2)(a) increase the complexity and difficulty of a lockout/tagout. Multiple energy sources as referenced in 120.2(D)(2)(a) are not limited to electric power; they could be compressed air, hydraulic, mechanical, pneumatic, natural gas, steam, thermal, or water. A process or machinery with multiple operations are examples of 120.2(D)(2)(a)(7).

(b) All complex lockout/tagout procedures shall require a written plan of execution that identifies the person in charge.

If one or more of the conditions listed in 120.2(D)(2)(a) exist, the lockout/tagout is defined as complex, and a person in charge must be assigned. The primary responsibility of the person in charge is to directly implement a complex lockout/tagout procedure.

(c) The complex lockout/tagout procedure shall vest primary responsibility in an authorized employee for a set number of employees working under the protection of a group lockout or tagout device (such as an operation lock). The person in charge shall be held accountable for safe execution of the complex lockout/tagout.

The person in charge must be both a qualified person and an authorized employee. The person in charge coordinates personnel and ensures continuity of lockout/tagout protection for all employees working under the protection of a group lockout/tagout device. The person in charge must understand that they are accountable for developing, implementing, and monitoring the plan.

(d) Each authorized employee shall affix a personal lockout or tagout device to the group lockout device, group lockbox, or comparable mechanism when he or she begins work and shall remove those devices when he or she stops working on the machine or equipment being serviced or maintained.

To adhere to the basic principle that each employee should be in control of the hazardous energy, each person involved in the work task must affix their personal lockout device to the group lockout device or to each individual lockout point. The person in charge is responsible for ensuring adherence to this principle. See Exhibits 120.8, 120.9, and 120.10 for examples of complex lockout/tagout devices.

EXHIBIT 120.8

Complex lockout/tagout devices.

EXHIBIT 120.9

Lockbox.

EXHIBIT 120.10

Disconnect switch with locks from more than one individual.

(e) The complex lockout/tagout procedure shall address all the concerns of employees who might be exposed. All complex lockout/tagout plans shall identify the method to account for all persons who might be exposed to electrical hazards in the course of the lockout/tagout.

The person in charge is responsible for ensuring that all employees are protected by the complex lockout/tagout. However, each employee must understand the process by which the person in charge will implement this responsibility.

(3) Coordination.

(a) The established electrical lockout/tagout procedure shall be coordinated with all other employer's procedures for control of exposure to electrical energy sources such that all employer's procedural requirements are adequately addressed on a site basis.

A project could involve contractors and employers other than the facility owner. Each employer is required to implement a lockout/tagout procedure, so different requirements can exist. To make sure that the requirements of each procedure are understood and observed, each employer might need to amend one or more procedures. When the lockout/tagout procedure for different employers must be implemented on the same work site, the procedures must be coordinated with each other and employees of each employer must be instructed about any unique aspect. The basic concerns of each employer must be addressed in the written lockout/tagout plan. Employees must understand the change and the necessity for the change.

(b) The procedure for control of exposure to electrical hazards shall be coordinated with other procedures for control of other hazardous energy sources such that they are based on similar/identical concepts.

Other standards may require an employer to implement a procedure for control of exposure to other hazardous energy sources. For instance, the general control of hazardous energy procedure and the electrical lockout/tagout procedure that is part of creating an electrically safe work condition could have similar or identical requirements for locks and tags. The electrical lockout/tagout procedure could be integrated into an overall control of hazardous energy procedure; however, that procedure needs to address all the issues identified in this standard.

(c) The electrical lockout/tagout procedure shall always include voltage testing requirements where there might be direct exposure to electrical hazards.

The lockout/tagout procedure must include the requirements necessary to ensure that all employees know whether they are exposed to an electrical hazard. The procedure should identify acceptable test instruments and contain a requirement to ensure that the test instrument is functioning properly before and after each use. Employee training must ensure that each qualified employee is familiar with the requirements for testing voltage. See Exhibits 110.4 and 120.4. Employees who use test instruments must understand how to use the device, how to protect themselves from any associated hazard, and how to interpret all possible indications provided by the test instrument.

(d) Electrical lockout/tagout devices shall be permitted to be similar to lockout/tagout devices for control of other hazardous energy sources, such as pneumatic, hydraulic, thermal, and mechanical, if such devices are used only for control of hazardous energy and for no other purpose.

Devices used for control of hazardous energy must be easily recognizable and be the only devices the employer uses in conjunction with the control hazardous energy.

Electrical lockout/tagout devices should have the same physical characteristics as devices used for control of other forms of energy so that employees are not confused.

-

(E) Equipment.

(1) Lock Application. Energy isolation devices for machinery or equipment installed after January 2, 1990, shall be capable of accepting a lockout device.

This requirement correlates with 29 CFR 1910.147(c)(2)(iii), which states the following: "After January 2, 1990, whenever replacement or major repair, renovation or modification of a machine or equipment is performed, and whenever new machines or equipment are installed, energy isolating devices for such machine or equipment shall be designed to accept a lockout device."

In addition, many *NEC* requirements for disconnecting means require that the equipment have provisions installed that allow it to be locked in the open position. Any disconnecting means that cannot be locked in the open position must not be used as an energy isolation device.

(2) Lockout/Tagout Device. Each employer shall supply, and employees shall use, lockout/tagout devices and equipment necessary to execute the requirements of 120.2(E). Locks and tags used for control of exposure to electrical hazards shall be unique, shall be readily identifiable as lockout/tagout devices, and shall be used for no other purpose.

Each employer must provide the necessary equipment — such as locks, tags, chains, wedges, key blocks, adaptor pins, and self-locking fasteners — for employees to control exposure to electrical hazards.

Lockout/tagout devices for control of exposure to electrical energy can be identical to lockout/tagout devices used for the control of hazardous energy from other energy sources, but these devices must not be used for any other purpose. See the commentary following 120.2(B)(7) for identification of lockout/tagout devices. Exhibit 120.6 is an example of a lockout station.

(3) Lockout Device.

(a) A lockout device shall include a lock (either keyed or combination).
(b) The lockout device shall include a method of identifying the individual who installed the lockout device.
(c) A lockout device shall be permitted to be only a lock, if the lock is readily identifiable as a lockout device, in addition to having a means of identifying the person who installed the lock.

A uniquely identified lock used as a lockout device may be used in conjunction with other components, but the basic lockout device is a lock. The method of locking (by key or combination) must prevent unauthorized removal of the lock, and the lock's installer must be in control of the key or combination. The lockout device must include information that identifies the person who installed the lock and be installed in a manner that prevents operation of the energy isolation device.

(d) Lockout devices shall be attached to prevent operation of the disconnecting means without resorting to undue force or the use of tools.
(e) Where a tag is used in conjunction with a lockout device, the tag shall contain a statement prohibiting unauthorized operation of the disconnecting means or unauthorized removal of the device.

(f) Lockout devices shall be suitable for the environment and for the duration of the lockout.

(g) Whether keyed or combination locks are used, the key or combination shall remain in the possession of the individual installing the lock or the person in charge, when provided by the established procedure.

One basic premise of lockout requirements is that the person who might be exposed to an electrical hazard is in control of the electrical energy. To accomplish that purpose, the key to the lock must remain in the possession of the person who installed it.

(4) Tagout Device.

(a) A tagout device shall include a tag together with an attachment means.

(b) The tagout device shall be readily identifiable as a tagout device and suitable for the environment and duration of the tagout.

(c) A tagout device attachment means shall be capable of withstanding at least 224.4 N (50 lb) of force exerted at a right angle to the disconnecting means surface. The tag attachment means shall be nonreusable, attachable by hand, self-locking, nonreleasable, and equal to an all-environmental tolerant nylon cable tie.

The tagout device must be unique and easily recognizable as a tagout device and must be securely attached to the energy isolation device to alert employees that equipment is not to be operated until the tag is removed. Lockout is a positive method to ensure that an energy isolation device will not be inadvertently operated. Tagout must not be used if it is possible to install a lockout device. When tagout is employed at least one additional safety measure used as indicated in 120.2(F)(2)(k)(4).

(d) Tags shall contain a statement prohibiting unauthorized operation of the disconnecting means or removal of the tag.

(e) A hold card tagging tool on an overhead conductor in conjunction with a hotline tool to install the tagout device safely on a disconnect that is isolated from the work(s) shall be permitted. Where a hold card is used, the tagout procedure shall include the method of accounting for personnel who are working under the protection of the hold card.

For many years, utility systems have successfully relied on a hold card to provide warning that operating a disconnecting means would place employees in danger. As indicated in the preceding commentary, an additional measure, such as removing the cutout should be used in addition to the use of the hold card. The work environment for utility employees means that the location of the disconnecting means might be several miles away from the work site. Utility workers are trained to respect the system associated with the hold card, which results in a positive and effective system of energy control. Employees of utility systems must be covered by a written policy that describes how the hold card system functions. When contract employees perform utility maintenance or construction, the contractor must provide a program that is at least as effective as the program of the utility authorizing the contractor's work.

(5) Electrical Circuit Interlocks. Up-to-date diagrammatic drawings shall be consulted to ensure that no electrical circuit interlock operation can result in reenergizing the circuit being worked on.

In many instances, a diagrammatic drawing is used to determine the possibility for a circuit to be re-energized from backfeeds, either directly or from a "sneak" circuit. It is crucial that the drawing be up-to-date and it accurately depict the installation. As an installed circuit or system changes, single-line diagrams on record, schematic diagrams, or similar drawings should be marked to record the change in the system.

(6) Control Devices. Locks/tags shall be installed only on circuit disconnecting means. Control devices, such as pushbuttons or selector switches, shall not be used as the primary isolating device.

Some control devices are fitted with a mechanism intended to accept a lock. However, the control device is not be used as the primary means for de-energizing circuits or equipment. Control devices do not render a circuit inoperative; therefore, installing a tag or a lock on a control device simply does not meet the objective of a primary isolating device. If the device being locked or tagged does not create a break in the conductors providing energy to the equipment or circuit, the device must not be used as a lockout or tagout point.

(F) Procedures. The employer shall maintain a copy of the procedures required by this section and shall make the procedures available to all employees.

Although an employer is responsible for providing the lockout/tagout procedure, employees should be involved in gathering the information to generate the procedure. The procedure required by this section must be implemented as a step in the process of establishing an electrically safe work condition. The lockout/tagout procedure can be included in an overall lockout/tagout procedure for an employer or site, or it can be a stand-alone procedure. In either instance, all the requirements of 120.2 must be addressed.

(1) Planning. The procedure shall require planning, including the requirements of 120.2(F)(1)(a) through 120.2(F)(2)(n).

The lockout/tagout procedure must include a detailed method of achieving the objectives of 120.2(F)(1) through 120.2(F)(2)(n). A complete procedure must indicate whether the plan must be in writing and whether any authorization is necessary.

(a) Locating Sources. Up-to-date single-line drawings shall be considered a primary reference source for such information. When up-to-date drawings are not available, the employer shall be responsible for ensuring that an equally effective means of locating all sources of energy is employed.

A crucial requirement in the lockout/tagout procedure is to determine all possible sources of electrical supply to the specific equipment. Although applicable up-to-date drawings must be the initial source of information, when they are not available other sources can include manuals, panel schedules, and information tags. The system should be visually inspected so that all hazards are properly identified.

(b) Exposed Persons. The plan shall identify persons who might be exposed to an electrical hazard and the PPE required during the execution of the job or task.

The plan must identify those who might be exposed to an unacceptable risk from an electrical hazard, either directly or indirectly. The plan must also identify what PPE employees must use while they implement the lockout or tagout and during the execution of the job or task.

(c) Person In Charge. The plan shall identify the person in charge and his or her responsibility in the lockout/tagout.

Responsibilities of the person in charge might include implementing the energy control procedures, communicating the purpose of the operation to the servicing and maintenance employees, coordinating the operation, and ensuring that all procedural steps have been properly completed.

(d) Simple Lockout/Tagout. Simple lockout/tagout procedure shall be in accordance with 120.2(D)(1).

Lockout/tagout procedures that involve a qualified person(s) de-energizing one set of conductors or one circuit part source for the sole purpose of safeguarding employees from exposure to electrical hazards are considered to be a simple lockout/tagout. Section 120.2(D)(1) does not require simple lockout/tagout plans to be written for each application. Each worker should be responsible for his or her own lockout/tagout.

(e) Complex Lockout/Tagout. Complex lockout/tagout procedure shall be in accordance with 120.2(D)(2).

Section 120.2(D)(2) requires the complex lockout/tagout procedure to have a written plan of execution that identifies the person in charge, and this person will have responsibility for directly implementing the complex lockout/tagout procedure.

(2) Elements of Control. The procedure shall identify elements of control.

(a) De-energizing Equipment (Shutdown). The procedure shall establish the person who performs the switching and where and how to de-energize the load.

(b) Stored Energy. The procedure shall include requirements for releasing stored electric or mechanical energy that might endanger personnel. All capacitors shall be discharged, and high capacitance elements shall also be short-circuited and grounded before the associated equipment is touched or worked on. Springs shall be released or physical restraint shall be applied when necessary to immobilize mechanical equipment and pneumatic and hydraulic pressure reservoirs. Other sources of stored energy shall be blocked or otherwise relieved.

Although the most common energy source is electrical energy, the procedure should consider all forms of energy that might be present, such as hydraulic energy (liquid under pressure), pneumatic energy (gas/air under pressure), or mechanical motion (kinetic energy, energy of motion). The procedure must indicate the appropriate steps to control these sources of energy, such as the blocking of springs.

Electrical energy can be generated electrical power, stored energy such as in a capacitor or static electricity. Generated electrical power can be turned off or on, while stored or static electricity can only be dissipated or controlled.

(c) Disconnecting Means. The procedure shall identify how to verify that the circuit is de-energized (open).

Creating an electrically safe work condition always involves lack of voltage verification. In some cases, verifying that the disconnecting means is open can involve looking through an observation port or opening a door and observing the position of the contacts in the disconnecting means. In other cases, the contacts are not readily visible, and observation of the contacts is not possible, such as where arc shoots are involved. Employees must be instructed how to determine that all conductors are disconnected from the source of energy.

(d) Responsibility. The procedure shall identify the person who is responsible for verifying that the lockout/tagout procedure is implemented and who is responsible for ensuring that the task is completed prior to removing locks/tags. A mechanism to accomplish lockout/

tagout for multiple (complex) jobs/tasks where required, including the person responsible for coordination, shall be included.

(e) Verification. The procedure shall verify that equipment cannot be restarted. The equipment operating controls, such as pushbuttons, selector switches, and electrical interlocks, shall be operated or otherwise it shall be verified that the equipment cannot be restarted.

The operation of control devices, such as pushbuttons, are not to be relied upon as the only means to verify that equipment cannot be restarted when creating an electrically safe work condition. Additional measures may be necessary.

(f) Testing. The procedure shall establish the following:

(1) Voltage detector to be used, the required PPE, and the person who will use it to verify proper operation of the voltage detector before and after use

(2) Requirement to define the boundary of the electrically safe work condition

(3) Requirement to test before touching every exposed conductor or circuit part(s) within the defined boundary of the work area

(4) Requirement to retest for absence of voltage when circuit conditions change or when the job location has been left unattended

(5) Planning considerations that include methods of verification where there is no accessible exposed point to take voltage measurements

The procedure must identify the appropriate test instrument for the task, and the tool should be visually inspected for external defects or damage before being used to determine that all conductors and circuit parts have been de-energized.

If the work task has no exposed point for measuring voltage, the procedure or plan must identify the method for verifying that all conductors and circuit parts have been de-energized.

(g) Grounding. Grounding requirements for the circuit shall be established, including whether the temporary protective grounding equipment shall be installed for the duration of the task or is temporarily established by the procedure. Grounding needs or requirements shall be permitted to be covered in other work rules and might not be part of the lockout/tagout procedure.

Applying temporary protective grounding devices is an important part of establishing an electrically safe work condition, but it may not be required in all cases. Where it is possible for conductors or circuit parts being de-energized to contact other exposed energized conductors or circuit parts, temporary protective ground devices rated for the available fault duty should be applied. Therefore, the procedure or plan must define any requirement for placement of temporary protective grounding equipment. If other work rules consider and address temporary protective grounding, the lockout/tagout procedure or plan is not required to address the same issue, but employees must be aware that the lockout/tagout procedure is augmented by other work rules that cover the use of temporary protective grounding equipment. Establishing general rules aids an employee who is attempting to determine if use of temporary protective grounding equipment is necessary.

(h) Shift Change. A method shall be identified in the procedure to transfer responsibility for lockout/tagout to another person or to the person in charge when the job or task extends beyond one shift.

When work extends beyond one shift and there are personnel changes, the procedure must ensure the continuity of lockout or tagout protection by providing for the orderly transfer of responsibilities between off-going and incoming employees. The person in charge of the off-going shift must transfer their responsibility to the incoming shift person in charge. Employees who installed locks and tags should remove their locks and tags, and the incoming employees must install their individual locks and tags.

The lockout/tagout procedure must define the method to be followed in transferring responsibility from one person in charge to another and any requirement for employees to remove locks and tags or to replace them should they be required to continue work on the task when they return to work.

(i) Coordination. The procedure shall establish how coordination is accomplished with other jobs or tasks in progress, including related jobs or tasks at remote locations, including the person responsible for coordination.

When more than one task or job is in progress, the actions of one employee might have an impact on the actions of others. One person must be assigned responsibility for ensuring that work tasks by others — such as different tradesmen or contractors — are coordinated, to minimize the possibility of one person's actions adversely impacting another's.

(j) Accountability for Personnel. A method shall be identified in the procedure to account for all persons who could be exposed to hazardous energy during the lockout/tagout.

Typically, the person in charge [see 120.2(F)(1)(c)] is assigned to account for everyone involved in the work task, including contract employees.

(k) Lockout/Tagout Application. The procedure shall clearly identify when and where lockout applies, in addition to when and where tagout applies, and shall address the following:

The procedure must define when and where lockout is applied and when tagout is acceptable. Tagout is permitted, without a lock, only in cases where equipment design precludes the installation of a lock. If tagout is permitted, the procedure or plan must define clearly and unambiguously individual responsibility and accountability for each person potentially exposed to an electrical hazard. In addition, all employees must receive training regarding the limitations of tags.

(1) Lockout shall be defined as installing a lockout device on all sources of hazardous energy such that operation of the disconnecting means is prohibited and forcible removal of the lock is required to operate the disconnecting means.

(2) Tagout shall be defined as installing a tagout device on all sources of hazardous energy, such that operation of the disconnecting means is prohibited. The tagout device shall be installed in the same position available for the lockout device.

Tagout devices are intended to ensure equipment is not re-energized inadvertently. The tagout device must be easily visible and recognizable, must be installed in the same position available for the lockout device, and must be suitable for the environment in which it is used.

(3) Where it is not possible to attach a lock to existing disconnecting means, the disconnecting means shall not be used as the only means to put the circuit in an electrically safe work condition.

A disconnecting means that does not accept a lockout device must not be the only means of controlling the energy source. When tagout is the authorized method of controlling exposure to electrical energy, the tagout must be supplemented by at least one additional safety measure.

(4) The use of tagout procedures without a lock shall be permitted only in cases where equipment design precludes the installation of a lock on an energy isolation device(s). When tagout is employed, at least one additional safety measure shall be employed. In such cases, the procedure shall clearly establish responsibilities and accountability for each person who might be exposed to electrical hazards.

Equipment installed prior to January 2, 1990, might not have a means to attach a lock. If the equipment cannot be locked, tagout is permitted when an additional safety measure is employed that is equivalent to the level of safety obtained by applying a lock, and the employer's procedure must clearly identify the acceptable methods that might be used as the additional safety measure. The informational note to this section provides examples of additional safety measures.

Informational Note: Examples of additional safety measures include the removal of an isolating circuit element such as fuses, blocking of the controlling switch, or opening an extra disconnecting device to reduce the likelihood of inadvertent energization.

(l) Removal of Lockout/Tagout Devices. The procedure shall identify the details for removing locks or tags when the installing individual is unavailable. When locks or tags are removed by someone other than the installer, the employer shall attempt to locate that person prior to removing the lock or tag. When the lock or tag is removed because the installer is unavailable, the installer shall be informed prior to returning to work.

The procedure for removal of a lockout/tagout under these circumstances should include measures to ensure that the lockout/tagout can be safely removed and that the employee who removes the lockout/tagout device documents its removal. The person who originally installed the device must be informed that their lockout/tagout device was removed before they return to work and should be briefed on the current status of the equipment.

(m) Release for Return to Service. The procedure shall identify steps to be taken when the job or task requiring lockout/tagout is completed. Before electric circuits or equipment are reenergized, appropriate tests and visual inspections shall be conducted to verify that all tools, mechanical restraints and electrical jumpers, short circuits, and temporary protective grounding equipment have been removed, so that the circuits and equipment are in a condition to be safely energized. Where appropriate, the employees responsible for operating the machines or process shall be notified when circuits and equipment are ready to be energized, and such employees shall provide assistance as necessary to safely energize the circuits and equipment. The procedure shall contain a statement requiring the area to be inspected to ensure that nonessential items have been removed. One such step shall ensure that all personnel are clear of exposure to dangerous conditions resulting from reenergizing the service and that blocked mechanical equipment or grounded equipment is cleared and prepared for return to service.

Inspection of the work area should also ensure that all guards and covers are installed properly and that everyone in the area is notified that the lockout/tagout devices will be removed and the equipment could start. The person in charge must make sure that all employees are safely positioned or removed from the area before removing the lockout devices.

(n) Temporary Release for Testing/Positioning. The procedure shall clearly identify the steps and qualified persons' responsibilities when the job or task requiring lockout/tagout is to be interrupted temporarily for testing or positioning of equipment; then the steps shall be identical to the steps for return to service.

If the equipment is not going to be returned to service after testing or repositioning, all lockout/tagout devices should be reinstalled immediately. Temporarily restoring electrical energy to reposition equipment or to facilitate testing is very hazardous and should be avoided.

> Informational Note: See 110.4(A) for requirements when using test instruments and equipment.

120.3 Temporary Protective Grounding Equipment.

The purpose of temporary grounding is to provide protection against electrical shock to personnel while working on de-energized circuits. Temporary protective grounds are critical safety devices used to create a circuit path so that the circuit overcurrent protective device can operate upon accidental energizing of the circuit. De-energized circuits could be accidentally energized from many energy sources, including human error (such as switching errors), equipment failure, mechanical failure, stored charges from capacitors, static buildup, or induced voltage feedback from adjacent circuits.

Grounding devices are available in many forms. Most commonly used are the clamp and cable type. Some equipment manufacturers provide grounding devices for switchgear, which are built on a frame for inserting into a space or cubicle that normally holds a circuit breaker or fusible switch. Selecting the appropriate temporary grounding device with an established fault-duty rating for the circuit is paramount. See Exhibit 120.5 for a typical set of safety grounds.

See Exhibit 250.1 for a manual grounding and test device.

(A) Placement. Temporary protective grounding equipment shall be placed at such locations and arranged in such a manner as to prevent each employee from being exposed to a shock hazard (hazardous differences in electrical potential). The location, sizing, and application of temporary protective grounding equipment shall be identified as part of the employer's job planning.

A worker must fully understand the system and the equipment to be worked on to select the appropriate temporary protective grounding device(s) and to determine their placement. The protective grounding device(s) must be installed on the conductor at a point between the person and the source of energy. Care must be taken to properly locate the conductors so that the movement of the conductors from mechanical forces does not cause harm to the employee. Equipment grounding conductors or equipment grounding terminals are a likely point of attachment for temporary protective grounding equipment. If the unexpected source of electricity could be from both directions in the electrical circuit, temporary protective grounding equipment must be installed on both sides of the worker.

(B) Capacity. Temporary protective grounding equipment shall be capable of conducting the maximum fault current that could flow at the point of grounding for the time necessary to clear the fault.

Temporary safety grounding equipment must also be capable of withstanding the high mechanical stress imposed when conducting fault current. The temporary protective grounding equipment must have a rating established by the manufacturer, and it must be applied within that rating. To establish the necessary rating of the temporary protective grounding equipment, the maximum available fault current must be determined and the equipment rating, which must be at least as great as the available fault current.

(C) Equipment Approval. Temporary protective grounding equipment shall meet the requirements of ASTM F855, *Standard Specification for Temporary Protective Grounds to be Used on De-energized Electric Power Lines and Equipment.*

(D) Impedance. Temporary protective grounding equipment and connections shall have an impedance low enough to cause immediate operation of protective devices in case of accidental energizing of the electric conductors or circuit parts.

The objective of temporary protective grounding is to create a circuit path so that the overcurrent protective device can operate upon accidental energizing of the conductor that has been temporarily grounded.

The impedance of the ground-fault current return path through earth should be verified on a frequency determined through use. If the impedance of the ground-fault return path is high, the overcurrent device is likely to not operate or might not operate rapidly enough to protect the worker and limit damage to the equipment. Therefore, creating a path for ground-fault return current other than through the earth is necessary to achieve the immediate overcurrent device operation upon accidental energizing of conductors or of equipment that is being worked on by personnel.

Cable size, cable length, and proper attachment of temporary protective grounding equipment are all important factors to be considered in creating a low impedance path. Cables must be of adequate length for the task but should not interfere with the worker within the defined boundary of the work area. Excessive cable length could increase cable impedance and worker exposure when conducting fault current and should be avoided to reduce possible injury to workers due to cable whipping action from high fault currents.

Work Involving Electrical Hazards   Article | 130

Summary of Article 130 Changes

130.1: Revised to clarify the scope of Article 130.

130.2(2): Changed *risk of injury* to *likelihood of injury*.

130.2(A)(1): Changed the title from *Greater Hazard* to *Additional Hazards or Increased Risk* to clarify and promote the consistent use of terminology associated with hazard and risk.

130.2(A)(4): Added to provide requirements where normal operation of electric equipment is permitted.

130.2(A)(4) Informational Note: Added to explain the meaning of the phrases *properly installed, properly maintained*, and *evidence of impending failure*.

130.2(B)(1): Changed *limited approach boundary* to *restricted approach boundary*. Revised to simplify the requirement for the need for an energized work permit.

130.2(B)(2)(4): Replaced the phrase *shock hazard analysis* with *shock risk assessment* and deleted reference to *prohibited approach boundary* as it no longer appears throughout this standard.

130.2(B)(2)(5): Replaced the phrase *arc flash hazard analysis* with *arc flash risk assessment*.

130.2(B)(3): Revised into a list format and added item referencing general housekeeping and miscellaneous non-electrical tasks.

130.4: Added the phrase *for Shock Protection* to the title.

130.4(A): Changed the phrase *hazard analysis* to *risk assessment* and *personal protective equipment* to *PPE* to promote consistent use of terminology.

130.4(B): Deleted the phrase *prohibited approach boundary,* as it no longer appears throughout this standard.

130.4(C): Relocated former 130.4(C) to 130.4(D) and former 130.4(D) to 130.4(C). Revised former 130.4(D) into three sections regarding limited approach boundary: (C)(1) Approach by Unqualified Persons, (C)(2) Working at or Close to the Limited Approach Boundary, and (C)(3) Entering the Limited Approach Boundary.

130.4(D): Relocated former 130.4(D) to 130.4(C) and former 130.4(C) to 130.4(D).

130.4(D)(1): Revised former 130.4(C)(1) to replace the phrase *cross the prohibited approach boundary* with *contact exposed energized conductors or circuit parts,* and updated the table references.

130.4(D)(3): Revised former 130.4(C)(2) to delete the phrase *bare-hand work.*

Table 130.4(D)(a): Several revisions were made to this table:

- Voltage range in the first row was adjusted from *50 V–300 V* to *50 V–150 V*
- Voltage range in the second row was adjusted from *300 V–750 V* to *151 V–750 V.*
- Revised Footnote "a" to specify single-phase systems above 250 V. Added a new Footnote "d" to clarify that the voltage range includes circuits where the exposure does not exceed 120 V.

130.5: Deleted first paragraph and exception. Replaced the phrase *arc flash hazard analysis* with *arc flash risk assessment.* Revised paragraphs into list format for usability. Updated subsections for (A) Documentation, (B) Arc Flash Boundary, (C) Arc Flash PPE, and (D) Equipment Labeling.

130.6(A)(1): Replaced the term *Hazardous* with the phrase *Electrical Hazards Might Exist* in the title to promote the consistent use of terminology.

130.6(C)(1): Replaced the term *containing* with *where* to promote the consistent use of terminology associated with hazard and risk.

130.6(D): Added the phrase *within the restricted approach boundary* to correlate with the standard.

130.6(H): Added new section *Clear Spaces* to clarify that working space requirements are found in other codes and standards and to provide information on maintaining clear working space in front of electrical equipment.

130.6(M): Added the phrase *from the design of the circuit and the overcurrent devices involved* to the second sentence. This change returns the requirement to the 2009 edition text.

130.6(N): Relocated former 130.3(B)(2) to new 130.6(N).

130.7(A) Informational Note No. 3: Revised to address electrical hazards.

130.7(C)(8) Informational Note: Replaced the term *shoes* with *footwear* to align with the ASTM family of standards.

130.7(C)(9)(a): Revised to clarify that nonmelting, flammable fiber garments used as underlayers do not contribute to the overall arc rating of the layering system. Added the term *coverall* and the phrase *shirts shall be tucked into pants.*

130.7(C)(10)(2)(a): Added the phrase *and the anticipated incident energy exposure is greater than 4 cal/cm²* to clarify when a balaclava is required to be worn.

130.7(C)(11) Exception: Updated ASTM standards to reflect current titles.

130.7(C)(12) Exception: Deleted the use of *Hazard/Risk Category 0* to clarify and promote the consistent use of terminology associated with hazard and risk.

130.7(C)(13): Revised to move reference to ASTM F1506 to the informational note, and added second informational note for both ASTM F2757 and ASTM F1449. Referencing these standards provides useful information on laundering flame, thermal, and arc resistant clothing.

Table 130.7(C)(14): Updated edition dates and titles.

Table 130.7(C)(15)(A): Combined former Table 130.7(C)(15)(a) and Table 130.7(C)(15)(b) to simplify use of the standard by eliminating redundant information. Deleted all references pertaining to rubber insulating gloves and insulating hands tools. Removed Notes 5 and 6. Expanded existing Note 4. New Note 1 includes information on current-limiting circuit breakers. Revised arc flash boundary distances to be in conformance with the *NEC Manual of Style.* Additional text and other editorial revisions were made for clarity.

Table 130.7(C)(16): Deleted the phrase *Protective Clothing and* from title and the reference to *hazard/risk category 0.* Revised Note 1 to correlate with the removal of requirements for rubber insulating gloves from new Table 130.7(C)(15)(b). Added the term *arc flash* before *PPE* for usability. Additional editorial changes made for clarity.

130.7(D)(1): Changed the term *limited* to *restricted* and the phrase *Fuse Holding* to *Fuseholder Handling.* Deleted references to Table 130.7(C)(15)(a) and Table 130.7(C)(15)(b) for tasks that require insulated and insulating hand tools.

130.7(E)(2): Replaced the term *cause* with the phrase *increase the likelihood of exposure to.* (Conductive objects do not cause injury or damage to health; rather, they increase the likelihood of exposure to injury or damage to health.)

130.7(F): Revised to update edition dates and titles.

130.8: Deleted the term *Uninsulated* from title.

130.8(E): Changed table references from Table 130.4(C)(a) to Table 130.4(D)(a) and Table 130.4(C)(b) to Table 130.4(D)(b).

130.9: Relocated former 110.5 to 130.9 for underground electrical lines. The phrase *risk of contracting* was replaced with *reasonable possibility of contracting,* and *personal protective equipment (PPE)* was simplified to *PPE.*

130.10: Added section *Cutting and Drilling* with new requirements regarding penetrating a floor, wall, or equipment to address penetration through walls or floors into a space that contains possible unknown electrical hazards.

While Article 120 describes the process of creating an electrically safe work condition (ESWC), Article 130 defines the situations under which an electrically safe work condition must be established. As the requirements in *NFPA 70E* and OSHA regulations indicate, users should work de-energized unless energized work can be justified. Section 130.2(A) explains the situations under which energized electrical work can be justified, as well as the necessary requirements for working safely with energized electrical equipment.

130.1 General.

Article 130 covers the following:

(1) When an electrically safe work condition must be established
(2) The electrical safety-related work practices when an electrically safe work condition cannot be established

All requirements of this article shall apply whether an incident energy analysis is completed or if Table 130.7(C)(15)(A)(a), Table 130.7(C)(15)(A)(b), Table 130.7(C)(15)(B), and Table 130.7(C)(16) are used in lieu of an incident energy analysis in accordance with 130.5.

There are two methods for determining the arc-rated PPE required to performing a task safely when energized electrical equipment is involved:

1. The arc flash PPE categories method uses Table 130.7(C)(15)(A)(a), 130.7(C)(15)(A)(b), Table 130.7(15)(B), and Table 130.7(C)(16) to determine when PPE is required, the category of PPE required, and the arc flash boundary. It is important to note that even when this method is used, all of the requirements included in Article 130 are still applicable. It is necessary to review all the applicable sections in Article 130 and determine if additional PPE not indicated in the tables is required.

2. The incident energy level method consists of a variety of possible calculation approaches. It is up to the user to determine which ones are applicable, under what conditions they are applicable, and which calculation method should be used. The information in Informative Annex D may be of assistance in this regard.

Each of the available methods should be applied within their designated parameters. Neither method should be applied when determined values are outside of the allowable parameter ranges.

It should be noted that Table 130.7(C)(15)(B) is only applicable to open-air situations, not to arc-in-a-box situations.

130.2 Electrically Safe Working Conditions.

Energized electrical conductors and circuit parts shall be put into an electrically safe work condition before an employee performs work if any of the following conditions exist:

(1) The employee is within the limited approach boundary.
(2) The employee interacts with equipment where conductors or circuit parts are not exposed but an increased likelihood of injury from an exposure to an arc flash hazard exists.

Under both *NFPA 70E* and OSHA, work is required to be performed in a verified de-energized state — referred to as "creating an electrical safe work condition (ESWC)" in *NFPA 70E*, unless energized work can be justified. The requirements for creating an ESWC are found in Article 120. If an employee is within the limited approach boundary or interacts with energized electrical equipment, conductors, or circuit parts in such a manner that there is a likelihood of injury from exposure to a potential arc flash event, the energized conductors and circuit parts are to be put into an ESWC unless the work falls under one of the exceptions allowed in 130.2(A), such as for diagnostics (testing, trouble-shooting, or voltage measurement). The act of creating an ESWC uses the safety control of elimination from the hierarchy of safety controls and is considered the most effective form of safety control available.

The risk of exposure to a possible arc flash incident increases where an employee is within the restricted approach boundary of exposed energized conductors and circuit parts. In 130.2, the boundary for establishing an ESWC is the limited approach boundary. However, the point at which an energized electrical work permit is required is the restricted approach boundary. Unqualified employees are allowed inside the limited approach boundary when they are aware of the hazards and continuously escorted by a qualified person. However, only qualified persons are allowed within the restricted approach boundary at any time.

There are three conditions given in 130.2(A) under which energized work is permitted. It is understood that under *NFPA 70E* and 29 CFR 1926.416, infeasibility is not a valid justification for working energized.

Exception: Where a disconnecting means or isolating element that has been properly installed and maintained is operated, opened, closed, removed, or inserted to achieve an electrically safe work condition for connected equipment or to return connected equipment to service that has been placed in an electrically safe work condition, the equipment supplying the disconnecting means or isolating element shall not be required to be placed in an electrically safe work condition provided a risk assessment is performed and does not identify unacceptable risks for the task.

If an enclosed fusible safety disconnect or an enclosure circuit breaker disconnect is nippled to an industrial control panel and used to create an ESWC, since the live conductors to the main lugs of the disconnect are not exposed when the industrial control panel doors are opened, the risk assessment should identify the risk as acceptable, unless there is another power source into the industrial control panel that has not been properly locked out. A similar situation is where the industrial control panel has two doors and the compartment containing the disconnect switch has a separate door and is completely barriered and sealed off from the other compartment in a manner similar to that of a nippled enclosed disconnect. Under either of these scenarios, the equipment supplying the disconnect would not be required to be put in an ESWC.

In OSHA's letter of interpretation to Mr. Rick Kante dated July 28, 2006, OSHA indicated in its answer to Question 1, where the disconnect is located inside a panelboard enclosure and is not barriered off (dead front) and the main lugs are energized, the panelboard is considered energized and not considered to be in a verified de-energized locked-out/tagged-out state (i.e., in an ESWC). The same would be true if it were an industrial control panel instead of a panelboard.

(A) Energized Work.
(1) Additional Hazards or Increased Risk. Energized work shall be permitted where the employer can demonstrate that de-energizing introduces additional hazards or increased risk.

The additional hazards do not have to be electrical hazards, and the increased risk does not have to be from electrical hazards; they can be chemical, mechanical, or environmental hazards — such as those associated with chemical plants, refineries, or ethanol production facilities. Informational Note No. 1 after 130.2(A)(3) provides some examples of equipment and processes that could justify energized electrical work due to new hazards or increased risk of exposure from existing potential hazards.

(2) Infeasibility. Energized work shall be permitted where the employer can demonstrate that the task to be performed is infeasible in a de-energized state due to equipment design or operational limitations.

Infeasible does not mean *impractical*. In many situations, due to equipment design limitations, diagnostic work such as voltage measurement, troubleshooting, and testing of electrical equipment is infeasible to perform without the employee being exposed to energized conductors and circuit parts. With the advent of remote programming and monitoring ports, however, what is considered to be infeasible may be changing. If it is feasible to install a remote programming port to program a motor control center bucket electronic overload relay, opening the door and exposing an employee to energized electrical conductors and circuit parts may no longer be justified. What is considered infeasible for older designed equipment may not be infeasible for newly designed equipment.

(3) Less Than 50 Volts. Energized electrical conductors and circuit parts that operate at less than 50 volts shall not be required to be de-energized where the capacity of the source and any overcurrent protection between the energy source and the worker are considered and it is determined that there will be no increased exposure to electrical burns or to explosion due to electric arcs.

Only those systems operating at 50 volts and less — where there is an appropriately sized overcurrent protective device (OCPD) located between the worker and the source, and the characteristics of the device and the capacity of the source considered together do not increase the risk of exposure to electrical burns or to explosions from electrical arcs — fall under this exception. Human skin provides fair to good protection from exposure to the normal elements, but it only offers modest protection from exposure to certain electrical incidents. Burn injuries from the use of electricity can be one of three types: an electrical burn that occurs when sufficient electrical current flows through the body and the I^2R heat from the resistance of the tissue and organs that it flows through cannot be dissipated fast enough; a thermal burn resulting from when the skin touches a hot surface or from when clothing catches on fire; and an arc flash burn from the radiation from an arc flash event. An arc flash event can occur provided there is a large enough battery (even if less than 50 volts), or as a result of discharge from a capacitor. If a regular screwdriver is left across the terminals of even a 12-volt battery for long enough, it can become red hot and can easily burn skin if touched without appropriate protection.

In most situations where the voltage to energized conductors and circuit parts is less than 50 volts, the risk of being injured from touching energized conductors or circuit parts is reduced to an acceptable level. However, where the skin is broken or the skin or clothing is wet, the resistance to electrical shock can be as low as 500 ohms. For an alternating current circuit, a shock lasting 1 second or longer at a level of about 40 mA can be fatal due to ventricular fibrillation. For a body resistance of 500 ohms, this would only take 20 volts rms $(40/100 \times 500)$ or greater. Based on the information in 340.5, an employer might consider using a value of 30 volts rms, instead of 50 volts for ac circuits where there is a potential for contact with exposed energized conductors and circuit parts.

> Informational Note No. 1: Examples of additional hazards or increased risk include, but are not limited to, interruption of life-support equipment, deactivation of emergency alarm systems, and shutdown of hazardous location ventilation equipment.
>
> Informational Note No. 2: Examples of work that might be performed within the limited approach boundary of exposed energized electrical conductors or circuit parts because of infeasibility due to equipment design or operational limitations include performing diagnostics and testing (for example, start-up or troubleshooting) of electric circuits that can only be performed with the circuit energized and work on circuits that form an integral part of a continuous process that would otherwise need to be completely shut down in order to permit work on one circuit or piece of equipment.

(4) Normal Operation. Normal operation of electric equipment shall be permitted where all of the following conditions are satisfied:

(1)   The equipment is properly installed.
(2)   The equipment is properly maintained.

Scenario and Causation.

Ivan, a 75-year old electrician and owner of an electrical contracting firm, had 50 years of experience as an electrician. Ivan was contracted to install conduit, branch circuit conductors, and a new circuit breaker for a new aggregate wash plant and water pump. Robert, the plant supervisor, asked Ivan if the old wash plant could be operated while he worked on the new electrical installation. Ivan replied that the old plant could be operated. A few minutes later, while feeding material into the old wash plant, Jake, the loader operator, observed that the old wash plant had unexpectedly shut down.

Results.

Jake went into the building housing the electrical equipment to investigate the cause of the wash plant shutdown. This building contained the 480-volt motor control center supplying the motors associated with the wash plant and is where Ivan was working to install the circuit and equipment for the new wash plant. Upon entering the room, Jake found Ivan leaning against exposed and energized 480-volt circuit parts contained within an electrical equipment enclosure. Jake realized that the electrical equipment was still energized and rushed to notify Robert, who immediately called for emergency medical assistance and contacted the fire department. Emergency rescue personnel arrived to find the electrical equipment building on fire. They first disconnected the power to the building and then proceeded to assist Ivan. By that time he was pronounced dead at the scene, and the cause of death was attributed to electrocution.

Analysis.

Accidents typically occur due to a chain of events or mistakes that lead up to the actual event. If one of the links in the chain of events had been eliminated, the accident most likely would not have happened.

The fundamental failure in this situation was that work was performed on energized electrical equipment. Ivan's death could have been prevented if he had not attempted to perform this task while the equipment was energized. There is no evidence to suggest that it was either infeasible or that a greater hazard would have been created if the power to the motor control center had been turned off and an electrically safe work condition was established in order to perform this task. This task could have been performed after the normal operating hours of the material wash plant.

Additionally, it is apparent that the host employer and the contract employee failed to hold a documented safety meeting, and the host employer did not provide any safety-related work rules that the contract employee was to follow or information about the host employer's installation. It is also possible that Ivan's life could have been saved if immediate action had been taken to release him from the electric circuit responsible for his death. In accordance with 110.2(C), employees responsible for taking action in case of a shock hazard emergency are to be trained in methods of release of victims from contact with energized electrical conductors or circuit parts.

Relevant *NFPA 70E* Requirements.

If Ivan and the company he was working for had followed all of the requirements of *NFPA 70E*, Ivan's death could have been prevented. Some of the steps they should have taken include the following:

• Create an electrically safe work condition by disconnecting the power to the motor control center per 130.2.

- Create an energized electrical work permit (EEWP) per 130.2(B).
- Perform shock and arc flash risk assessments per 130.4 and 130.5.
- Use proper PPE for protection against shock and arc flash hazards per 130.7.
- Install protective shields or barriers, rubber insulating equipment, or voltage-rated plastic guard equipment per 130.7(D)(1)(f), (g), or (h).
- Provide employees with emergency response training per 110.2(C).

(3)   The equipment doors are closed and secured.

(4)   All equipment covers are in place and secured.

(5)   There is no evidence of impending failure.

> Informational Note: The phrase *properly installed* means that the equipment is installed in accordance with applicable industry codes and standards and the manufacturer's recommendations. The phrase *properly maintained* means that the equipment has been maintained in accordance with the manufacturer's recommendations and applicable industry codes and standards. The phrase *evidence of impending failure* means that there is evidence such as arcing, overheating, loose or bound equipment parts, visible damage, or deterioration.

When equipment is considered to be operating normally, the risk associated with normal operation is generally considered to be acceptable. There may be some unusual cases, such as where rodents or snakes can get into equipment through unsealed openings.

(B) Energized Electrical Work Permit.

No matter how simple, there is no such thing as routine work where electrical safety is involved. Before issuing an energized electrical work permit (EEWP) for simple, repetitive, or similar tasks for a period of time such as 1 to 3 months, due consideration must be given to the fact that, where exposed energized conductors and circuit parts are involved, no work should be considered as "routine" work — even if the employee is trained to understand the risk associated with exposure to the potential electrical hazards that can be involved and is wearing appropriate PPE, as situations can easily change.

(1) When Required. When energized work is permitted in accordance with 130.2(A), an energized electrical work permit shall be required under the following conditions:

(1)   When work is performed within the restricted approach boundary

(2)   When the employee interacts with the equipment when conductors or circuit parts are not exposed but an increased likelihood of injury from an exposure to an arc flash hazard exists

When the conditions of this section are met, an energized electrical work permit (EEWP) is only required when the work is to be performed within the restricted approach boundary or when the qualified employee is interacting with the equipment in a manner that increases the likelihood of injury from exposure to an arc flash incident, such as when racking equipment in-or-out, or when opening up doors or removing covers that expose energized conductors or circuit parts.

It is a good idea to consider the use of an EEWP even when one is not required, because this demonstrates that proper consideration was given to all aspects required to protect an employee when diagnostic work is being done.

(2) Elements of Work Permit. The energized electrical work permit shall include, but not be limited to, the following items:

(1) Description of the circuit and equipment to be worked on and their location
(2) Justification for why the work must be performed in an energized condition *[see 130.2(A)]*
(3) Description of the safe work practices to be employed *[see 130.3]*
(4) Results of the shock risk assessment *[see 130.4(A)]*

    a. Voltage to which personnel will be exposed
    b. Limited approach boundary *[see 130.4(B), Table 130.4(D)(a), and Table 130.4(D)(b)]*
    c. Restricted approach boundary *[see 130.4(B) and Table 130.4(D)(a) and Table 130.4(D)(b)]*

    d. Necessary personal and other protective equipment to safely perform the assigned task *[see 130.4(C), 130.7(C)(1) through (C)(16), Table 130.7(C)(15)(A)(a), Table 130.7(C)(16), and 130.7(D)]*

(5) Results of the arc flash risk assessment *[see 130.5]*

    a. Available incident energy at the working distance or arc flash PPE category *[see 130.5]*
    b. Necessary PPE to protect against the hazard *[see 130.5(C), 130.7(C)(1) through (C)(16), Table 130.7(C)(15)(A)(a), Table 130.7(C)(16), and 130.7(D)]*
    c. Arc flash boundary *[see 130.5(B)]*

(6) Means employed to restrict the access of unqualified persons from the work area *[see 130.3]*
(7) Evidence of completion of a job briefing, including a discussion of any job-specific hazards *[see 130.3]*
(8) Energized work approval (authorizing or responsible management, safety officer, or owner, etc.) signature(s)

    Informational Note: For an example of an acceptable energized work permit, see Figure J.1.

This section has been revised in this edition to clarify that the task to be performed needs to be clearly listed on the permit. The term *prohibited approach boundary* was deleted throughout the standard, including here. The phrase *voltage to which personnel will be exposed* was added to item (4) to clarify the shock boundaries and PPE to be used for shock protection. An example of an energized electrical work permit is given in Figure J.1 in Informative Annex J.

(3) Exemptions to Work Permit. An energized electrical work permit shall not be required if a qualified person is provided with and uses appropriate safe work practices and PPE in accordance with Chapter 1 under any of the following conditions:

(1) Testing, troubleshooting, and voltage measuring
(2) Thermography and visual inspections if the restricted approach boundary is not crossed
(3) Access to and egress from an area with energized electrical equipment if no electrical work is performed and the restricted approach boundary is not crossed
(4) General housekeeping and miscellaneous non-electrical tasks if the restricted approach boundary is not crossed

This section has been revised to improve understanding of when an energized electrical work permit energized electrical work permit (EEWP) is not required — such as for diagnostic tasks. However it might be a good idea to use an EEWP even when it is not required. Should the work task change to include removing, installing, or maintaining conductors or circuit parts, the work is no longer considered diagnostic, and an EEWP is required. A work function is diagnostic when it is used to identify the condition of equipment or its parts to assure the condition of the equipment or to identify and eliminate potential problems.

When energized electrical equipment is operating normally, the risk of injury is generally considered acceptable. Under such conditions, access and egress, or the performance of housekeeping and miscellaneous non-electrical tasks in areas where energized electrical equipment is located, is generally considered to be acceptable without the requirement of an EEWP, even for unqualified persons. Where a normal operating condition is not considered to exist — such as when there are energized conductors and circuit parts that are exposed — access and egress or the performance of housekeeping and miscellaneous non-electrical tasks in areas where energized electrical equipment is located is acceptable, even for unqualified persons, provided they are outside the limited approach boundary and any arc flash boundary that is considered to exist, and an EEWP is not required.

If an unqualified person is inside the limited approach boundary or in an area where an arc flash boundary is considered to exist but is outside the restricted approach boundary, this person needs to be continuously escorted by a qualified person to perform such activities and wear appropriate PPE if an arc flash boundary is considered to exist, and an EEWP is not required. For a qualified person performing such activity, they do not need to be escorted, and an EEWP is not required.

130.3 Working While Exposed to Electrical Hazards.

The terminology in 130.3 has been revised to be consistent with the new definitions of *hazard* and *risk*. The term *hazard analysis* has been replaced with the term *risk assessment*. These changes make the terminology consistent with other revised sections of this standard. This section has been revised to refer to *shock risk assessment* and *arc flash risk assessment* in accordance with the other revisions.

Under normal conditions, electrical conductors energized at a voltage less than 50 volts do not present an electrical shock hazard. However, a thermal hazard can exist in circuits that have a significant capacity to deliver energy, even when the voltage level is less than 50 volts. For instance, battery installations can be connected so that arcing resulting from a short circuit could present a significant thermal hazard. Many control circuits operate at a voltage level less than 50 volts. Creating an open circuit or short circuit in one of these control circuits could result in a different type of hazard. An interruption or other unintended action within an industrial process can result in exposure to a chemical hazard or creation of an unacceptable environmental condition.

When an installation is in compliance with the *NEC* and the manufacturer's installation instructions, it is considered to be operating normally. Equipment that is considered to be operating normally is generally considered safe, barring extremely unusual circumstances. However, when electrical equipment changes state — such as when equipment is switched from energized to de-energized, an overload relay is reset, a door is opened or closed, a circuit breaker is reset, or some other physical movement occurs — the result might be an initiation of an arcing fault. Depending on the state and condition of the equipment and the circuit protective devices, an employee could be exposed to this arcing fault.

Safety-related work practices must be consistent with the parameters of the hazard. The PPE used by the employee must be consistent with the characteristics of the hazard and exposure. Selecting and using PPE is considered a work practice. Protective equipment must be selected based on the degree of the hazard. The boundaries determine whether risk of injury exists; if no risk of injury exists because there is no exposure, then PPE is not necessary.

It should be presumed that there is an unacceptable risk of injury from shock or thermal hazards — arc flash, electrical burn, or thermal burn — from exposure to energized conductors and circuit parts operating at greater than or equal to 50 volts, unless the required risk assessment finds otherwise. Employers are required to provide procedures for employees to prevent injury. Work that is performed on exposed energized electrical conductors or a circuit part is dangerous. Work on exposed conductors that does not take place within an electrically safe work condition exposes employees to potential injury. Only qualified persons should perform work on energized electrical conductors or circuit parts, and only where live work can be justified.

Safety-related work practices shall be used to safeguard employees from injury while they are exposed to electrical hazards from electrical conductors or circuit parts that are or can become energized. The specific safety-related work practices shall be consistent with the electrical hazards and the associated risk. Appropriate safety-related work practices shall be determined before any person is exposed to the electrical hazards involved by using both shock risk assessment and arc flash risk assessment. Only qualified persons shall be permitted to work on electrical conductors or circuit parts that have not been put into an electrically safe work condition.

•

Employers are required to provide procedures that prevent injury to employees. Work that is performed on exposed energized electrical conductors or circuit parts is dangerous. Work on conductors that are not in an electrically safe work condition (ESWC) exposes employees to potential injury. Only qualified persons may perform work within the restricted approach boundary and the arc flash protection boundary of these conductors. Employees must use work procedures that are consistent with the degree of the potential hazard. For instance, if the potential hazard includes possible exposure to an arc flash, the employee must select arc-rated clothing and other PPE that is at least as protective as the potential hazard is dangerous.

The primary protective strategy must be to establish an ESWC, which is a form of the safety control of elimination from the hierarchy of safety controls. After this strategy is executed, all electrical energy has been removed from all conductors and circuit parts to which the employee could be exposed. After the ESWC has been established, no PPE is required, and unqualified workers are permitted to execute the remainder of the work. The only exception to this requirement is where working on exposed energized electrical conductors or circuit parts can be justified as described under one of the two conditions listed in this section. Except for voltage measuring, troubleshooting, and diagnostic tasks, justification for work within the limited approach or arc flash boundary of exposed energized electrical conductors or circuit parts must be in writing.

Some functions require a circuit to be energized. A qualified person can perform work on exposed energized conductors or circuit parts under one of the following two conditions:

1. De-energizing the conductors or equipment could result in an increased hazard. For instance, a hazardous area ventilation system may depend on the continuity of electrical power.
2. De-energizing the conductors or equipment could require a complete shutdown of a continuous process. For instance, the design of the electrical circuit is such that a continuous processing facility must be taken completely out of production. Employees are often reluctant to question a decision by a supervisor that a work task must be conducted while the circuit remains energized. Section 130.2(A) requires the employer to demonstrate that de-energizing introduces additional hazards or increased risk. Often the tendency is for employees to err on the side of accepting exposure to electrical hazards, while managers and supervisors tend to be reluctant to accept increased risk

of exposure to hazards, particularly if their authorization for employees to work on energized equipment has to be in writing. There is a distinct safety benefit provided to employees and their employers of a properly communicated and administered energized electrical work permit (EEWP) covered by 130.2(B).

The risk assessment procedure must determine whether any conductor will remain energized for the duration of the work task. Hazard identification is now part of risk assessment. This process must be used to determine the shock protection boundaries and the arc flash boundary.

Both a shock risk assessment and an arc flash risk assessment are required before any person is permitted to approach the exposed energized electrical conductors or circuit parts. These risk assessments must answer the following questions:

- Is there unacceptable risk of injury from exposure to a potential shock hazard?
- Will the employee be exposed to the shock hazard at any point during the work task?
- What is the degree of the hazard?
- What protective equipment is necessary to minimize the exposure?
- Is there unacceptable risk of injury from exposure to a potential arc flash hazard?
- Will the employee be exposed to a thermal hazard at any point during the work task?
- What is the degree of the arc flash hazard?
- What protective equipment is necessary to minimize exposure to the thermal hazard?
- Are there special considerations due to a host employer contract employer relationship?
- Will energized electrical work impact other work?
- What measures will be taken to minimize the impact of other work?
- Will other employees be exposed to an electrical hazard because of the work task?
- Will the employee be exposed to any other electrical hazard while executing the work task?
- What authorization is necessary to justify executing the work task while the exposed conductor(s) is (are) energized?
- What employees are required to be within an approach boundary?
- Are unqualified employees required?
- How will the voltage on the conductor (or nearby conductors) be determined?
- Are there situations where the risk of injury is unacceptable?
- Is the risk of injury acceptable?

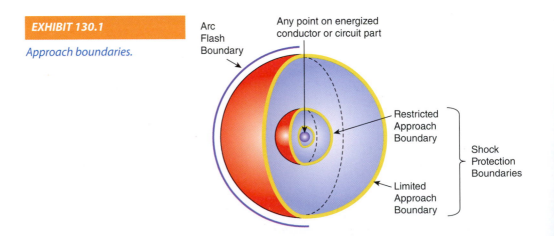

EXHIBIT 130.1

Approach boundaries.

Arc Flash Boundary

Any point on energized conductor or circuit part

Restricted Approach Boundary

Shock Protection Boundaries

Limited Approach Boundary

130.4 Approach Boundaries to Energized Electrical Conductors or Circuit Parts for Shock Protection.

Within *NFPA 70E* are two types of approach boundaries: the shock protection boundary and the arc flash boundary. Each type is determined by the kind of hazard associated with it. Safe approach boundaries are established on the basis of potential exposure to injury. Although the boundaries are related to the source of energy, they are each considered in different risk assessments — shock risk assessment and arc flash hazard risk assessment. Exhibit 130.1 illustrates the shock protection boundaries (restricted approach and limited approach) and arc flash boundary.

(A) Shock Risk Assessment. A shock risk assessment shall determine the voltage to which personnel will be exposed, the boundary requirements, and the PPE necessary in order to minimize the possibility of electric shock to personnel.

In order to understand shock risk assessment, it is helpful to be familiar with the new definition of *risk assessment* in Article 100. Risk assessment is the overall process that identifies hazards, estimates the potential severity of injury or damage to health, estimates the likelihood of occurrence of injury or damage to health, and determines if protective measures are required. Shock risk assessment is the process that identifies exposure to the potential electrical shock hazards, estimates the potential severity of a shock injury, estimates the likelihood of occurrence of this injury, then determines if protective measures are required and determines the appropriate protective measure (PPE) to use. The severity of the injury depends on the current flow through the body, the path it takes, and whether the heart is in a vulnerable point in the cardiac cycle. See 340.5 for more on the effects of electricity on the human body.

As the approach distance to an exposed energized electrical conductor decreases, the risk of direct contact with the conductor increases. Employees must determine the closest approach to a conductor that is necessary to perform the work task. The shock approach boundary is determined on that necessary approach distance. When conducting the shock risk assessment, the employee must determine the voltage of all conductors in the vicinity of the employee's body or of tool that will be used. Qualified persons are expected to have the knowledge and skill to be able to approach an exposed energized electrical conductor closer than would unqualified persons. The shock approach boundaries provide a trigger for added protection for all employees.

(B) Shock Protection Boundaries. The shock protection boundaries identified as limited approach boundary and restricted approach boundary shall be applicable where approaching personnel are exposed to energized electrical conductors or circuit parts. Table 130.4(D)(a) shall be used for the distances associated with various ac system voltages. Table 130.4(D)(b) shall be used for the distances associated with various dc system voltages.

The electrical safety program must include a procedure that provides workers with guidance related to required protection when the approach distance is less than the appropriate shock protection boundary. There are two protection boundaries: the limited approach boundary and the restricted approach boundary.

The dimension associated with each of these boundaries depends on the maximum voltage to which an employee might be exposed. The dimensions are given in Table 130.4(D)(a) for ac systems and Table 130.4(D)(b) for dc systems. For ac systems, the voltages given in Table 130.4(D)(a) are line-to-line voltages and refer to the table's footnote where single-phase systems are involved.

The limited approach boundary is the closest approach distance for an *unqualified* employee, unless additional protective measures are used. The restricted approach boundary is the closest approach distance for a *qualified* employee, unless additional protective measures are used. The prohibited approach boundary must not be crossed unless the

EXHIBIT 130.2

Working on equipment inside the shock boundary. (Courtesy of Arnold Air Force Base)

work task is guided by the measures identified in 130.1. Exhibit 130.2 shows an employee working inside the shock boundary.

Shock approach boundaries are related to direct contact with energized electrical conductors and circuit parts only and do not consider exposure to arc flash.

> Informational Note: In certain instances, the arc flash boundary might be a greater distance from the energized electrical conductors or circuit parts than the limited approach boundary. The shock protection boundaries and the arc flash boundary are independent of each other.

•

(C) Limited Approach Boundary.

Unqualified persons have not been trained to recognize and react to risk of injury of shock or electrocution from exposure to potential electrical shock hazards. The limited approach boundary is the approach limit for unqualified persons, unless the conditions of supervision specified in 130.4(C)(3) are met. Only qualified persons should be permitted to be within the space defined by the limited approach boundary. If the conductors are placed in an electrically safe work condition (ESWC), approach boundaries no longer exist and unqualified persons can approach the conductor without risk of injury.

(1) Approach by Unqualified Persons. Unless permitted by 130.4(C)(3), no unqualified person shall be permitted to approach nearer than the limited approach boundary of energized conductors and circuit parts.

Employees who are not associated with an electrical work task or tasks could be exposed to an electrical hazard. For instance, a painter could be working in the same room with an exposed energized electrical conductor or circuit part. The electrical supervisor and the painter's supervisor must establish a communication path so that the painter is advised of the location of the exposed energized electrical conductor, as well as how to avoid exposure to any associated electrical hazard(s). Signs or barricades might be necessary.

(2) Working at or Close to the Limited Approach Boundary. Where one or more unqualified persons are working at or close to the limited approach boundary, the designated person in charge of the work space where the electrical hazard exists shall advise the unqualified person(s) of the electrical hazard and warn him or her to stay outside of the limited approach boundary.

When unqualified persons working near a limited approach boundary cross the shock protection boundary, they are at an elevated risk of injury from exposure to potential electrical shock hazards. They need to be made aware of the potential hazards they can face by the designated person in charge of the work space where the electrical hazards are found. If there is a limited approach boundary, there is an arc flash boundary. If the arc flash boundary is outside of the limited approach boundary, the unqualified persons also have to be kept outside of the arc flash boundary. Where alerting techniques, such as safety signs and tags, barricades, and attendants — as identified in 130.7(E) — are used, the unqualified persons in the area where the work is being performed also need to know what the alerting techniques mean, so that they stay a safe distance away from where the work is being performed.

(3) Entering the Limited Approach Boundary. Where there is a need for an unqualified person(s) to cross the limited approach boundary, a qualified person shall advise him or her of the possible hazards and continuously escort the unqualified person(s) while inside the limited approach boundary. Under no circumstance shall the escorted unqualified person(s) be permitted to cross the restricted approach boundary.

In some instances, an unqualified person might be required to perform one or more tasks within the limited approach boundary. If this becomes necessary, a qualified person must ensure that the unqualified person is advised about the location of all exposed energized electrical conductors. The unqualified person must be advised that the risk of shock or electrocution exists, and a qualified person must escort the unqualified person at all times. The unqualified person must not cross the restricted approach boundary under any circumstances.

(D) Restricted Approach Boundary. No qualified person shall approach or take any conductive object closer to exposed energized electrical conductors or circuit parts operating at 50 volts or more than the restricted approach boundary set forth in Table 130.4(D)(a) and Table 130.4(D)(b), unless one of the following conditions applies:

(1) The qualified person is insulated or guarded from the energized electrical conductors or circuit parts operating at 50 volts or more. Insulating gloves or insulating gloves and sleeves are considered insulation only with regard to the energized parts upon which work is being performed. If there is a need for an uninsulated part of the qualified person's body to contact exposed energized electrical conductors or circuit parts, a combination of 130.4(D)(1), 130.4(D)(2), and 130.4(D)(3) shall be used to protect the uninsulated body parts.
(2) The energized electrical conductors or circuit part operating at 50 volts or more are insulated from the qualified person and from any other conductive object at a different potential.
(3) The qualified person is insulated from any other conductive object.

The restricted approach boundary is the closest approach distance for a qualified person without the use of personal and other shock protective equipment. If it is necessary to cross the restricted approach boundary, the qualified person must take additional precautionary measures. If a qualified person must approach an exposed energized electrical conductor closer than the restricted approach boundary, insulating materials with a defined voltage rating must be placed between the person and the conductor. The insulating material can take several forms. The insulating material can be installed so that the conductor is insulated from possible contact. The employee can be insulated by wearing appropriately rated PPE or can be insulated from ground, as in live-line bare-hand work.

In some cases, using more than one protective scheme is desirable. For instance, appropriately rated rubber blankets might be installed to cover or partially cover one or more conductors, and the employee also might wear appropriate voltage-rated PPE.

Insulating the qualified person from ground as the sole protective measure requires special training and qualifications. This alternative is not recommended except when the work is located on an overhead conductor at an elevated position.

Section 130.4(D) establishes requirements that must be observed before a qualified person is permitted to cross the restricted approach boundary. This section recognizes that a tool or other object is considered to be an extension of the person's body. If a person is holding an object, the requirements of this section apply to both the person and the object. The effect of these requirements is that the employee (or extended body part) is prevented from being exposed to any difference of potential 50 volts or more.

The purpose of columns 2 and 3 of Table 130.4(D)(a) and Table 130.4(D)(b) is to recognize that an electrical conductor can move. If the conductor is fixed into position, the distance between the employee and the conductor is under the control of the employee. If that distance can vary because the conductor can move (such as a bare overhead conductor or a

Table 130.4(D)(a) *Approach Boundaries to Energized Electrical Conductors or Circuit Parts for Shock Protection for Alternating-Current Systems (All dimensions are distance from energized electrical conductor or circuit part to employee.)*

(1)	(2)	(3)	(4)
	Limited Approach Boundary[b]		**Restricted Approach Boundary[b]; Includes Inadvertent Movement Adder**
Nominal System Voltage Range, Phase to Phase[a]	**Exposed Movable Conductor[c]**	**Exposed Fixed Circuit Part**	
<50 V	Not specified	Not specified	Not specified
50 V–150 V[d]	3.0 m (10 ft 0 in.)	1.0 m (3 ft 6 in.)	Avoid contact
151 V–750 V	3.0 m (10 ft 0 in.)	1.0 m (3 ft 6 in.)	0.3 m (1 ft 0 in.)
751 V–15 kV	3.0 m (10 ft 0 in.)	1.5 m (5 ft 0 in.)	0.7 m (2 ft 2 in.)
15.1 kV–36 kV	3.0 m (10 ft 0 in.)	1.8 m (6 ft 0 in.)	0.8 m (2 ft 7 in.)
36.1 kV–46 kV	3.0 m (10 ft 0 in.)	2.5 m (8 ft 0 in.)	0.8 m (2 ft 9 in.)
46.1 kV–72.5 kV	3.0 m (10 ft 0 in.)	2.5 m (8 ft 0 in.)	1.0 m (3 ft 3 in.)
72.6 kV–121 kV	3.3 m (10 ft 8 in.)	2.5 m (8 ft 0 in.)	1.0 m (3 ft 4 in.)
138 kV–145 kV	3.4 m (11 ft 0 in.)	3.0 m (10 ft 0 in.)	1.2 m (3 ft 10 in.)
161 kV–169 kV	3.6 m (11 ft 8 in.)	3.6 m (11 ft 8 in.)	1.3 m (4 ft 3 in.)
230 kV–242 kV	4.0 m (13 ft 0 in.)	4.0 m (13 ft 0 in.)	1.7 m (5 ft 8 in.)
345 kV–362 kV	4.7 m (15 ft 4 in.)	4.7 m (15 ft 4 in.)	2.8 m (9 ft 2 in.)
500 kV–550 kV	5.8 m (19 ft 0 in.)	5.8 m (19 ft 0 in.)	3.6 m (11 ft 10 in.)
765 kV–800 kV	7.2 m (23 ft 9 in.)	7.2 m (23 ft 9 in.)	4.9 m (15 ft 11 in.)

Note (1): For arc flash boundary, see 130.5(A).
Note (2): All dimensions are distance from exposed energized electrical conductors or circuit part to employee.
[a] For single-phase systems above 250V, select the range that is equal to the system's maximum phase-to-ground voltage multiplied by 1.732.
[b] See definition in Article 100 and text in 130.4(D)(2) and Informative Annex C for elaboration.
[c] *Exposed movable conductors* describes a condition in which the distance between the conductor and a person is not under the control of the person. The term is normally applied to overhead line conductors supported by poles.
[d] This includes circuits where the exposure does not exceed 120V.

Table 130.4(D)(b) Approach Boundaries to Energized Electrical Conductors or Circuit Parts for Shock Protection, Direct-Current Voltage Systems

(1)	(2)	(3)	(4)
	Limited Approach Boundary		Restricted Approach Boundary; Includes Inadvertent Movement Adder
Nominal Potential Difference	Exposed Movable Conductor*	Exposed Fixed Circuit Part	
<100 V	Not specified	Not specified	Not specified
100 V–300 V	3.0 m (10 ft 0 in.)	1.0 m (3 ft 6 in.)	Avoid contact
301 V–1 kV	3.0 m (10 ft 0 in.)	1.0 m (3 ft 6 in.)	0.3 m (1 ft 0 in.)
1.1 kV–5 kV	3.0 m (10 ft 0 in.)	1.5 m (5 ft 0 in.)	0.5 m (1 ft 5 in.)
5 kV–15 kV	3.0 m (10 ft 0 in.)	1.5 m (5 ft 0 in.)	0.7 m (2 ft 2 in.)
15.1 kV–45 kV	3.0 m (10 ft 0 in.)	2.5 m (8 ft 0 in.)	0.8 m (2 ft 9 in.)
45.1 kV– 75 kV	3.0 m (10 ft 0 in.)	2.5 m (8 ft 0 in.)	1.0 m (3 ft 2 in.)
75.1 kV–150 kV	3.3 m (10 ft 8 in.)	3.0 m (10 ft 0 in.)	1.2 m (4 ft 0 in.)
150.1 kV–250 kV	3.6 m (11 ft 8 in.)	3.6 m (11 ft 8 in.)	1.6 m (5 ft 3 in.)
250.1 kV–500 kV	6.0 m (20 ft 0 in.)	6.0 m (20 ft 0 in.)	3.5 m (11 ft 6 in.)
500.1 kV–800 kV	8.0 m (26 ft 0 in.)	8.0 m (26 ft 0 in.)	5.0 m (16 ft 5 in.)

Note: All dimensions are distance from exposed energized electrical conductors or circuit parts to worker.
Exposed movable conductor describes a condition in which the distance between the conductor and a person is not under the control of the person. The term is normally applied to overhead line conductors supported by poles.

conductor installed on racks in a manhole), or if the distance can vary because the platform (articulating arm) on which the employee is standing can move, then the distance is not under the employee's control, and column 2 of Table 130.4(D)(a) and Table 130.4(D)(b) applies. That column would also apply if a large size conductor is being disconnected from a fixed part (such as a connection point on an equipment bus) and it is anticipated that the act of disconnection could cause the conductor to swing away from the connection point. If the arc flash boundary is outside the limited approach boundary or restricted approach boundary, appropriate PPE must be worn for protection when an employee is within the boundary.

130.5 Arc Flash Risk Assessment.

An arc flash risk assessment shall be performed and shall:
(1) Determine if an arc flash hazard exists. If an arc flash hazard exists, the risk assessment shall determine:
   a.   Appropriate safety-related work practices
   b.   The arc flash boundary
   c.   The PPE to be used within the arc flash boundary

Whenever there are energized electrical conductors or circuit parts operating at 50 volts or greater that are considered to be exposed and an employee is within the limited approach boundary, a potential arc flash hazard is presumed to exist. A potential arc flash hazard can also be presumed to exist when an employee interacts with energized electrical equipment when conductors or circuit parts are not exposed.

The severity of a potential arc flash injury depends upon various factors, including whether or not personal and other protective equipment is used, the incident energy level that is impinged upon the body, whether the enclosure is open or not, and most importantly, where the person is standing in relation to the arc flash — that is, whether or not the person is within the arc flash boundary.

| Incident Energy and the Inverse Square Law | It is important to know |

whether or not the worker is within the arc flash boundary, because different measures are required depending on whether the worker is inside or outside this boundary. As a rule of thumb, the incident energy value increases, roughly, based on an inverse square law — meaning if the distance is halved, the incident energy increases by a factor of four (4).

Alerting techniques, in accordance with 130.7(E), are used to keep unqualified persons and those without appropriate PPE outside of the arc flash boundary. Appropriate procedures are used by those within the arc flash boundary to work safely, such as the selection and use of correct work procedures including the selection of appropriate personal and other protective equipment, which has been verified and confirmed by means of demonstrated ability.

(2) Be updated when a major modification or renovation takes place. It shall be reviewed periodically, at intervals not to exceed 5 years, to account for changes in the electrical distribution system that could affect the results of the arc flash risk assessment.

In some facilities the electrical distribution system changes infrequently, while in others it can change constantly; changes could result in constant upgrading of the arc flash risk assessment for particular pieces of electrical distribution equipment. *NFPA 70E* does not define what is meant by a major modification or renovation. Where there is exposure to a possible arc flash hazard, *a major modification or renovation* could be interpreted to mean when the data on the arc flash hazard equipment label is no longer applicable, or when new equipment is added that requires an arc flash label.

The arc flash risk assessment has to be reviewed at least every 5 years, or whenever a major modification or renovation takes place. Based on the frequency of changes to the electrical distribution system in some facilities, the period between updates to an arc flash risk assessment could be quite frequent. The modification that causes a new risk assessment to be required does not have to be a change of conditions on the premises itself. It could result from changes made by the utility company to their electrical distribution system feeding the premises, resulting in a higher or lower short circuit current to the facility.

(3) Take into consideration the design of the overcurrent protective device and its opening time, including its condition of maintenance.

•

Informational Note No. 1: Improper or inadequate maintenance can result in increased opening time of the overcurrent protective device, thus increasing the incident energy. Where equipment is not properly installed or maintained, PPE selection based on incident energy analysis or the PPE category method may not provide adequate protection from arc flash hazards.

Informational Note No. 2: Both larger and smaller available short-circuit currents could result in higher available arc flash energies. If the available short-circuit current increases without a decrease in the opening time of the overcurrent protective device, the arc flash energy will increase. If the available short-circuit current decreases, resulting in a longer opening time for the overcurrent protective device, arc flash energies could also increase.

The arc flash risk assessment process should take into consideration the characteristics of the upstream overcurrent device from an arcing event and the condition of maintenance of this overcurrent protective device.

The first question that should be asked regarding the maintenance condition of an overcurrent protective device is whether the overcurrent protective device has been maintained in accordance with the manufacturer's instructions and applicable industry codes and standards, such as NFPA 70B, *Recommended Practice for Electrical Equipment Maintenance.* The second question is whether there is evidence of impending failure — such as arcing, overheating, loose or bound equipment parts, unusual vibration, unusual smell, visible damage, or deterioration. See 130.2(A)(4). Where equipment has not been properly installed, or where there a signs of impending failure, an overcurrent protective device may fail to clear a fault, or fail to clear the fault in accordance with the manufacturer's published time–current characteristic curve. Under this particular circumstance, the PPE selection based on the incident energy value given on the arc flash hazard label, or the level of PPE given on the label, may not provide adequate protection from the arc flash hazard. See Exhibit 130.3.

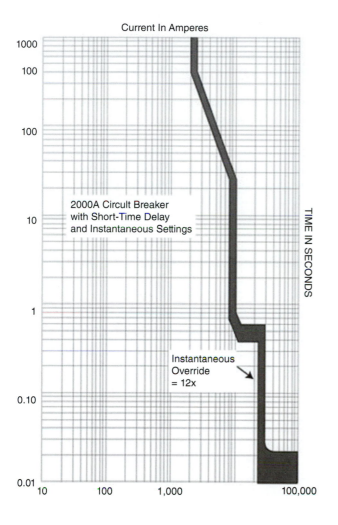

EXHIBIT 130.3

Circuit breaker with overload protection, short-time delay, and instantaneous trip override. (Courtesy of IAEI)

Informational Note No. 2 above deals with the situation where an overcurrent protective device is operating correctly, but the parameters of the circuit in which the device is installed has changed. Under this circumstance, there may be a larger or smaller available short-circuit current, both of which can result in higher available arc flash energies. Where the available short-circuit current increases without a decrease in the opening time of the overcurrent protective device, arc flash energy can increase; and where the available short-circuit current decreases, resulting in a longer clearing time for the overcurrent protective device, arc flash energies can also increase. See also the commentary on 130.5(2).

Informational Note No. 3:   The occurrence of an arcing fault inside an enclosure produces a variety of physical phenomena very different from a bolted fault. For example, the arc energy resulting from an arc developed in the air will cause a sudden pressure increase and localized overheating. Equipment and design practices are available to minimize the energy levels and the number of procedures that could expose an employee to high levels of incident energy.

Proven designs such as arc-resistant switchgear, remote racking (insertion or removal), remote opening and closing of switching devices, high-resistance grounding of low-voltage and 5000 volts (nominal) systems, current limitation, and specification of covered bus or covered conductors within equipment are available to reduce the risk associated with an arc flash incident. See Informative Annex O for Safety-Related Design Requirements.

Informational Note No. 4: For additional direction for performing maintenance on overcurrent protective devices, see Chapter 2, Safety-Related Maintenance Requirements.

The serviceable decal from the test or calibration decal system described in 11.27 of NFPA 70B, *Recommended Practice for Electrical Equipment Maintenance,* can be used as an indication that the electrical equipment under consideration has been properly maintained.

Informational Note No. 5: See IEEE 1584, *Guide for Performing Arc Flash Calculations,* for more information regarding arc flash hazards for three-phase systems.

The National Fire Protection Association (NFPA) and the Institute of Electrical and Electronic Engineers (IEEE) are sponsoring a joint endeavor, the IEEE/NFPA Arc Flash Phenomena Collaborative Research Project. The project will produce the data necessary to further understanding of the arc flash phenomena, which will help better protect worker safety. The results of this project have not yet been issued, but could have a significant impact on hazard assessment methods used today.

(A) Documentation. The results of the arc flash risk assessment shall be documented.

The use of an energized electrical work permit (EEWP), even when it is not required, is one of the best methods for documenting that the required risk assessments have been performed. See Figure J.2 in Informative Annex J for an example of an EEWP.

(B) Arc Flash Boundary.
(1) The arc flash boundary shall be the distance at which the incident energy equals 5 J/cm² (1.2 cal/cm²).

An arc fault incident can happen as a result of an open air event. However, arc fault events most often happen inside electrical equipment enclosures, and an enclosure can actually concentrate an arc flash event. Usually, an arc fault event occurs as a result of some movement, such as when a power circuit breaker is racked-in or -out, when an equipment door is opened or a cover is removed exposing energized conductors, when a conductor springs loose, a wrench drops, an animal contacts an energized circuit part, a disconnect is opened or closed, or a starter or contractor operates. Most arcing faults are started as a result of human interaction — such as when a person blindly reaches into a cubicle containing energized conductors or circuit parts with a conductive object.

Informational Note: For information on estimating the arc flash boundary, see Informative Annex D.

Informative Annex D provides information on various methods presented for estimating the available incident energy and the arc flash boundary, and illustrates several of the various techniques that can be used. Other appropriate techniques may be available. When using any of the techniques identified in Informative Annex D, it is important to be familiar with the content in the source documents, which are listed in Table D.1.

(2) The arc flash boundary shall be permitted to be determined by Table 130.7(C)(15)(A)(b) or Table 130.7(C)(15)(B), when the requirements of these tables apply.

Where the arc flash PPE categories method is used, the arc flash boundary can be determined by the use of Table 130.7(C)(15)(A)(b) for ac systems, or Table 130.7(C)(15)(B) for dc systems, which is based on open-air incident energy calculations.

(C) Arc Flash PPE. One of the following methods shall be used for the selection of PPE. Either, but not both, methods shall be permitted to be used on the same piece of equipment. The results of an incident energy analysis to specify an arc flash PPE Category in Table 130.7(C)(16) shall not be permitted.

If an employee's body (or part of the employee's body) must be within the arc flash boundary in order for the employee to perform a task, an arc flash risk assessment must be performed to determine the amount of incident energy that might be impinged upon the employee's body or clothing. The employee then must select and use PPE having a rating at least as high as the predicted incident energy that is available from the energy source. The level of the potential arc flash hazard must be determined in calories per square centimeter (cal/cm²) or converted to that measure. There are several ways to do the arc flash hazard analysis and determine the PPE. The employer's procedure that defines how to perform the arc flash risk assessment must also specify which method is preferred.

Section 130.5(C) requires that arc-rated clothing be worn within the arc flash boundary, which is generally considered to be the proximity to energized electrical equipment at which the employee is exposed to an incident energy of 1.2 cal/cm² or higher. Hazard/risk categories no longer exist in the 2015 edition of *NFPA 70E*. They have been replaced with PPE categories, which only deal with the PPE required to protect an employee from a potential arc flash exposure.

There is no PPE category 0. If there is an unacceptable risk of injury from exposure to a potential arc flash hazard and the worker is inside the arc flash boundary, appropriate arc-rated PPE is required. The arc-rated PPE required for each PPE category is described in Table 130.7(C)(16). If an incident energy analysis is performed as part of an arc flash hazard risk assessment instead of using the arc flash hazard PPE categories method — Table 130.7(C)(15)(A)(a), Table 130.7(C)(15)(A)(b) or Table 130.7(C)(15)(B), and Table 130.7(C)(16) — the arc flash PPE categories in Table 130.7(C)(16) should not be used in accordance with the requirements of this section.

(1) Incident Energy Analysis Method. The incident energy exposure level shall be based on the working distance of the employee's face and chest areas from a prospective arc source for the specific task to be performed. Arc-rated clothing and other PPE shall be used by the employee based on the incident energy exposure associated with the specific task. Recognizing that incident energy increases as the distance from the arc flash decreases, additional PPE shall be used for any parts of the body that are closer than the distance at which the incident energy was determined.

Informational Note: For information on estimating the incident energy, see Informative Annex D. For information on selection of arc-rated clothing and other PPE, see Table H.3(b) in Informative Annex H.

An incident energy analysis can be used to predict the amount of incident energy that might be available from the energy source at the point in the circuit where the work task is to be performed. If the distance between a person and the potential arc source is different from the dimension used by the analysis method, the actual thermal energy might be greater or smaller than that tested for in the analysis. The selected PPE should be determined by this actual thermal energy. Any body parts, such as the hands and arms, might require greater protection than the person's chest and torso. These parts should be protected accordingly.

Arc-Rated versus Flame-Resistant Protective Clothing and Equipment

Protective equipment and clothing that is intended for protection from an arcing fault must be rated by the manufacturer for use in an environment influenced by an electrical arc. Although the term *flame resistant* (FR) has been used in previous editions of *NFPA 70E*, an FR marking on a garment does not necessarily mean it is arc rated. The term *flame resistant* is no longer used in *NFPA 70E*; this change has been made to clarify that only clothing or other PPE with an arc rating is acceptable for protecting persons against arc flash hazards. The term *flame resistant* could indicate exposure to other events such as flames from a fire. As an example, FR clothing for race car drivers or pilots is not suitable for arc flash protection. Arc-rated clothing is the only clothing marked with the cal/cm² rating.

(2) Arc Flash PPE Categories Method. The requirements of 130.7(C)(15) and 130.7(C)(16) shall apply when the arc flash PPE category method is used for the selection of arc flash PPE.

The sections referenced here identify protective garments and equipment by PPE categories. Table 130.7(C)(16) illustrates arc flash PPE categories and assigns maximum protection afforded by each PPE category. To clarify the word *maximum*, if an equipment category in Table 130.7(C)(15)(A)(b) or Table 130.7(C)(15)(B) requires PPE category 2, then PPE with a rating no less than 8 cal/cm² per Table 130.7(C)(16) needs to be used.

Table 130.7(C)(15)(A)(a) provides a list of work tasks and condition parameters and, based on the work task and the specific condition parameters, defines whether or not arc flash PPE is required for the work task.

Inherent in the arc flash PPE categories method is that no equipment category in Table 130.7(C)(15)(A)(b) or Table 130.7(C)(15)(B) identifies a PPE category over 4, so no category requires a protection rating greater than 40 cal/cm². Therefore, where Table 130.7(C)(15)(A)(a) identifies that arc-rated PPE is not required and the risk has been determined to be acceptable, arc-rated PPE is not required to perform the task safely.

The employer can choose to use these tables to determine the amount of protection that is necessary. When these tables are used, however, the currents and clearing times included in the equipment category/rating headings must be checked to confirm that use of the tables is permitted. If a task is not in Table 130.7(C)(15)(a), or the working distance is closer than those that used in the tables or the clearing time of the arcing current, then the arc flash PPE categories method should not be used, and the incident energy analysis method should be used instead.

(D) Equipment Labeling. Electrical equipment such as switchboards, panelboards, industrial control panels, meter socket enclosures, and motor control centers that are in other than dwelling units and that are likely to require examination, adjustment, servicing, or maintenance while energized shall be field-marked with a label containing all the following information:

(1)   Nominal system voltage
(2)   Arc flash boundary
(3)   At least one of the following:
    a.   Available incident energy and the corresponding working distance, or the arc flash PPE category in Table 130.7(C)(15)(A)(b) or 130.7(C)(15)(B) for the equipment, but not both
    b.   Minimum arc rating of clothing
    c.   Site-specific level of PPE

•

The average employee should not be expected to calculate an incident energy value or to determine whether a job complies with the *NFPA 70E* arc flash PPE categories method. However, the employee should be capable of reading and interpreting the information included on an arc flash hazard equipment label. It is the employer's responsibility to provide employees performing work tasks with the information they need to work safely. Part of this is ensuring that arc flash hazard equipment labels are in place and that the employee has the information necessary to select the appropriate PPE.

The first edition of *NFPA 70E* was published in 1979. The requirements for arc flash personal and other protective equipment have been in effect since the 2000 edition of this standard. The requirement for the installation of field-marked arc flash hazard equipment labels first appeared in the 2009 edition of *NFPA 70E*.

The information on the label required by this section is vital for employees performing tasks where exposure to electrical hazards is possible, because it helps the user perform the shock risk and arc flash hazard risk assessments required by this standard. See Exhibit 130.4.

This standard does not require homeowners to install field-marked arc flash hazard equipment labels on the equipment installed in their premises.

Retroactive Requirements in *NFPA 70E*   Each edition of *NFPA 70E*, as of its issue date, supersedes all previous editions. The only requirement in *NFPA 70E* that can be considered retroactive is the Exception to 130.5(D) for existing field-marked arc flash hazard equipment labels. See the commentary below that exception for further information.

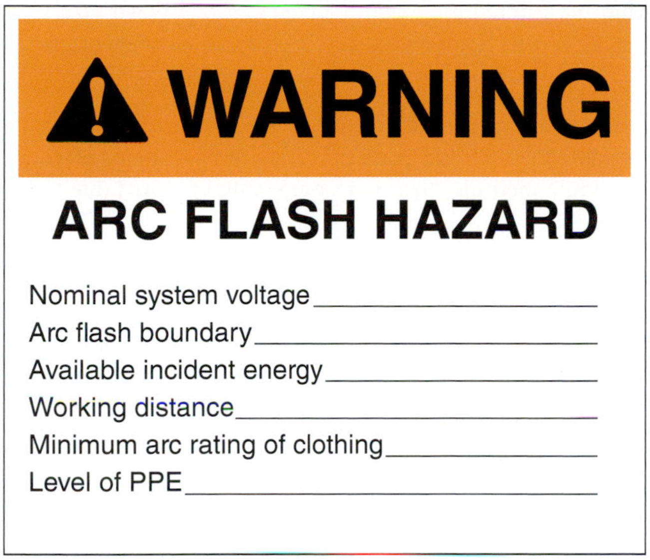

EXHIBIT 130.4

Warning label.

⚠ WARNING

ARC FLASH HAZARD

Nominal system voltage_____

Arc flash boundary_____

Available incident energy_____

Working distance_____

Minimum arc rating of clothing_____

Level of PPE_____

However, anyone performing energized work in dwelling units where electrical conductors and circuits are considered exposed is subject to a possible injury. So, a shock risk assessment and an arc flash risk assessment must be performed. The service and distribution equipment for most dwelling units is rated 120/240 volts nominal. Generally the upstream overcurrent protective device for the service and distribution equipment is the utility company's fuse on the primary side of the pole-mounted step-down distribution transformer. The fuse is generally a class T fuse and is not considered current limiting. Therefore in most situations, the new arc flash categories PPE method tables identified in 130.5(C)(2) cannot be used for dwelling services and distribution equipment.

> *Exception: Labels applied prior to September 30, 2011 are acceptable if they contain the available incident energy or required level of PPE.*

The method of calculating and the data to support the information for the label shall be documented. Where the review of the arc flash hazard risk assessment identifies a change that renders the label inaccurate, the label shall be updated.

If the nominal system voltage value is not given on the electrical equipment by the manufacturer, the method of determining the nominal system voltage must also be documented.

Where the values used on the label were determined by a power system analysis study, the power system analysis study report should be kept. This report must include information on the calculation method used, which may be documented by identifying the software manufacturer and the version of the software that was used. Where the arc flash PPE categories method is used to determine the required level of PPE, the manner in which the level of PPE was determined needs to be documented. Additional information, such as the party determining the values and the date the values where determined, can also be documented.

The owner of the electrical equipment shall be responsible for the documentation, installation, and maintenance of the field-marked label.

130.6 Other Precautions for Personnel Activities.

(A) Alertness.

(1) When Electrical Hazards Might Exist. Employees shall be instructed to be alert at all times when they are working within the limited approach boundary of energized electrical conductors or circuit parts operating at 50 volts or more and in work situations when electrical hazards might exist.

(2) When Impaired. Employees shall not be permitted to work within the limited approach boundary of energized electrical conductors or circuit parts operating at 50 volts or more, or where other electrical hazards exist, while their alertness is recognizably impaired due to illness, fatigue, or other reasons.

The employer or employer's agent has a duty to observe the physical and mental condition of the employees. At the time of the job briefing, the person in charge must evaluate whether any employee is impaired for any reason. In addition to illness and fatigue, impairment could be the result of drug or alcohol influence or the employee's emotional state.

(3) Changes in Scope. Employees shall be instructed to be alert for changes in the job or task that may lead the person outside of the electrically safe work condition or expose the person to additional hazards that were not part of the original plan.

When the job briefing is given, employees must be instructed about the details and limits of the expected work task. When the work task cannot be executed as planned, the work task must be stopped until a new job plan has been provided. A change in the condition or work environment that was not a part of the original plan might result in unanticipated exposure to an electrical hazard.

(B) Blind Reaching. Employees shall be instructed not to reach blindly into areas that might contain exposed energized electrical conductors or circuit parts where an electrical hazard exists.

Blind reaching is when an employee reaches into an area where lack of illumination or an obstruction impedes direct visual observation of the work to be performed. The use of a mirror to see behind an obstruction is also considered blind reaching.

(C) Illumination.
(1) General. Employees shall not enter spaces where electrical hazards exist unless illumination is provided that enables the employees to perform the work safely.

The employer should ensure that adequate lighting is provided to perform the work safely and should be included in the risk assessment procedure. Additional illumination could be required due to darkness of a face shield. Prior to start of work, the work area should be viewed wearing the face shield to see if additional illumination is necessary.

(2) Obstructed View of Work Area. Where lack of illumination or an obstruction precludes observation of the work to be performed, employees shall not perform any task within the limited approach boundary of energized electrical conductors or circuit parts operating at 50 volts or more or where an electrical hazard exists.

An obstruction could be a barrier installed by the equipment manufacturer to protect specific points within the equipment, or a barrier that has been installed to isolate an energized component.

(D) Conductive Articles Being Worn. Conductive articles of jewelry and clothing (such as watchbands, bracelets, rings, key chains, necklaces, metalized aprons, cloth with conductive thread, metal headgear, or metal frame glasses) shall not be worn within the restricted approach boundary or where they present an electrical contact hazard with exposed energized electrical conductors or circuit parts.

Employees must be aware that their jewelry or clothing could present an electrical hazard. This requirement also applies to conductive body piercing jewelry, which must be removed before entering the restricted approach boundary.

(E) Conductive Materials, Tools, and Equipment Being Handled.
(1) General. Conductive materials, tools, and equipment that are in contact with any part of an employee's body shall be handled in a manner that prevents accidental contact with energized electrical conductors or circuit parts. Such materials and equipment shall include, but are not limited to, long conductive objects, such as ducts, pipes and tubes, conductive hose and rope, metal-lined rules and scales, steel tapes, pulling lines, metal scaffold parts, structural members, bull floats, and chains.

Objects that do not have a voltage rating must be considered conductive. Testing to determine the effective insulation level and resultant voltage rating for electrical insulating equipment is typically performed by manufacturers and testing laboratories.

Long objects that are difficult to control, such as ladders, should be handled by assigning an employee to each end of the object. This work practice allows the object to be handled safely without crossing the limited approach boundary. Conductive fish tapes should not be used for a work task associated with an exposed energized electrical conductor or circuit part. Nonconductive pulling and fishing equipment is recommended.

(2) Approach to Energized Electrical Conductors and Circuit Parts. Means shall be employed to ensure that conductive materials approach exposed energized electrical conductors or circuit parts no closer than that permitted by 130.2.

(F) Confined or Enclosed Work Spaces. When an employee works in a confined or enclosed space (such as a manhole or vault) that contains exposed energized electrical conductors or circuit parts operating at 50 volts or more, or where an electrical hazard exists, the employer shall provide, and the employee shall use, protective shields, protective barriers, or insulating materials as necessary to avoid inadvertent contact with these parts and the effects of the electrical hazards.

Confined spaces are areas that are not necessarily designed for people but are large enough for workers to enter and perform certain tasks. Another characteristic of a confined space is limited or restricted means for entry or exit. Large equipment housing is an example of a confined space that could contain exposed energized electrical conductors or circuit parts. Only authorized qualified employees should be permitted to enter these areas.

Confined spaces might contain a hazardous atmosphere — due to toxic gases, low oxygen, flammability, or excessive heat — that could expose an employee to a life-threatening situation. Ventilation, emergency evacuation, and atmospheric testing should be included in the risk assessment procedure. If the assessment determines that the risks could be reduced to an acceptable level by installing barriers, shields, or other isolating devices, the task can be performed, provided all hazards identified in the risk assessment procedure are mitigated. An energized electrical work permit must cover the work task if the task could expose an employee to injury from an electrical hazard.

(G) Doors and Hinged Panels. Doors, hinged panels, and the like shall be secured to prevent their swinging into an employee and causing the employee to contact exposed energized electrical conductors or circuit parts operating at 50 volts or more or where an electrical hazard exists if movement of the door, hinged panel, and the like is likely to create a hazard.

(H) Clear Spaces. Working space required by other codes and standards shall not be used for storage. This space shall be kept clear to permit safe operation and maintenance of electrical equipment.

Where working space is used for storage, it can create a hazardous situation for employees performing work on or around electrical distribution equipment. See *NEC* 110.26 for working space requirements for equipment 600 volts, nominal or less, and see *NEC* 110.32 and 110.34 for equipment over 600 volts, nominal.

(I) Housekeeping Duties. Employees shall not perform housekeeping duties inside the limited approach boundary where there is a possibility of contact with energized electrical conductors or circuit parts, unless adequate safeguards (such as insulating equipment or barriers) are provided to prevent contact. Electrically conductive cleaning materials (including conductive solids such as steel wool, metalized cloth, and silicone carbide, as well as conductive liquid solutions) shall not be used inside the limited approach boundary unless procedures to prevent electrical contact are followed.

Unqualified persons must remain outside the limited approach boundary unless they meet the conditions of 130.4(C)(3). In applying this requirement, it is important to understand that the limited approach boundary only exists where there are exposed energized electrical conductors and circuit parts. This requirement does not apply where mitigation strategies have been employed to prevent employees from accidentally contacting exposed electrical conductors and circuit parts.

(J) Occasional Use of Flammable Materials. Where flammable materials are present only occasionally, electric equipment capable of igniting them shall not be permitted to be used, unless measures are taken to prevent hazardous conditions from developing. Such materials shall include, but are not limited to, flammable gases, vapors, or liquids; combustible dust; and ignitible fibers or flyings.

> Informational Note: Electrical installation requirements for locations where flammable materials are present on a regular basis are contained in *NFPA 70, National Electrical Code.*

Many cleaning materials and the propellant used in some spray cans are flammable. Liquids sometimes vaporize readily and flow into recesses and crevices of electrical equipment. An electrical arc caused by the operation of the equipment could ignite any cleaning liquid that remains. If the use of flammable liquids or gases for any purpose is necessary in the immediate vicinity of an exposed electrical conductor or circuit part, an electrically safe work condition must be created before the material is used.

(K) Anticipating Failure. When there is evidence that electric equipment could fail and injure employees, the electric equipment shall be de-energized, unless the employer can demonstrate that de-energizing introduces additional hazards or increased risk or is infeasible because of equipment design or operational limitation. Until the equipment is de-energized or repaired, employees shall be protected from hazards associated with the impending failure of the equipment by suitable barricades and other alerting techniques necessary for safety of the employees.

> Informational Note: See 130.7(E) for alerting techniques.

Electrical equipment frequently offers indications that failure is impending, and workers should be trained to recognize these indications. Possible indications of impending failure are hot enclosures, unusual noises or sounds, and unfamiliar smells. If any of these indications is observed, the equipment must be isolated through the use of barricades or similar protective measures to protect employees against accidental contact with the equipment while in failure mode. After the equipment has been isolated, it should be de-energized from a remote location. Disconnecting means located in the equipment should not be operated unless the employee is protected from the effects of equipment failure.

(L) Routine Opening and Closing of Circuits. Load-rated switches, circuit breakers, or other devices specifically designed as disconnecting means shall be used for the opening, reversing, or closing of circuits under load conditions. Cable connectors not of the load-break type, fuses, terminal lugs, and cable splice connections shall not be permitted to be used for such purposes, except in an emergency.

Where non–load-break-rated devices are used to break electrical load currents in an emergency situation, the party breaking the load needs to be aware of the risk of injury from exposure to potential electrical hazards and must carefully consider the use of appropriate PPE.

(M) Reclosing Circuits After Protective Device Operation. After a circuit is de-energized by the automatic operation of a circuit protective device, the circuit shall not be manually reenergized until it has been determined that the equipment and circuit can be safely energized. The repetitive manual reclosing of circuit breakers or reenergizing circuits through replaced fuses shall be prohibited. When it is determined from the design of the circuit and the overcurrent devices involved that the automatic operation of a device was caused by an overload rather than a fault condition, examination of the circuit or connected equipment shall not be required before the circuit is reenergized.

Only equipment that has been rated to serve as load-break equipment should be used for routine control of electrical equipment or circuits. The load-break rating is established by the manufacturer after testing to ensure that the equipment can safely interrupt the full-load current of the circuit. Unless the rating equals or exceeds the full-load current of the circuit, the disconnecting device must not be operated unless the load has been interrupted or the circuit has been de-energized.

Unless specifically designed, fuses should not be removed from an energized circuit. Removing a fuse or opening a cable connector that is conducting current can initiate an arc that could escalate into an arcing fault.

(N) Safety Interlocks. Only qualified persons following the requirements for working inside the restricted approach boundary as covered by 130.4(C) shall be permitted to defeat or bypass an electrical safety interlock over which the person has sole control, and then only temporarily while the qualified person is working on the equipment. The safety interlock system shall be returned to its operable condition when the work is completed.

If an overcurrent device (such as a fuse, circuit breaker, relay, overload, or similar device) operates, it can be assumed that the rated current has been exceeded or that a short-circuit or ground-fault condition exists. Reclosing a circuit into a short circuit or ground fault could have a disastrous result and must be avoided.

If it has been determined that the device was exposed to an overloaded condition, the circuit can be returned to its operating position. For instance, a motor overload relay or an internal thermal protector in an appliance can be reset.

130.7 Personal and Other Protective Equipment.

Personal and other protective equipment is normally fabricated to provide protection from a particular electrical hazard for a particular body part, but it may also provide protection from another electrical hazard. A classic example is the use of rubber insulated gloves and leather protectors that provide shock protection for the hands and forearms, which are also considered to provide a fair amount of arc flash protection for the hands and forearms.

Ensembles are put together, which are considered to provide protection from more than one hazard. The ensemble should also include arc-rated pants, nonmelting or arc-rated underwear and socks, and heavy duty work shoes or boots with metatarsal protection. The underwear may include a small amount of elastic bands for support, and a leather belt may be used to support the pants. Such an ensemble is considered to provide protection up to the arc-rating of the lowest rated outer piece. However, it should be noted that using arc-rated PPE at its rating provides only a 50 percent probability of protection from a second degree burn. PPE does not have an arc blast rating; it is only rated for arc flash or shock protection, not for arc blast protection.

(A) General. Employees working in areas where electrical hazards are present shall be provided with, and shall use, protective equipment that is designed and constructed for the specific part of the body to be protected and for the work to be performed.

The performance of a proper analysis involves identifying each electrical hazard involved, the body parts that are within the specific protection boundary, the severity of the hazard where the body part is to be located, and comparing the level of the hazard with the protection offered by the PPE selected. Where a worker's face (but not the back of the worker's head) is going to be within the arc flash boundary, a hard hat and face shield combination that is rated for the anticipated available incident energy level is required. Where the whole head is going to be within the arc flash boundary, a properly rated arc flash hood, or hard hat and face shield combination with arc-rated balaclava, is required. Properly rated hard hat and face shield combinations with arc-rated balaclava are only considered to provide protection up to 12 cal/cm². [See 130.7(C)(10)(b)(2).] When the back of the head is within the arc flash boundary it has to be properly protected, as with any other body part.

> Informational Note No. 1: The PPE requirements of 130.7 are intended to protect a person from arc flash and shock hazards. While some situations could result in burns to the skin, even with the protection selected, burn injury should be reduced and survivable. Due to the explosive effect of some arc events, physical trauma injuries could occur. The PPE requirements of 130.7 do not address protection against physical trauma other than exposure to the thermal effects of an arc flash.

When an arcing fault occurs, it creates different results than those from a short circuit or bolted fault, which mainly result from magnetic forces and stresses. During an arcing fault, electrical energy is converted into various forms of energy. Electrical energy can vaporize metal, which can change from solid state to gas vapor, in an expanding plasma ball with explosive concussive force. When copper vaporizes, it expands by a factor of 67,000 times in volume and can create a superheated plasma.

An arc flash event can result in arc temperatures as high as 19,500°C (35,000°F) creating molten metal, aluminum or copper vapor or superheated air. The threshold for a second degree burn is 80°C (175°F) for 0.1 seconds), and the threshold for a third degree burn is 96°C (205°F) for 0.1 seconds. Eye damage can result from the radiation associated with the intense light. Unprotected eardrums can rupture from the pressures resulting from expansion of metals by superheated air. Eardrum rupture occurs at a threshold of 34,475 Pa (720 lbs/ft²), and lung damage occurs at a threshold of 82,737 to 103,421 Pa (1,728 to 2,160 lbs/ft²). Shrapnel can reach speeds in excess of 1100 kph (700 mph). Objects and personnel can be flung across a room, from a ladder, or out of the bucket of a bucket truck.

Informational Note No. 2: It is the collective experience of the Technical Committee on Electrical Safety in the Workplace that normal operation of enclosed electrical equipment, operating at 600 volts or less, that has been properly installed and maintained by qualified persons is not likely to expose the employee to an electrical hazard.

Proper installation involves compliance with the applicable installation code. In the United States, the *NEC* is adopted and used in an overwhelming majority of jurisdictions. Compliance with the manufacturer's instructions is also a key component of ensuring a safe initial installation. Electrical equipment is subject to stresses both electrically and environmentally. Maintaining electrical equipment per the manufacturer's instructions or in accordance with consensus-based documents such as NFPA 70B, *Recommended Practice for Electrical Equipment Maintenance,* helps ensure that the equipment operates properly without posing an electrical hazard to employees who interface with the equipment as part of their job task(s).

Informational Note No. 3: When incident energy exceeds 40 cal/cm² at the working distance, greater emphasis may be necessary with respect to de-energizing when exposed to electrical hazards.

In addition to the thermal energy generated from an arc flash event, an arcing fault can generate a significant pressure wave. The level of a pressure wave is related to the amount of arcing current that is generated. There is an equation identified in the commentary on Informative Annex K, but it has not been generally adopted. Therefore, the degree of the pressure wave, and possible results from it, are not currently considered to be predictable with reasonable reliability. Where high arcing currents are involved and the thermal energy exceeds 40 cal/cm², greater emphasis needs to be put on creating an electrically safe work condition and minimizing the potential for injury by minimizing the time of exposure to a potential arc flash incident.

(B) Care of Equipment. Protective equipment shall be maintained in a safe, reliable condition. The protective equipment shall be visually inspected before each use. Protective equipment shall be stored in a manner to prevent damage from physically damaging conditions and from moisture, dust, or other deteriorating agents.

Informational Note: Specific requirements for periodic testing of electrical protective equipment are given in 130.7(C)(14) and 130.7(F).

Following the manufacturer's instructions as well as the guidance and requirements contained in consensus standards assures that personal and other protective equipment is kept in a suitable condition ready for use when needed.

Personal and other protective equipment has a life cycle that can include many use cycles, such as donning and doffing, cleaning, and laundering. To be assured that personal and other protective equipment is maintained in a reliable condition ready for use and will be effective when needed, it must be cared for and inspected at regular intervals and prior to use. Table 130.7(C)(7)(c) *Rubber Insulating Equipment, Maximum Test Intervals* and Table 130.7(F) *Standards on Other Protective Equipment* may be useful in this regard.

Specific guidance regarding some of the various types of personal and other protective equipment may be found in the following national consensus standards, as well as in international consensus standards:

- ASTM F496, *Standard Specification for In-Service Care of Insulating Gloves and Sleeves*
- ASTM F478, *Standard Specification for In-Service Care of Insulating Line Hose and Covers*
- ASTM F479, *Standard Specification for In-Service Care of Insulating Blankets*
- ASTM F1449, *Standard Guide for Industrial Laundering of Flame, Thermal, and Arc Resistant Clothing*
- ASTM F1505, *Standard Specification for Insulated and Insulating Hand Tools*
- ASTM F2249, *Standard Specification for In-Service Test Methods for Temporary Grounding Jumper Assemblies Used on De-Energized Electric Power Lines and Equipment*
- ASTM F2757, *Standard Guide for Home Laundering Care and Maintenance of Flame, Thermal and Arc Resistant Clothing*

ASTM F478, ASTM F479, and ASTM F496 include information on the periodic testing requirements for specific types of rubber insulating equipment. For instance, rubber insulating gloves are to be tested before their first issue and then every 6 months thereafter. If the rubber insulating gloves have been electrically tested but not issued for service, they can be issued within 12 months from the date they were tested. After that point they have to be retested before being issued. It has to be determined what type of identification or marking system is going to be used to keep track of the rubber insulating gloves, such as the use of color coding by stamping with a test date and issue date, and so forth.

Guidance on commercial and home laundering can be found in ASTM F1449 and ASTM F2757, identified above. These standards may be helpful in determining the types of cleaning agents to be used, whether bleach-type additives can be used, and which cleaning processes and procedures are allowed to be used, including the temperature of the water. The entity needs to determine whether commercial or home laundering is to be used. If arc-rated PPE is to be shared, sanitary considerations need to be addressed, such as cleaning between each use.

(C) Personal Protective Equipment (PPE).

(1) General. When an employee is working within the restricted approach boundary, the worker shall wear PPE in accordance with 130.4. When an employee is working within the arc flash boundary, he or she shall wear protective clothing and other PPE in accordance with 130.5. All parts of the body inside the arc flash boundary shall be protected.

(2) Movement and Visibility. When arc-rated clothing is worn to protect an employee, it shall cover all ignitible clothing and shall allow for movement and visibility.

When protective apparel and equipment are selected and appropriately sized, the employee's movement is not restricted. All parts of the employee's ignitable clothing must be protected from the available incident energy to avoid igniting flammable apparel. See Informational Note No. 3 to 130.7(C)(16) to determine the arc-rating of a protective clothing system. Any clothing that melts is not considered as providing thermal insulation and should not be used as a component of an arc-rated protective clothing system.

Rehearsing a work task on similar equipment in an electrically safe work condition and wearing PPE is one way to determine if PPE is non-restricting.

(3) Head, Face, Neck, and Chin (Head Area) Protection. Employees shall wear nonconductive head protection wherever there is a danger of head injury from electric shock or burns due to contact with energized electrical conductors or circuit parts or from flying objects resulting from electrical explosion. Employees shall wear nonconductive protective equipment for the face, neck, and chin whenever there is a danger of injury from exposure to electric arcs or flashes or from flying objects resulting from electrical explosion. If employees use hairnets or beard nets, or both, these items must be arc rated.

Informational Note: See 130.7(C)(10)(b) and (c) for arc flash protective requirements.

When working within the arc flash boundary, protection above the shoulders is as critical as the protection for the torso and lower body. Depending on the level of the hazard, an arc-rated face shield with balaclava, or an arc-rated hood, is required to provide protection to the head, face, neck, and chin. The face shield or hood window should be impact resistant per ANSI Z87.1, *Practice for Occupational and Educational Eye and Face Protection,* because of the potential for flying debris. Where required by the risk assessment, a Class E or G hard hat should be used to provide protection against impact or from contact with the electrical energy. Additional PPE such as hairnets, beard nets, hard cap liners, as well as other devices worn on or about the head and neck should not melt or drip and need to be arc-rated.

(4) Eye Protection. Employees shall wear protective equipment for the eyes whenever there is danger of injury from electric arcs, flashes, or from flying objects resulting from electrical explosion.

Safety spectacles or goggles protect the eyes from impact that exceeds the protection offered by face shields and hood windows. Face shields or hoods in accordance with ANSI Z87.1 are considered as a "secondary eye protective device" that must be used with a "primary eye protective device" (spectacle or goggle) underneath.

(5) Hearing Protection. Employees shall wear hearing protection whenever working within the arc flash boundary.

The sound pressure of an arc flash incident could exceed 140 decibels. Therefore, employees are required to wear hearing protection when they perform tasks within the arc flash boundary. Ear canal inserts are the identified form of hearing protection recognized in Table 130.7(C)(16). They interfere less with the proper fit and application of other PPE intended to protect the head and face. If the arc flash PPE categories method is not used, compliance with the requirement of 130.7(C)(5) can be achieved through the use of ear canal inserts. Hearing protection should not adversely impact the performance of other items of PPE.

(6) Body Protection. Employees shall wear arc-rated clothing wherever there is possible exposure to an electric arc flash above the threshold incident energy level for a second degree burn [5 J/cm^2 (1.2 cal/cm^2)].

The arc flash boundary is defined in Article 100 as the "distance from a prospective arc source within which a person could receive a second degree burn if an electrical arc were to occur." The level of thermal energy at which a second degree burn would occur is defined in research by Alice Stoll and Maria Chianta. They conducted burn injury research in the late 1950s and early 1960s at the Aerospace Medical Research Department of the Naval Air Development Center. Their research led to the development of the Stoll Curve, which quantifies the level of heat and time required for a second degree burn. The Stoll Curve is used as the benchmark for determining the level of arc-rated clothing and PPE that should be worn for protection against the thermal effects of an arc flash. Based on this research it was determined that employees exposed to greater than 1.2 cal/cm^2 need to wear arc-rated clothing for protection from the thermal effects of an arcing event. This clothing could consist of a shirt and pants combination, coveralls, or any other assembly that provides protection for exposed skin or flammable clothing.

(7) Hand and Arm Protection. Hand and arm protection shall be provided in accordance with 130.7(C)(7)(a), (b), and (c).

(a) Shock Protection. Employees shall wear rubber insulating gloves with leather protectors where there is a danger of hand injury from electric shock due to contact with energized electrical conductors or circuit parts. Employees shall wear rubber insulating gloves with leather protectors and rubber insulating sleeves where there is a danger of hand and arm injury from electric shock due to contact with energized electrical conductors or circuit parts. Rubber insulating gloves shall be rated for the voltage for which the gloves will be exposed.

Exception: Where it is necessary to use rubber insulating gloves without leather protectors, the requirements of ASTM F496, Standard Specification for In-Service Care of Insulating Gloves and Sleeves, shall be met.

•

Where unacceptable risk involves contact of the hands or arms with exposed energized conductors or circuit parts, the protection should minimize the possibility that the employee might be shocked or electrocuted. This can involve the use of rubber insulating gloves (with leather protectors) and insulating sleeves.

(b) Arc Flash Protection. Hand and arm protection shall be worn where there is possible exposure to arc flash burn. The apparel described in 130.7(C)(10)(d) shall be required for protection of hands from burns. Arm protection shall be accomplished by the apparel described in 130.7(C)(6).

If a task has been identified to expose the employee to thermal energy up to 40 cal/cm^2, the PPE selected for body as well as the arms and hands must be rated to provide at least 40 cal/cm^2 of protection. Where the hands and arms are closer than the working distance, a greater level of protection is required. Sleeves must not be shortened or rolled up, exposing skin or flammable undergarments. Voltage-rated gloves with leather protectors are considered to provide significant thermal protection.

Where the employee's hands are within the arc flash boundary, rubber insulating gloves should not be worn without leather protectors. If shock protection is not required, heavy-duty leather gloves (gloves made entirely of leather with a minimum thickness of 0.03 in.) are considered acceptable for arc flash protection up to 10 cal/cm^2. Arc-rated gloves are not voltage rated and, as such, provide offer no protection from shock or electrocution.

(c) Maintenance and Use. Electrical protective equipment shall be maintained in a safe, reliable condition. Insulating equipment shall be inspected for damage before each day's use and immediately following any incident that can reasonably be suspected of having caused damage. Insulating gloves shall be given an air test, along with the inspection. Electrical protective equipment shall be subjected to periodic electrical tests. Test voltages and the maximum intervals between tests shall be in accordance with Table 130.7(C)(7)(c).

Informational Note: See OSHA 1910.137 and ASTM F496, *Standard Specification for In-Service Care of Insulating Gloves and Sleeves.*

Manufacturers of arc-rated clothing and other arc-rated protective equipment provide instructions for cleaning and care of their products. The electrical safety program must

Table 130.7(C)(7)(c)   Rubber Insulating Equipment, Maximum Test Intervals

Rubber Insulating Equipment	When to Test	Governing Standard for Test Voltage*
Blankets	Before first issue; every 12 months thereafter[†]	ASTM F479
Covers	If insulating value is suspect	ASTM F478
Gloves	Before first issue; every 6 months thereafter[†]	ASTM F496
Line hose	If insulating value is suspect	ASTM F478
Sleeves	Before first issue; every 12 months thereafter[†]	ASTM F496

*ASTM F478, *Standard Specification for In-Service Care of Insulating Line Hose and Covers*; ASTM F479, *Standard Specification for In-Service Care of Insulating Blankets*; ASTM F496, *Standard Specification for In-Service Care of Insulating Gloves and Sleeves.*
[†]If the insulating equipment has been electrically tested but not issued for service, it is not permitted to be placed into service unless it has been electrically tested within the previous 12 months.

describe the maintenance and cleaning process or processes that ensure the integrity of the apparel. Employees should inspect their arc-rated clothing and ensure that the apparel is not soiled with a flammable contaminant. Any patches or labels attached to arc-rated clothing should also be of arc-rated material. Shock-protective equipment must be physically inspected immediately prior to use.

The insulating integrity of this equipment must be verified by tests conducted in accordance with the manufacturer's instructions and the appropriate standards as shown in Table 130.7(C)(7)(c). This table describes the maximum testing intervals for various products and identifies the appropriate test standard.

(8) Foot Protection. Where insulated footwear is used as protection against step and touch potential, dielectric footwear shall be required. Insulated soles shall not be used as primary electrical protection.

Informational Note: Electrical hazard footwear meeting ASTM F2413, *Standard Specification for Performance Requirements for Protective (Safety) Toe Cap Footwear*, can provide a secondary source of electric shock protection under dry conditions.

Heavy-duty leather footwear should cover the employee's feet. If shock protection is warranted, the employee must wear dielectric overshoes (boots), use rubber insulating mats, or a combination of these two protective items. Other protective techniques such as equipotential bonding can also be used for shock protection. The integrity of the insulating quality of footwear with insulated soles cannot be established easily after they have been worn in a work environment. Electrical hazard (EH) rated footwear with insulated soles must not serve as the primary protection from touch and step potential in a dry environment.

(9) Factors in Selection of Protective Clothing. Clothing and equipment that provide worker protection from shock and arc flash hazards shall be used. If arc-rated clothing is required, it shall cover associated parts of the body as well as all flammable apparel while allowing movement and visibility.

Clothing and equipment required for the degree of exposure shall be permitted to be worn alone or integrated with flammable, nonmelting apparel. Garments that are not arc rated shall not be permitted to be used to increase the arc rating of a garment or of a clothing system.

Informational Note: Protective clothing includes shirts, pants, coveralls, jackets, and parkas worn routinely by workers who, under normal working conditions, are exposed to momentary electric arc and related thermal hazards. Arc-rated rainwear worn in inclement weather is included in this category of clothing.

Arc-rated clothing must be rated for use in an environment that is or could be influenced by the presence of electrical energy. The nature of an arcing fault is that thermal energy is generated very rapidly. The protective clothing must be capable of providing the necessary protection in that environment. To avoid injury from a thermal hazard, the employee's clothing must ensure that his or her skin surface does not receive more than 1.2 cal/cm² of incident energy, the level of energy that would produce the onset of a second degree burn. The protective clothing must prevent any flammable underclothing from igniting and burning. Additionally, the protective clothing should ensure that any underclothing with limited meltable fabric does not melt onto the employee's skin. See the exception in 130.7(C)(11) for limited meltable fabric explanation.

Materials that have an established arc rating can be used either alone or in combination with materials without an established rating. Materials from one manufacturer also can be used with materials from a different manufacturer, provided both sets of clothing have an established rating. The employee's clothing should be viewed as a system that avoids injury from the thermal hazard. The system can consist of multiple clothing items, but the employee must understand that the system includes all garments being worn.

Arc-rated equipment must contain a label or other mark that describes the maximum incident energy rating. Garments that have been tested and marked with an arc-rating can be used to increase the arc rating of an individual garment or as part of a clothing system with an overall arc rating. Combination systems have to be tested, and the overall rating of the system cannot be determined by simply adding the arc rating of the two garments together. Other thermodynamic factors, such as the air gap between the garments, can impact the overall arc rating of the system and have to be considered.

(a) Layering. Nonmelting, flammable fiber garments shall be permitted to be used as underlayers in conjunction with arc-rated garments in a layered system. If nonmelting, flammable fiber garments are used as underlayers, the system arc rating shall be sufficient to prevent breakopen of the innermost arc-rated layer at the expected arc exposure incident energy level to prevent ignition of flammable underlayers. Garments that are not arc rated shall not be permitted to be used to increase the arc rating of a garment or of a clothing system.

> Informational Note: A typical layering system might include cotton underwear, a cotton shirt and trouser, and an arc-rated coverall. Specific tasks might call for additional arc-rated layers to achieve the required protection level.

Layering can increase the overall protective characteristics of arc-rated clothing. If ignitable fabrics are used as underlayers, the arc rating of the clothing system must not permit the outer layer to break open and directly expose the ignitable material to the heat of the arcing fault. The outer layers must limit the temperature rise of the ignitable underlayers to no more than 1.2 cal/cm². Air is a good thermal insulator. Wearing multiple layers of clothing traps air in between the clothing layers. The layering effect tends to improve the thermal insulating efficiency of the overall protective system. However, the increase in efficiency does not necessarily increase the arc rating of the protective clothing system. The clothing manufacturer should be consulted to determine the arc rating of the overall system of protective clothing; the combined rating of two arc-rated garments is not linear and can only be determined through testing. In order to have a combined rating, both garments must be arc-rated.

(b) Outer Layers. Garments worn as outer layers over arc-rated clothing, such as jackets or rainwear, shall also be made from arc-rated material.

When the protective clothing system consists of arc-rated protective clothing and clothing that is flammable, the employee must ensure that all flammable material is not exposed to the arcing fault. Flammable garments worn over arc-rated clothing, such as windbreakers, jackets, or rainwear, can act as a flame source themselves. This provides an additional ignition source for the arc-rated clothing underneath, which potentially exceeds the insulating capacity of the arc-rated clothing, exposing the user to sufficient thermal energy to cause a burn injury. Arc-rated rainwear made from arc-rated material that will enhance protection against an arc flash will be certified to ASTM F1506, *Standard Performance Specification for Flame Resistant and Arc Rated Textile Materials for Wearing Apparel for Use by Electrical Workers Exposed to Momentary Electric Arc and Related Thermal Hazards,* and bear an arc rating. See Exhibit 130.5 for examples of this type of clothing.

EXHIBIT 130.5

Arc-rated rainwear. (Courtesy of National Safety Apparel)

(c) Underlayers. Meltable fibers such as acetate, nylon, polyester, polypropylene, and spandex shall not be permitted in fabric underlayers (underwear) next to the skin.

Exception: An incidental amount of elastic used on nonmelting fabric underwear or socks shall be permitted.

Informational Note No. 1: Arc-rated garments (e.g., shirts, trousers, and coveralls) worn as underlayers that neither ignite nor melt and drip in the course of an exposure to electric arc and related thermal hazards generally provide a higher system arc rating than nonmelting, flammable fiber underlayers.

Informational Note No. 2: Arc-rated underwear or undergarments used as underlayers generally provide a higher system arc rating than nonmelting, flammable fiber underwear or undergarments used as underlayers.

Significant injuries occur when fabrics melt onto an employee's skin. However, an incidental quantity of these fabrics, such as those used in the elastic bands in underwear, is permitted. Exhibit 130.6 shows examples of arc-rated underlayers for women.

EXHIBIT 130.6

Arc-rated underlayers for women. (Courtesy of DRIFIRE, LLC)

(d) Coverage. Clothing shall cover potentially exposed areas as completely as possible. Shirt and coverall sleeves shall be fastened at the wrists, shirts shall be tucked into pants, and shirts, coveralls, and jackets shall be closed at the neck.

When required by the risk assessment, arc-rated clothing must completely cover all body areas that are within the arc flash boundary. Shirt sleeves must be fastened at the wrists, and the top button of shirts and/or jackets must be fastened to minimize the chance that heated air could reach below the arc-rated clothing. Shirt sleeves should fit under the gauntlet of the protective gloves to minimize the chance that thermal energy could enter under the shirt sleeves.

(e) Fit. Tight-fitting clothing shall be avoided. Loose-fitting clothing provides additional thermal insulation because of air spaces. Arc-rated apparel shall fit properly such that it does not interfere with the work task.

The fit of arc-rated clothing is important to the safety of the employee. When the surface of arc-rated clothing is heated, heat is conducted through the material. If the arc-rated clothing is touching skin, the heat energy that is conducted through the arc-rated clothing could result in a burn. To minimize this chance, arc-rated clothing must fit loosely to provide additional thermal insulation. However, arc-rated clothing must not be so loose that it interferes with the employee's movements. See Exhibit 130.7 for examples of arc-rated clothing sized specifically for women.

(f) Interference. The garment selected shall result in the least interference with the task but still provide the necessary protection. The work method, location, and task could influence the protective equipment selected.

The plan for the work task must define the protective garments to be worn by the employee. As the plan is developed, the location and position of the employee must be considered to provide the best chance that the PPE does not interfere with the person's move-

ments as he or she executes the work task. A dry run of the task wearing PPE will identify whether there are PPE restrictions for the task.

(10) Arc Flash Protective Equipment.

(a) Arc Flash Suits. Arc flash suit design shall permit easy and rapid removal by the wearer. The entire arc flash suit, including the hood's face shield, shall have an arc rating that is suitable for the arc flash exposure. When exterior air is supplied into the hood, the air hoses and pump housing shall be either covered by arc-rated materials or constructed of nonmelting and nonflammable materials.

The protection provided by an arc flash suit is more than simply the arc rating of the fabric or the fabric system. The design and construction of the suits are of great importance. A poorly or improperly designed suit may provide for the easy passage of energy through the holes or ports in the garment, thus bypassing the protection offered by the fabric. In the past, garments were constructed of double layers of fabric on the chest and single layers of fabric on the arms and back, but they were rated to the greater protection offered by the chest, not the protection provided by the single layer on the arms and back. Also, how the garment is constructed can determine its ease of donning and doffing.

Details provided in ASTM F1506 dictate that an arc-rated garment be constructed of flame resistant trim, such as zipper tape and hook-and-loop fasteners. ASTM F2178, *Standard Test Method for Determining the Arc Rating and Standard Specification for Eye or Face Protective Products,* dictates that the arc rating of an arc flash face shield or hood is not the lesser of the window and the fabric. It mandates that a manufacturer conduct actual testing on their product as it is sold to the marketplace, since the shape and design of the hood and face shield often has more to do with the performance of the product than the fabric and plastic qualities. Face shields and hood windows also must meet the impact requirements of ANSI Z87.1. Safety glasses, as primary eye protective devices, must be worn under the face shield or hood window. Neither goggles nor spectacles are recognized by *NFPA 70E* as arc flash protective devices.

> Paragraph 130.7(C)(10)(b)(1) was revised by a tentative interim amendment (TIA).

(b) Head Protection.

    (1) An arc-rated balaclava shall be used with an arc-rated face shield when the back of the head is within the arc flash boundary. An arc-rated hood shall be permitted to be used instead of an arc-rated face shield and balaclava.

    (2) An arc-rated hood shall be used when the anticipated incident energy exposure exceeds 12 cal/cm^2.

Complete protection for the head is required when the entire head is within the arc flash boundary. This protection can be achieved through the use of an arc-rated balaclava in conjunction with an arc-rated face shield. Each item must be individually rated for the incident energy exposure, unless a tested combination is available. Arc-rated hoods can be used in lieu of the balaclava and face shield combination and must be used if the incident energy exceeds 12 cal/cm^2.

(c) Face Protection. Face shields shall have an arc rating suitable for the arc flash exposure. Face shields with a wrap-around guarding to protect the face, chin, forehead, ears, and neck area shall be used. Face shields without an arc rating shall not be used. Eye protection (safety glasses or goggles) shall always be worn under face shields or hoods.

    Informational Note: Face shields made with energy-absorbing formulations that can provide higher levels of protection from the radiant energy of an arc flash are available, but these shields are tinted and can reduce visual acuity and color perception. Additional illumination of the task area might be necessary when these types of arc-protective face shields are used.

It is critical that the face and head be protected equal to or greater than the potential thermal energy exposure as determined the by risk assessment, as detailed in 130.7(C)(3). If the back of the head is within the arc flash protection boundary, the entire head must be protected, including the back of the head, by a balaclava conjunction with an arc-rated face shield or an arc-rated hood. ASTM F2178 stipulates that the entire face shield or the entire arc flash hood must be tested, as sold by the manufacturer, to determine its arc rating. The arc rating is not determined by the rating of the window alone or the fabric alone.

Recently, new developments have dramatically improved the visible light transmission and clarity of colors visible through arc flash face shields and hood windows. However, many have a green tint that can negatively impact visibility and color perception. Additional illumination may be required to enhance the employee's ability to see the task to be performed.

(d) Hand Protection.

    (1) Heavy-duty leather gloves or arc-rated gloves shall be worn where required for arc flash protection.

    Informational Note: Heavy-duty leather gloves are made entirely of leather with minimum thickness of 0.03 in. (0.7 mm) and are unlined or lined with nonflammable, nonmelting fabrics. Heavy-duty leather gloves meeting this requirement have been shown to have ATPV values in excess of 10 cal/cm^2.

    (2) Where insulating rubber gloves are used for shock protection, leather protectors shall be worn over the rubber gloves.

Informational Note: The leather protectors worn over rubber insulating gloves provide additional arc flash protection for the hands for arc flash protection exposure.

Normally, the hands are the most exposed part of a person's body, since they are used to perform work. Generally, an arc flash risk assessment is based on an exposure distance of 18 in. or 36 in. Because a person's hands are closer to the hazard than their body, the thermal exposure will likely be much greater to the hands. Additional thermal protection is warranted, since the employee's hands may be within the actual arc's plasma.

Care should be taken when evaluating the arc rating of rubber gloves in conjunction with leather protectors. Because of the wide variation in the thickness of leather protectors, correlation of the test value could prove difficult.

Rehearsal of the work task while wearing layers of hand protection will determine if the dexterity for the task is sufficient. If the hands do not extend beyond the restricted approach boundary within which there is the risk of shock (and are only being protected against exposure to an arc flash), then heavy-duty leather gloves or arc-rated gloves can be used without voltage-rated rubber gloves where the incident energy does not exceed 8 cal/cm². The leather protectors worn over insulating rubber gloves provide additional arc flash protection for the hands, but during high arc flash exposures the leather can shrink and cause a decrease in protection.

(e) Foot Protection. Heavy-duty leather footwear provide some arc flash protection to the feet and shall be used in all exposures greater than 4 cal/cm².

Footwear with an arc rating is not available. Normally, the employee's feet are less exposed than the hands or head due to the proximity of most tasks. However, employees should not wear footwear made from lightweight materials or materials that are flammable or may melt and drip. In most cases, heavy-duty leather work shoes with integral steel toes are satisfactory. If the risk assessment procedure indicates that arc flash protection is necessary while the task is performed, then heavy-duty leather work footwear must be worn. Where there is a risk of contact with energized systems, refer to 130.7(C)(8).

(11) Clothing Material Characteristics. Arc-rated clothing shall meet the requirements described in 130.7(C)(14) and 130.7(C)(12).

Section 130.7(C)(11) describes requirements that must be met by arc-rated clothing. Arc-rated clothing that has an arc rating based on testing defined by ASTM F1506 and ASTM F1959, *Standard Test Method for Determining the Arc Thermal Performance Value of Materials for Clothing,* adheres to these requirements.

Informational Note No. 1: Arc-rated materials, such as flame-retardant-treated cotton, meta-aramid, para-aramid, and poly-benzimidazole (PBI) fibers, provide thermal protection. These materials can ignite but will not continue to burn after the ignition source is removed. Arc-rated fabrics can reduce burn injuries during an arc flash exposure by providing a thermal barrier between the arc flash and the wearer.

Informational Note No. 2: Non–arc-rated cotton, polyester-cotton blends, nylon, nylon-cotton blends, silk, rayon, and wool fabrics are flammable. Fabrics, zipper tapes, and findings made of these materials can ignite and continue to burn on the body, resulting in serious burn injuries.

Informational Note No. 3: Rayon is a cellulose-based (wood pulp) synthetic fiber that is a flammable but nonmelting material.

Clothing consisting of fabrics, zipper tapes, and findings made from flammable synthetic materials that melt at temperatures below 315°C (600°F), such as acetate, acrylic, nylon, polyester, polyethylene, polypropylene, and spandex, either alone or in blends, shall not be used.

Informational Note: These materials melt as a result of arc flash exposure conditions, form intimate contact with the skin, and aggravate the burn injury.

Exception: Fiber blends that contain materials that melt, such as acetate, acrylic, nylon, polyester, polyethylene, polypropylene, and spandex, shall be permitted if such blends in fabrics meet the requirements of ASTM F1506, Standard Performance Specification for Flame Resistant and Arc Rated Textile Materials for Wearing Apparel for Use by Electrical Workers Exposed to Momentary Electric Arc and Related Thermal Hazards, and if such blends in fabrics do not exhibit evidence of a melting and sticking hazard during arc testing according to ASTM F1959/F1959M, Standard Test Method for Determining the Arc Rating of Materials for Clothing.

Some fabrics, such as flame-resistant treated cotton/nylon blends, contain quantities of material (in this case nylon) that may be flammable, or that may melt and drip within the construction of the fabric, that would otherwise not be permitted by itself. If the fabric does not demonstrate melting, dripping, or flammability, the fabric would be considered to be compliant with the ASTM F1506 standard and may be arc-rated.

Small amounts of flammable or meltable materials can be used in the elastic bands of underwear. See the exception to 130.7(C)(9)(c).

(12) Clothing and Other Apparel Not Permitted. Clothing and other apparel (such as hard hat liners and hair nets) made from materials that do not meet the requirements of 130.7(C)(11) regarding melting or made from materials that do not meet the flammability requirements shall not be permitted to be worn.

Informational Note: Some flame-resistant fabrics, such as non-flame-resistant modacrylic and nondurable flame-retardant treatments of cotton, are not recommended for industrial electrical or utility applications.

Apparel made from materials that are not arc-rated must not be worn. For instance, hair nets could melt onto an employee's hair and head (although arc-rated hair nets are available). Ear warmers and head covers must not be worn unless they are made of arc-rated material.

Exception No. 1:   Nonmelting, flammable (non–arc-rated) materials shall be permitted to be used as underlayers to arc-rated clothing, as described in 130.7(C)(11).

Provided that the innermost arc-rated layer above the non–arc-rated layer does not break open, nonmelting flammable materials are permitted to be used as underlayers to arc-rated clothing (such as underwear). However, there may be some additional benefits to wearing arc-rated undergarments.

Exception No. 2:   Where the work to be performed inside the arc flash boundary exposes the worker to multiple hazards, such as airborne contaminants, and the risk assessment identifies that the level of protection is adequate to address the arc flash hazard, non–arc-rated PPE shall be permitted.

(13) Care and Maintenance of Arc-Rated Clothing and Arc-Rated Arc Flash Suits.

(a) Inspection. Arc-rated apparel shall be inspected before each use. Work clothing or arc flash suits that are contaminated or damaged to the extent that their protective qualities are impaired shall not be used. Protective items that become contaminated with grease, oil, or flammable liquids or combustible materials shall not be used.

Any flammable soiling or contamination on the surface of the arc-rated clothing can reduce or even void the arc rating of the apparel. The clothing must also be free from tears, cuts, or rips. Garments that are not suitable for use must be removed from service. If corrections can be made, such as laundering or repair using appropriate techniques, the garment can be returned to service; otherwise, the garment should be disposed of.

(b) Manufacturer's Instructions. The garment manufacturer's instructions for care and maintenance of arc-rated apparel shall be followed.

The manufacturer's laundry care instructions are a required element of the labeling of any arc flash garment, per the ASTM F1506 standard. The information is typically present in pictorial form, using symbols from industry groups such as Textile Industry Affairs (www.textileaffairs.com/lguide.htm), which demonstrate how the garment should be cared for. The care and laundering information may also be present in writing, with additional detail beyond the symbology.

Improper laundering or failure to follow the instruction provided by the manufacturer in the garment labeling or additional printed guides may diminish the longevity of the products useful life as well as reduce the protective qualities of the product itself.

(c) Storage. Arc-rated apparel shall be stored in a manner that prevents physical damage; damage from moisture, dust, or other deteriorating agents; or contamination from flammable or combustible materials.

Arc-rated clothing is intended to protect employees from the thermal hazard associated with an arcing fault. For the protective clothing to perform as intended, it must be protected when in use and in storage. Contaminants such as grease and oil must be avoided, since contamination reduces the thermal protection provided by the clothing. Exposure to flammable materials also must be avoided.

(d) Cleaning, Repairing, and Affixing Items. When arc-rated clothing is cleaned, manufacturer's instructions shall be followed to avoid loss of protection. When arc-rated clothing is repaired, the same arc-rated materials used to manufacture the arc-rated clothing shall be used to provide repairs.

Manufacturers provide cleaning instructions for their products used for thermal protection. These instructions should be followed closely to ensure that the performance of the garment is not compromised. When arc-rated clothing is shared by employees, the electrical safety program should consider health aspects of shared PPE when determining the cleaning frequency.

> Informational Note No. 1: Additional guidance is provided in ASTM F1506, *Standard Performance Specification for Flame Resistant and Arc Rated Textile Materials for Wearing Apparel for Use by Electrical Workers Exposed to Momentary Electric Arc and Related Thermal Hazards*, when trim, name tags, logos, or any combination thereof are affixed to arc-rated clothing.

> Informational Note No. 2: Additional guidance is provided in ASTM F1449, *Standard Guide for Industrial Laundering of Flame, Thermal, and Arc Resistant Clothing*, and ASTM F2757, *Standard Guide for Home Laundering Care and Maintenance of Flame, Thermal, and Arc Resistant Clothing*.

(14) Standards for Personal Protective Equipment (PPE). PPE shall conform to the standards listed in Table 130.7(C)(14).

Informational Note: Non–arc-rated or flammable fabrics are not covered by any of the standards in Table 130.7(C)(14). See 130.7(C)(11) and 130.7(C)(12).

This table only encompasses protective equipment that is normally considered as PPE. For instance, voltmeters provide information that enables employees to protect themselves from electrical shock, but because they are not generally considered PPE, voltmeters are not covered in Table 130.7(C)(14).

(15) Selection of Personal Protective Equipment (PPE) When Required for Various Tasks.

In Article 100, risk is defined as the combination of the likelihood of occurrence of injury and the severity of injury that results from a hazard. Where electrical safety is involved, protection from injury associated with exposure to possible electrical hazards is the concern. For the 2015 edition, there is a definition for the term *risk assessment* in Article 100. Risk assessment is generally defined as the overall process that identifies hazards, estimates the potential severity of injury, estimates the likelihood of occurrence of injury, and determines if protective measures are required, such as the appropriate personal and other protective equipment required to perform a task safely. With the replacement of the

Table 130.7(C)(14)   *Standards on Protective Equipment*

Subject	Document Title	Document Number
Apparel — Arc Rated	Standard Performance Specification for Flame Resistant and Arc Rated Textile Materials for Wearing Apparel for Use by Electrical Workers Exposed to Momentary Electric Arc and Related Thermal Hazards	ASTM F1506
	Standard Guide for Industrial Laundering of Flame, Thermal, and Arc Resistant Clothing	ASTM F1449
	Standard Guide for Home Laundering Care and Maintenance of Flame, Thermal and Arc Resistant Clothing	ASTM F2757
Aprons — Insulating	Standard Specification for Electrically Insulating Aprons	ASTM F2677
Eye and Face Protection-General	Practice for Occupational and Educational Eye and Face Protection	ANSI Z87.1
Face — Arc Rated	Standard Test Method for Determining the Arc Rating and Standard Specification for Eye or Face Protective Products	ASTM F2178
Fall Protection	Standard Specification for Personal Climbing Equipment	ASTM F887
Footwear — Dielectric Specification	Standard Specification for Dielectric Footwear	ASTM F1117
Footwear — Dielectric Test Method	Standard Test Method for Determining Dielectric Strength of Dielectric Footwear	ASTM F1116
Footwear — Standard Performance Specification	Standard Specification for Performance Requirements for Protective (Safety) Toe Cap Footwear	ASTM F2413
Footwear — Standard Test Method	Standard Test Methods for Foot Protections	ASTM F2412
Gloves — Leather Protectors	Standard Specification for Leather Protectors for Rubber Insulating Gloves and Mittens	ASTM F696
Gloves — Rubber Insulating	Standard Specification for Rubber Insulating Gloves	ASTM D120
Gloves and Sleeves — In-Service Care	Standard Specification for In-Service Care of Insulating Gloves and Sleeves	ASTM F496
Head Protection — Hard Hats	Requirements for Protective Headwear for Industrial Workers	ANSI Z89.1
Rainwear — Arc Rated	Standard Specification for Arc and Flame Resistant Rainwear	ASTM F1891
Rubber Protective Products — Visual Inspection	Standard Guide for Visual Inspection of Electrical Protective Rubber Products	ASTM F1236
Sleeves — Insulating	Standard Specification for Rubber Insulating Sleeves	ASTM D1051

hazard/risk categories method with the arc flash PPE categories method, there is now a single methodology used for risk assessment; the approach used for the incident energy method is in harmony with the approach used for the arc flash PPE categories method. If the risk is found to be unacceptable, either method now requires the use of PPE for the level of the anticipated hazard — the maximum possible hazard.

The arc flash PPE category for equipment classifications given in Table 130.7(C)(15)(A)(b), for ac systems, or Table 130.7(C)(15)(B), for dc systems, generally is based on a determination of the estimated exposure level for the arc flash hazard involved. No equipment classification in Table 130.7(C)(15)(A)(b) or Table 130.7(C)(15)(B) results in an arc flash PPE category over 4, for which the maximum anticipated exposure level is 40 cal/cm². Thus, where the arc flash PPE categories method is used, the maximum anticipated exposure level for an equipment category is 40 cal/cm² or less. In fact, for most equipment classifications it is less. When using the arc flash PPE categories method, the severity of injury results from an incident energy level that does not exceed 40 cal/cm² — 40 cal/cm² sets the upper boundary.

As with any standard, the requirements included in the arc flash PPE categories method tables are minimum requirements only. The user of the standard is free to determine that arc flash PPE is required for a task, where the table indicates it is not. When it has been determined that arc flash PPE is required from Table 130.7(C)(15)(A)(a), the level of PPE required is determined from either Table 130.7(C)(15)(A)(b) for ac systems or Table 130.7(C)(15)(B) for open-air dc systems. Exhibits 130.8 through 130.12 show various PPE and tools used when performing various jobs.

EXHIBIT 130.8

Using Category 1 PPE to remove or install a circuit breaker in a panelboard. (Reproduced with Permission, Fluke Corporation)

(A) Alternating Current (ac) Equipment. When selected in lieu of the incident energy analysis of 130.5(B)(1), Table 130.7(C)(15)(A)(a) shall be used to identify when arc flash PPE is required. When arc flash PPE is required, Table 130.7(C)(15)(A)(b) shall be used to determine the arc flash PPE category. The estimated maximum available short-circuit current, maximum fault clearing times, and minimum working distances for various ac equipment types or classifications are listed in Table 130.7(C)(15)(A)(b). An incident energy analysis shall be required in accordance with 130.5 for the following:

(1) Tasks not listed in Table 130.7(C)(15)(A)(a)

(2) Power systems with greater than the estimated maximum available short-circuit current

EXHIBIT 130.9

Measuring voltage in a 480-V
motor control panel center using
Category 2 PPE.

EXHIBIT 130.10

Current measurement in a
480-V motor control center also
requires Category 2 PPE.

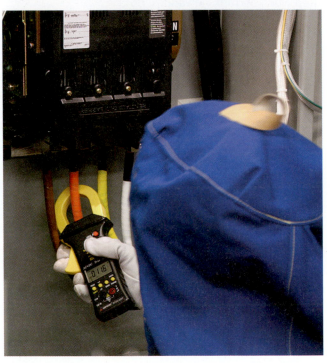

(3) Power systems with longer than the maximum fault clearing times
(4) Tasks with less than the minimum working distance

Generally, a determination of estimated incident energy involves determining or examining the system voltage, the working distance to be used, the probable arcing current
involved, and the total clearing time. Similarly, a determination of whether the arc flash
PPE categories method is applicable involves deciding whether or not the available fault
current and total fault clearing time fall within the maximum available short-circuit current and fault clearing times parameters, which are indicated in the applicable equipment
classification (category) for the appropriate arc flash PPE categories method table — Table
130.7(C)(15)(A)(b) for ac systems and Table 130.7(C)(15)(B) for dc systems.

EXHIBIT 130.11

Breathing pack. (Courtesy of Chicago Protective Apparel, Inc.)

EXHIBIT 130.12

Insertion of a 480-V drawout breaker with switchgear doors open, using Category 4 PPE.

One of the generally accepted methods for estimating total clearing time for an overcurrent protective device where arcing currents are involved is to determine the estimated arcing current and then to plot that arcing current on the protective device's time-current characteristic curve, thereby determining its tripping time. Where power circuit breakers are involved, the device opening time needs to be added to the tripping time determined from the time–current characteristic curve for the device to obtain the total fault clearing time.

Another method for estimating total clearing time for 480-volt nominal systems is to estimate the short-circuit current with reasonable accuracy, and then use 38 percent of that value to determine the total clearing time. For medium voltage systems, a reasonable approach for estimating the tripping time may involve using a value of 90 to 95 percent of the short-circuit current for the arcing current, and then using 85 percent of that value to determine the total clearing time.

The determination of a value for arcing current is made based on a determination of the probable available three-phase symmetrical fault current. If the task is not listed in Table 130.7(C)(15)(A)(a) or the values for working distance, available fault current, or total fault clearing time are determined to fall outside of the parameters given in the equipment category for the appropriate arc flash PPE categories method table [Table 130.7(C)(15)(A)(b) for ac systems and Table 130.7(C)(15)(B) for dc systems], then an incident energy analysis is required.

Table 130.7(C)(15)(A)(a) *Arc Flash Hazard Identification for Alternating Current (ac) and Direct Current (dc) Systems*

Task	Equipment Condition*	Arc Flash PPE Required
Reading a panel meter while operating a meter switch	Any	No
Normal operation of a circuit breaker (CB), switch, contactor, or starter	All of the following: The equipment is properly installed The equipment is properly maintained All equipment doors are closed and secured All equipment covers are in place and secured There is no evidence of impending failure	No
	One or more of the following: The equipment is not properly installed The equipment is not properly maintained Equipment doors are open or not secured Equipment covers are off or not secured There is evidence of impending failure	Yes
For ac systems: Work on energized electrical conductors and circuit parts, including voltage testing	Any	Yes
For dc systems: Work on energized electrical conductors and circuit parts of series-connected battery cells, including voltage testing	Any	Yes
Voltage testing on individual battery cells or individual multi-cell units	All of the following: The equipment is properly installed The equipment is properly maintained Covers for all other equipment are in place and secured There is no evidence of impending failure	No
	One or more of the following: The equipment is not properly installed The equipment is not properly maintained Equipment doors are open or not secured Equipment covers are off or not secured There is evidence of impending failure	Yes

Table 130.7(C)(15)(A)(a) *Arc Flash Hazard Identification for Alternating Current (ac) and Direct Current (dc) Systems*

Task	Equipment Condition*	Arc Flash PPE Required
Removal or installation of CBs or switches	Any	Yes
Removal or installation of covers for equipment such as wireways, junction boxes, and cable trays that does not expose bare energized electrical conductors and circuit parts	All of the following: The equipment is properly installed The equipment is properly maintained There is no evidence of impending failure	No
	Any of the following: The equipment is not properly installed The equipment is not properly maintained There is evidence of impending failure	Yes
Removal of bolted covers (to expose bare energized electrical conductors and circuit parts). For dc systems, this includes bolted covers, such as battery terminal covers.	Any	Yes
Removal of battery intercell connector covers	All of the following: The equipment is properly installed. The equipment is properly maintained Covers for all other equipment are in place and secured There is no evidence of impending failure	No
	One or more of the following: The equipment is not properly installed The equipment is not properly maintained Equipment doors are open or not secured Equipment covers are off or not secured There is evidence of impending failure	Yes
Opening hinged door(s) or cover(s) (to expose bare energized electrical conductors and circuit parts)	Any	Yes
Perform infrared thermography and other noncontact inspections outside the restricted approach boundary. This activity does not include opening of doors or covers.	Any	No
Application of temporary protective grounding equipment after voltage test	Any	Yes
Work on control circuits with exposed energized electrical conductors and circuit parts, 120 volts or below without any other exposed energized equipment over 120 V including opening of hinged covers to gain access	Any	No
Work on control circuits with exposed energized electrical conductors and circuit parts, greater than 120 V	Any	Yes
Insertion or removal of individual starter buckets from motor control center (MCC)	Any	Yes
Insertion or removal (racking) of CBs or starters from cubicles, doors open or closed	Any	Yes
Insertion or removal of plug-in devices into or from busways	Any	Yes
Insulated cable examination with no manipulation of cable	Any	No
Insulated cable examination with manipulation of cable	Any	Yes

(continued)

Table 130.7(C)(15)(A)(a) *Arc Flash Hazard Identification for Alternating Current (ac) and Direct Current (dc) Systems*

Task	Equipment Condition*	Arc Flash PPE Required
Work on exposed energized electrical conductors and circuit parts of equipment directly supplied by a panelboard or motor control center	Any	Yes
Insertion and removal of revenue meters (kW-hour, at primary voltage and current)	Any	Yes
For dc systems, insertion or removal of individual cells or multi-cell units of a battery system in an enclosure	Any	Yes
For dc systems, insertion or removal of individual cells or multi-cell units of a battery system in an open rack	Any	No
For dc systems, maintenance on a single cell of a battery system or multi-cell units in an open rack	Any	No
For dc systems, work on exposed energized electrical conductors and circuit parts of utilization equipment directly supplied by a dc source	Any	Yes
Arc-resistant switchgear Type 1 or 2 (for clearing times of <0.5 sec with a prospective fault current not to exceed the arc-resistant rating of the equipment) and metal enclosed interrupter switchgear, fused or unfused of arc resistant type construction, tested in accordance with IEEE C37.20.7: • Insertion or removal (racking) of CBs from cubicles • Insertion or removal (racking) of ground and test device • Insertion or removal (racking) of voltage transformers on or off the bus	All of the following: The equipment is properly installed The equipment is properly maintained All equipment doors are closed and secured All equipment covers are in place and secured There is no evidence of impending failure	No
	One or more of the following: The equipment is not properly installed The equipment is not properly maintained Equipment doors are open or not secured Equipment covers are off or not secured There is evidence of impending failure	Yes
Opening voltage transformer or control power transformer compartments	Any	Yes
Outdoor disconnect switch operation (hookstick operated) at 1 kV through 15 kV	Any	Yes
Outdoor disconnect switch operation (gang-operated, from grade) at 1 kV through 15 kV	Any	Yes

Note: Hazard identification is one component of risk assessment. Risk assessment involves a determination of the likelihood of occurrence of an incident, resulting from a hazard that could cause injury or damage to health. The assessment of the likelihood of occurrence contained in this table does not cover every possible condition or situation. Where this table indicates that arc flash PPE is not required, an arc flash is not likely to occur.

*The phrase *properly installed*, as used in this table, means that the equipment is installed in accordance with applicable industry codes and standards and the manufacturer's recommendations. The phrase *properly maintained*, as used in this table, means that the equipment has been maintained in accordance with the manufacturer's recommendations and applicable industry codes and standards. The phrase *evidence of impending failure*, as used in this table, means that there is evidence of arcing, overheating, loose or bound equipment parts, visible damage, deterioration, or other damage.

Table 130.7(C)(15)(A)(b) Arc-Flash Hazard PPE Categories for Alternating Current (ac) Systems

Equipment	Arc Flash PPE Category	Arc-Flash Boundary
Panelboards or other equipment rated 240 V and below Parameters: Maximum of 25 kA short-circuit current available; maximum of 0.03 sec (2 cycles) fault clearing time; working distance 455 mm (18 in.)	1	485 mm (19 in.)
Panelboards or other equipment rated >240 V and up to 600 V Parameters: Maximum of 25 kA short-circuit current available; maximum of 0.03 sec (2 cycles) fault clearing time; working distance 455 mm (18 in.)	2	900 mm (3 ft)
600-V class motor control centers (MCCs) Parameters: Maximum of 65 kA short-circuit current available; maximum of 0.03 sec (2 cycles) fault clearing time; working distance 455 mm (18 in.)	2	1.5 m (5 ft)
600-V class motor control centers (MCCs) Parameters: Maximum of 42 kA short-circuit current available; maximum of 0.33 sec (20 cycles) fault clearing time; working distance 455 mm (18 in.)	4	4.3 m (14 ft)
600-V class switchgear (with power circuit breakers or fused switches) and 600 V class switchboards Parameters: Maximum of 35 kA short-circuit current available; maximum of up to 0.5 sec (30 cycles) fault clearing time; working distance 455 mm (18 in.)	4	6 m (20 ft)
Other 600-V class (277 V through 600 V, nominal) equipment Parameters: Maximum of 65 kA short circuit current available; maximum of 0.03 sec (2 cycles) fault clearing time; working distance 455 mm (18 in.)	2	1.5 m (5 ft)
NEMA E2 (fused contactor) motor starters, 2.3 kV through 7.2 kV Parameters: Maximum of 35 kA short-circuit current available; maximum of up to 0.24 sec (15 cycles) fault clearing time; working distance 910 mm (36 in.)	4	12 m (40 ft)
Metal-clad switchgear, 1 kV through 15 kV Parameters: Maximum of 35 kA short-circuit current available; maximum of up to 0.24 sec (15 cycles) fault clearing time; working distance 910 mm (36 in.)	4	12 m (40 ft)
Arc-resistant switchgear Type 1 or 2 [for clearing times of < 0.5 sec (30 cycles) with a perspective fault current not to exceed the arc-resistant rating of the equipment], and metal-enclosed interrupter switchgear, fused or unfused of arc-resistant-type construction, tested in accordance with IEEE C37.20.7, 1 kV through 15 kV	N/A (doors closed)	N/A (doors closed)
Parameters: Maximum of 35 kA short-circuit current available; maximum of up to 0.24 sec (15 cycles) fault clearing time; working distance 910 mm (36 in.)	4 (doors open)	12 m (40 ft)
Other equipment 1 kV through 15 kV Parameters: Maximum of 35 kA short-circuit current available; maximum of up to 0.24 sec (15 cycles) fault clearing time; working distance 910 mm (36 in.)	4	12 m (40 ft)

Note: For equipment rated 600 volts and below, and protected by upstream current-limiting fuses or current-limiting circuit breakers sized at 200 amperes or less, the arc flash PPE category can be reduced by one number but not below arc flash PPE category 1.

(B) Direct Current (dc) Equipment. When selected in lieu of the incident energy analysis of 130.5(C)(1), Table 130.7(C)(15)(A)(a) shall be used to identify when arc flash PPE is required. When arc flash PPE is required, Table 130.7(C)(15)(B) shall be used to determine the arc flash PPE category. The estimated maximum available short circuit current, maximum arc duration and working distances for dc equipment are listed in 130.7(C)(15)(B). An incident energy analysis shall be required in accordance with 130.5 for the following:

(1)   Tasks not listed in Table 130.7(C)(15)(A)(a)

(2)   Power systems with greater than the estimated maximum available short circuit current

(3) Power systems with longer than the maximum fault clearing times

(4) Tasks with less than the minimum working distance

The commentary following 130.7(C) and the commentary regarding alternating current (ac) equipment should prove helpful in comprehending how the arc flash PPE categories method works for direct current (dc) systems. Note that Table 130.7(C)(15)(B) is only applicable to open-air arc situations. The principle involved in estimating the total clearing time for dc arcs is similar to those of ac arcs, but involves an additional level of complexity.

> Informational Note No. 1: The arc flash PPE category, work tasks, and protective equipment provided in Table 130.7(C)(15)(A)(a), Table 130.7(C)(15)(A)(b), and Table 130.7(C)(15)(B) were identified and selected, based on the collective experience of the NFPA 70E Technical Committee. The arc flash PPE category of the protective clothing and equipment is generally based on determination of the estimated exposure level.

As indicated in the informational note, the risk assessments given in 130.7(C)(15) are based on the collective experience of the technical committee for *NFPA 70E,* the Technical Committee on Electrical Safety in the Workplace. As with any standard, the requirements included in the arc flash PPE categories method tables are minimum requirements only. Users of the standard are free to determine that arc flash PPE is required for a task where the table indicates it is not, and they are free to use a higher level of PPE if they determine that it is appropriate.

> Informational Note No. 2: The collective experience of the NFPA 70E Technical Committee is that, in most cases, closed doors do not provide enough protection to eliminate the need for PPE in situations in which the state of the equipment is known to readily change (e.g., doors open or closed, rack in or rack out).

See the commentary following 130.2(A)(4) regarding normal operation. In most cases where normal operation is not involved and the state of the equipment is known to readily change — such as when taking off covers and exposing energized conductors and circuit parts, or racing-in or racking-out circuit breakers — PPE for the maximum possible hazard or other protective techniques such as remote racking should be used.

> Informational Note No. 3: The premise used by the NFPA 70E Technical Committee in developing the criteria discussed in Informational Note No. 1 and Informational Note No. 2 is considered to be reasonable, based on the consensus judgment of the committee.

Where dc equipment is involved, the arc flash PPE categories method is only suitable for use where open-air arcs are involved, such as those found in many battery systems or other dc process systems that are in open areas or rooms. The method does not cover arc-in-a-box type situations where the incident energy is focused and a multiplier effect is involved.

(16) Protective Clothing and Personal Protective Equipment (PPE). Once the arc flash PPE category has been identified from Table 130.7(C)(15)(A)(b) or Table 130.7(C)(15)(B), Table 130.7(C)(16) shall be used to determine the required PPE for the task. Table 130.7(C)(16) lists the requirements for PPE based on arc flash PPE categories 1 through 4. This clothing and equipment shall be used when working within the arc flash boundary.

In the 2015 edition of *NFPA 70E,* Table 130.7(C)(16) deals only with arc flash protection. As a result, hazard/risk categories (HRCs) are no longer represented there. Instead, there are now arc flash PPE categories. And since this table only deals with arc flash protection, HRC 0 has been eliminated (and there is no equivalent arc flash PPE category 0).

Table 130.7(C)(15)(B)   Arc-Flash Hazard PPE Categories for Direct Current (dc) Systems

Equipment	Arc Flash PPE Category	Arc-Flash Boundary
Storage batteries, dc switchboards, and other dc supply sources 100 V> Voltage < 250 V Parameters: Voltage: 250 V Maximum arc duration and working distance: 2 sec @ 455 mm (18 in.)		
Short-circuit current <4 kA	1	900 mm (3 ft)
4 kA ≤ short-circuit current < 7 kA	2	1.2 m (4 ft)
7 kA ≤ short-circuit current < 15 kA	3	1.8 m (6 ft)
Storage batteries, dc switchboards, and other dc supply sources 250 V ≤ Voltage ≤ 600 V Parameters: Voltage: 600 V Maximum arc duration and working distance: 2 sec @ 455 mm (18 in.)		
Short-circuit current < 1.5 kA	1	900 mm (3 ft)
1.5 kA ≤ short-circuit current < 3 kA	2	1.2 m (4 ft)
3 kA ≤ short-circuit current < 7 kA	3	1.8 m (6 ft.)
7 kA ≤ short-circuit current < 10 kA	4	2.5 m (8 ft)

Note: Apparel that can be expected to be exposed to electrolyte must meet both of the following conditions:
(1) Be evaluated for electrolyte protection in accordance with ASTM F1296, *Standard Guide for Evaluating Chemical Protective Clothing*
(2) Be arc-rated in accordance with ASTM F1891, *Standard Specification for Arc Rated and Flame Resistant Rainwear*, or equivalent

Informational Note No. 1: "Short-circuit current," as used in this table, is determined from the dc power system maximum available short-circuit, including the effects of cables and any other impedances in the circuit. Power system modeling is the best method to determine the available short-circuit current at the point of the arc. Battery cell short-circuit current can be obtained from the battery manufacturer. See Informative Annex D.5 for the basis for table values and alternative methods to determine dc incident energy. Methods should be used with good engineering judgment.

Informational Note No. 2: The methods for estimating the dc arc flash incident energy that were used to determine the categories for this table are based on open-air incident energy calculations. Open-air calculations were used because many battery systems and other dc process systems are in open areas or rooms. If the specific task is within in an enclosure, it would be prudent to consider additional PPE protection beyond the value shown in this table. Research with ac arc flash has shown a multiplier of as much as 3x for arc-in-a-box [508 mm (20 in.) cube] versus open air. Engineering judgment is required when reviewing the specific conditions of the equipment and task to be performed, including the dimensions of the enclosure and the working distance involved.

The arc-rated clothing and protective equipment shown in Table 130.7(C)(16) is intended to be used with Table 130.7(C)(15)(A)(a), and Table 130.7(C)(15)(A)(b) for ac systems and Table 130.7(C)(15)(B) for dc systems to protect only from an arc flash hazard. Arc-rated clothing is available in many constructions. Using arc-rated clothing in layers is one way

A face shield with (a) a balaclava sock hood and (b) a flash suit hood. (Courtesy of Salisbury of Honeywell International, Inc.)

to achieve a higher level of protection. Table 130.7(C)(16) suggests acceptable combinations of clothing items to achieve a desired arc flash PPE category. Other combinations are possible. See Exhibit 130.13 for two examples of arc flash PPE category 2 head protection.

Table 130.7(C)(16) provides general information that can help an employee understand the process for selecting clothing based on an arc flash PPE category designation, but it does not describe any required combination or construction of a protective system. The manufactured system can differ from the content of the table; the clothing manufacturer needs to be consulted. See Informational Note No. 3 to 130.7(C)(16).

PPE rated in cal/cm² is suitable for that incident energy level. The rating of layered PPE can only be determined by test. The rating of a clothing system recommended by the manufacturer of an arc-rated garment is only determined by testing those specific layers together. See Informational Note No. 3 and No. 4 to the definition of *arc rating* in Article 100.

> Informational Note No. 1: See Informative Annex H for a suggested simplified approach to ensure adequate PPE for electrical workers within facilities with large and diverse electrical systems.

To simplify administration of arc-rated clothing, an employer might implement a procedure that has two or three different assemblies of arc flash protective equipment. An employer could define a general requirement for workers to wear a minimum level of protective clothing. The procedure could define a secondary level of protection that is easily recognized by employees and, still, a third level of protection that is required in special situations. Two key factors in this type of requirement are that the protection is adequate for the greatest exposure for each level of protective clothing and that each employee can recognize when a higher level of protection is necessary to wear.

> Informational Note No. 2:   The PPE requirements of this section are intended to protect a person from arc flash hazards. While some situations could result in burns to the skin, even with the protection described in Table 130.7(C)(16), burn injury should be reduced and survivable. Due to the explosive effect of some arc events, physical trauma injuries could occur. The PPE requirements of this section do not address protection against physical trauma other than exposure to the thermal effects of an arc flash.

The nature of thermal energy dictates the wearing of arc-rated clothing and protective equipment to reduce the chance of thermal injury from an arc flash event. However, because of the nature of an arcing fault, determining the degree of each hazard associated with an arcing fault can be difficult. There currently is no agreed upon method for determining the pressure wave associated with an arc blast from an arc flash event. Any work

Table 130.7(C)(16) Personal Protective Equipment (PPE)

PPE Category	PPE
•	
1	**Arc-Rated Clothing, Minimum Arc Rating of 4 cal/cm² (see Note 1)** Arc-rated long-sleeve shirt and pants or arc-rated coverall Arc-rated face shield (see Note 2) or arc flash suit hood Arc-rated jacket, parka, rainwear, or hard hat liner (AN) **Protective Equipment** Hard hat Safety glasses or safety goggles (SR) Hearing protection (ear canal inserts) Heavy duty leather gloves (see Note 3) Leather footwear (AN)
2	**Arc-Rated Clothing, Minimum Arc Rating of 8 cal/cm² (see Note 1)** Arc-rated long-sleeve shirt and pants or arc-rated coverall Arc-rated flash suit hood or arc-rated face shield (see Note 2) and arc-rated balaclava Arc-rated jacket, parka, rainwear, or hard hat liner (AN) **Protective Equipment** Hard hat Safety glasses or safety goggles (SR) Hearing protection (ear canal inserts) Heavy duty leather gloves (see Note 3) Leather footwear
3	**Arc-Rated Clothing Selected so That the System Arc Rating Meets the Required Minimum Arc Rating of 25 cal/cm²** (see Note 1) Arc-rated long-sleeve shirt (AR) Arc-rated pants (AR) Arc-rated coverall (AR) Arc-rated arc flash suit jacket (AR) Arc-rated arc flash suit pants (AR) Arc-rated arc flash suit hood Arc-rated gloves (see Note 3) Arc-rated jacket, parka, rainwear, or hard hat liner (AN) **Protective Equipment** Hard hat Safety glasses or safety goggles (SR) Hearing protection (ear canal inserts) Leather footwear

(continued)

Table 130.7(C)(16) **Personal** Protective Equipment (PPE)

PPE Category	PPE
4	**Arc-Rated Clothing Selected so That the System Arc Rating Meets the Required Minimum Arc Rating of 40 cal/cm²** (see Note 1) Arc-rated long-sleeve shirt (AR) Arc-rated pants (AR) Arc-rated coverall (AR) Arc-rated arc flash suit jacket (AR) Arc-rated arc flash suit pants (AR) Arc-rated arc flash suit hood Arc-rated gloves (see Note 3) Arc-rated jacket, parka, rainwear, or hard hat liner (AN) **Protective Equipment** Hard hat Safety glasses or safety goggles (SR) Hearing protection (ear canal inserts) Leather footwear

AN: as needed (optional). AR: as required. SR: selection required.

Notes:

(1) *Arc rating* is defined in Article 100.

(2) Face shields are to have wrap-around guarding to protect not only the face but also the forehead, ears, and neck, or, alternatively, an arc-rated arc flash suit hood is required to be worn.

(3) If rubber insulating gloves with leather protectors are used, additional leather or arc-rated gloves are not required. The combination of rubber insulating gloves with leather protectors satisfies the arc flash protection requirement.

performed on or near exposed energized electrical conductors exposes an employee to an elevated risk of shock or arc flash injury.

Informational Note No. 3: The arc rating for a particular clothing system can be obtained from the arc-rated clothing manufacturer.

When layered systems are involved, the only way to determine the arc rating of the system is by testing.

(D) Other Protective Equipment.

(1) Insulated Tools and Equipment. Employees shall use insulated tools or handling equipment, or both, when working inside the restricted approach boundary of exposed energized electrical conductors or circuit parts where tools or handling equipment might make accidental contact. Insulated tools shall be protected from damage to the insulating material.

Informational Note: See 130.4(B), Shock Protection Boundaries.

(a) Requirements for Insulated Tools. The following requirements shall apply to insulated tools:

(1) Insulated tools shall be rated for the voltages on which they are used.

(2) Insulated tools shall be designed and constructed for the environment to which they are exposed and the manner in which they are used.

(3) Insulated tools and equipment shall be inspected prior to each use. The inspection shall look for damage to the insulation or damage that can limit the tool from performing its intended function or could increase the potential for an incident (e.g., damaged tip on a screwdriver).

When working inside the restricted approach boundary, employees must select and use work procedures, including using insulated tools that provide maximum protection from a release of energy. If contact with the exposed energized electrical conductor or circuit part is likely, the employee must use insulated tools. An unqualified person within the limited approach boundary is considered likely to contact an exposed energized electrical conductor. Also, a qualified person performing a work task within the restricted approach boundary is considered likely to contact an exposed energized electrical conductor.

The term *insulated* means that the tool manufacturer has assigned a voltage rating to the insulating material. If the task requires the worker to cross or work in the vicinity of the restricted approach boundary, insulated tools are required. Qualified persons are expected to be competent to inspect an insulated tool for potential damage. Qualified persons also must be able to determine whether the voltage rating remains intact.

Exhibit 130.14(a) shows a marked insulated tool with cutaway, and Exhibit 130.14 (b) shows a covered tool, which is not insulated. It is important for employees to look for the markings on the tool before using it.

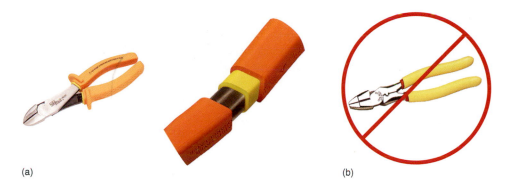

(a)   (b)

EXHIBIT 130.14

Example of (a) insulated tool with cutaway and (b) non-insulated tool. (Courtesy of Ideal Industries, Inc.)

(b) Fuse or Fuseholder Handling Equipment. Fuse or fuseholder handling equipment, insulated for the circuit voltage, shall be used to remove or install a fuse if the fuse terminals are energized.

Authorized and qualified persons can remove and install fuses routinely with the fuse terminals energized using a live-line tool (hot stick), provided they remain outside of the shock boundaries or the arc flash boundary. If they are within either type of boundary, appropriate PPE must be used. Replacing fuses live is energized work and, as such, must be justified.

(c) Ropes and Handlines. Ropes and handlines used within the limited approach boundary of exposed energized electrical conductors or circuit parts operating at 50 volts or more, or used where an electrical hazard exists, shall be nonconductive.

If the employees feel that ropes and handlines are necessary to control the lift, they should reevaluate the lift to determine whether it could be performed from a different position, making ropes or handlines unnecessary. However, where ropes and handlines are used around energized conductors and circuit parts, they need to be made of nonconductive material to protect the employees from potential shock hazards.

(d) Fiberglass-Reinforced Plastic Rods. Fiberglass-reinforced plastic rod and tube used for live-line tools shall meet the requirements of applicable portions of electrical codes and standards dealing with electrical installation requirements.

Informational Note: For further information concerning electrical codes and standards dealing with installation requirements, refer to ASTM F 711, *Standard Specification for Fiberglass-Reinforced Plastic (FRP) Rod and Tube Used in Live Line Tools.*

Case Study

Scenario and Causation.

David, aged 54, was a contract electrician. He arrived at a work site to check the fuses in a three-phase, 480-volt distribution panel that supplied power to a lighting transformer and other loads throughout the facility. The distribution panelboard was approximately 3½ ft wide and 5 ft tall. The panelboard was main lug only and had six cubicles (buckets), three on each side accessible from the front of the panelboard. Each bucket had an individual hinged door with a switching (disconnect) mechanism that was equipped with a safety interlock that prevented the opening of the cubicle door with the switch in the closed (on) position. Within each bucket were three horizontal positioned cartridge fuses, which were held in place with spring fuse holder clips located on each end of each fuse.

A thermal photo image performed as part of a preventative maintenance program indicated that an area in one of the buckets was overheating. To check the fuse clips and conductor terminations in the bucket, David opened the bucket door by first placing the disconnect switch in the open (off) position. He did not bypass the safety interlock. David was not wearing shock or arc flash personal protective garments. David removed the three 100-amp fuses in the bucket. It is unknown whether David tested for power after he opened the bucket door.

Daniel, the plant supervisor, came over to check on what David was doing. Daniel stated that David stuck his screwdriver into the top fuse clip on the right side of the opening and spread the clip to demonstrate that the tension of the clip was fine and it did not need to be changed. It is not known if David was using an insulated screwdriver. David then removed the screwdriver from the top clip and proceeded to move down to the next clip, and suddenly there was an explosion. An arc flash incident ensued.

Results.

Both Daniel and David sustained burn injuries. However, David's injuries were so severe that he died four weeks later; he had third-degree burns over 60 percent of his body, as well as other injuries, sustained at the time of the incident.

Analysis.

Accidents typically occur due to a chain of events or mistakes that lead up to the actual event. If one of the links in the chain of events is eliminated, the accident most likely would not have happened.

Although opening (turning off) the disconnecting means (switch) of the bucket de-energized the load terminals and load side of the fuse clips, the line side fuse clips remained energized. This distribution panelboard was supplied from upstream distribution equipment.

Sticking a screw driver into the fuse clips as a measure of the clip's tension is not likely to be a maintenance practice recommended by the equipment manufacturer and proved to be a lethal mistake. Even if this was a sanctioned method, the fact that David was not wearing any personal protective equipment is contrary to the requirements in Article 130 of *NFPA 70E*.

Employees performing inspection and diagnostic and testing work must wear appropriate personal protective equipment. Furthermore, after inspection work is performed and prior to performing repair or maintenance work, employees should create an electrically safe work condition. By their actions we can infer that both David and Daniel were not qualified to perform energized electrical work, including diagnostic or testing tasks, nor did David recognize that his interface with the energized fuse clip was not a testing or diagnostic task. Only testing, troubleshooting, and voltage-measuring tasks are exempt from the *energized electrical work permit* process specified in 130.2(B). An energized electrical work permit is required for maintenance type tasks, and more importantly, energized work has to be justified based on at least one of the criteria established in 130.2(A)(1), (2), and (3).

Relevant *NFPA 70E* Requirements.

The host employer and the contract employer failed to hold a documented safety meeting. The host employer did not provide any safety-related rules that the contract employer was to follow. The contract employer failed to follow the work practices required by *NFPA 70E*. If David had installed a barrier or warned Daniel to stay outside the limited appropriate boundary and arc flash boundary, Daniel might not have sustained his injuries. If David had used insulated tools with less metal exposed, the incident might not have happened. If David, Daniel, and the companies they work for had followed all of the requirements of *NFPA 70E*, David would be alive today and Daniel would not have had to experience the painful recovery from his severe burns. Some of the steps they should have taken include the following:

- Create an electrically safe work condition by disconnecting the power to the panelboard per 130.2.
- Perform risk assessment (risk assessment includes hazard identification) of the task per 110.1(G).
- Train employees to be qualified to recognize and avoid the electrical hazard per 110.2.
- Create an energized electrical work permit (EEWP) per 130.2(B).
- Perform shock and arc flash risk assessment per 130.4 and 130.5.
- Use proper personal protective equipment for protection against shock and arc flash hazards per 130.7.
- Follow all procedures established by the electrical safety program for working within the approach boundaries of electrical equipment per 110.1(F).

(e) Portable Ladders. Portable ladders shall have nonconductive side rails if they are used where an employee or ladder could contact exposed energized electrical conductors or circuit parts operating at 50 volts or more or where an electrical hazard exists. Nonconductive ladders shall meet the requirements of ANSI standards for ladders listed in Table 130.7(F).

The use of ladders with conductive side rails in the vicinity of exposed energize conductors or circuit parts creates an unacceptable risk of injury from an electric shock. It may also create an unacceptable risk of injury from a potential arc flash hazard, depending on the circumstances.

(f) Protective Shields. Protective shields, protective barriers, or insulating materials shall be used to protect each employee from shock, burns, or other electrically related injuries while an employee is working within the limited approach boundary of energized conductors or circuit parts that might be accidentally contacted or where dangerous electric heating or arcing might occur. When normally enclosed energized conductors or circuit parts are exposed for maintenance or repair, they shall be guarded to protect unqualified persons from contact with the energized conductors or circuit parts.

The use of these protective techniques is required to protect the employee from injuries from potential exposure to electrical hazards, including electrically associated burns or other electrically related injuries that might result from accidental contact with energized conductors or circuit parts, from hot electrical equipment, or where arcing might occur. Only qualified persons are allowed to work within the restricted approach boundary of energized electrical conductors and circuit parts that are considered exposed. Generally, protective shields, protective barriers, and insulating material are not considered to be full protection from an arc flash event.

When qualified persons are performing maintenance or repair activities within the restricted approach boundary and unqualified persons are allowed within the limited approach boundary, the energized conductors and circuit parts are to be guarded to protect the unqualified persons from contacting them and thereby initiating an event. For equipment to be considered guarded, the unqualified person must be isolated from, kept away from, or protected from the energized conductors or circuit parts by means of protective barriers, protective shields, or insulating materials. It should be noted that an unqualified person cannot be within the limited approach boundary unless the person aware of the potential hazards and is continuously escorted by a qualified person. Since protective barriers, protective shields, or insulating materials are not considered fully protective against an arc flash event, anyone within the arc flash boundary, qualified or not, must use appropriate arc flash personal and other protective equipment to be effectively protected from a potential arc flash event.

Prior to the 2000 edition of *NFPA 70E*, and before the advent of arc flash personal protective clothing and equipment, this section was the only means available for protecting employees from an arc flash event. However, the requirements in this section are still enforced.

(g) Rubber Insulating Equipment. Rubber insulating equipment used for protection from accidental contact with energized conductors or circuit parts shall meet the requirements of the ASTM standards listed in Table 130.7(F).

(h) Voltage-Rated Plastic Guard Equipment. Plastic guard equipment for protection of employees from accidental contact with energized conductors or circuit parts, or for protection of employees or energized equipment or material from contact with ground, shall meet the requirements of the ASTM standards listed in Table 130.7(F).

(i) Physical or Mechanical Barriers. Physical or mechanical (field-fabricated) barriers shall be installed no closer than the restricted approach boundary distance given in Table 130.4(D)(a) and Table 130.4(D)(b). While the barrier is being installed, the restricted approach boundary distance specified in Table 130.4(D)(a) and Table 130.4(D)(b) shall be maintained, or the energized conductors or circuit parts shall be placed in an electrically safe work condition.

(E) Alerting Techniques.

People who are not involved in the work task can be exposed to an electrical hazard when the work task is being executed. To avoid unnecessary exposure to electrical hazards, people must be provided with a warning that an electrical hazard exists.

(1) Safety Signs and Tags. Safety signs, safety symbols, or accident prevention tags shall be used where necessary to warn employees about electrical hazards that might endanger them. Such signs and tags shall meet the requirements of ANSI Z535, *Series of Standards for Safety Signs and Tags*, given in Table 130.7(F).

Informational Note: Safety signs, tags, and barricades used to identify energized "look-alike" equipment can be employed as an additional preventive measure.

(2) Barricades. Barricades shall be used in conjunction with safety signs where it is necessary to prevent or limit employee access to work areas containing energized conductors or circuit parts. Conductive barricades shall not be used where it might increase the likelihood of exposure to an electrical hazard. Barricades shall be placed no closer than the limited approach boundary given in Table 130.4(D)(a) and Table 130.4(D)(b). Where the arc flash boundary is greater than the limited approach boundary, barricades shall not be placed closer than the arc flash boundary.

Barricades are not intended to prevent approach to an area. Instead, a barricade is intended to act as a warning device. When installed, the barricade should enclose the area containing the electrical hazard. The barricade must not be closer to the exposed energized electrical conductor or circuit part than the limited approach boundary. The barricade should be placed so as not to impede the exit of employees within the boundary. See Exhibit 130.15 for an example of a barricade.

EXHIBIT 130.15

Barricade used to keep unqualified persons from an area. (Courtesy of The T-CAP / www.TheTCap.com)

(3) Attendants. If signs and barricades do not provide sufficient warning and protection from electrical hazards, an attendant shall be stationed to warn and protect employees. The primary duty and responsibility of an attendant providing manual signaling and alerting shall be to keep unqualified employees outside a work area where the unqualified employee might be exposed to electrical hazards. An attendant shall remain in the area as long as there is a potential for employees to be exposed to the electrical hazards.

When an attendant is necessary to deliver the warning, the attendant should have no other duty.

(4) Look-Alike Equipment. Where work performed on equipment that is de-energized and placed in an electrically safe condition exists in a work area with other energized equipment that is similar in size, shape, and construction, one of the alerting methods in 130.7(E)(1), (2), or (3) shall be employed to prevent the employee from entering look-alike equipment.

When an installation involves multiple similar processes, similar electrical equipment is likely to exist in the same physical area, with different labels as the only visible difference. Employees must be aware that equipment similar to the one that is being maintained exists. Employees should consider installing some temporary identifying mark to reduce the chance of opening the wrong equipment.

Table 130.7(F) *Standards on Other Protective Equipment*

Subject	Document	Document Number
Arc Protective Blankets	Standard Test Method for Determining the Protective Performance of an Arc Protective Blanket for Electric Arc Hazards	ASTM F2676
Blankets	Standard Specification for Rubber Insulating Blankets	ASTM D1048
Blankets — In-service Care	Standard Specification for In-Service Care of Insulating Blankets	ASTM F479
Covers	Standard Specification for Rubber Covers	ASTM D1049
Fiberglass Rods — Live Line Tools	Standard Specification for Fiberglass-Reinforced Plastic (FRP) Rod and Tube Used in Live Line Tools	ASTM F711
Insulated Hand Tools	Standard Specification for Insulated and Insulating Hand Tools	ASTM F1505
Ladders	American National Standard for Ladders — Wood — Safety Requirements	ANSI/ASC A14.1
	American National Standard for Ladders — Fixed — Safety Requirements	ANSI/ASC A14.3
	American National Standard Safety Requirements for Job Made Ladders	ANSI ASC A14.4
	American National Standard for Ladders-Portable Reinforced- Safety Requirements	ANSI ASC A14.5
Line Hose	Standard Specification for Rubber Insulating Line Hoses	ASTM D1050
Line Hose and Covers — In-service Care	Standard Specification for In-Service Care of Insulating Line Hose and Covers	ASTM F478
Plastic Guard	Standard Test Methods and Specifications for Electrically Insulating Plastic Guard Equipment for Protection of Workers	ASTM F712
Sheeting	Standard Specification for PVC Insulating Sheeting	ASTM F1742
	Standard Specification for Rubber Insulating Sheeting	ASTM F2320
Safety Signs and Tags	Series of Standards for Safety Signs and Tags	ANSI Z535
Shield Performance on Live Line Tool	Standard Test Method for Determining the Protective Performance of a Shield Attached on Live Line Tools or on Racking Rods for Electric Arc Hazards	ASTM F2522
Temporary Protective Grounds — In-service Testing	Standard Specification for In-Service Test Methods for Temporary Grounding Jumper Assemblies Used on De-energized Electric Power Lines and Equipment	ASTM F2249
Temporary Protective Grounds — Test Specification	Standard Specification for Temporary Protective Grounds to Be Used on De-energized Electric Power Lines and Equipment	ASTM F855

(F) **Standards for Other Protective Equipment.** Other protective equipment required in 130.7(D) shall conform to the standards given in Table 130.7(F).

Table 130.7(F) identifies standards that define requirements for specific protective equipment. All equipment listed in the table impacts safe work procedures that a qualified person should implement. Equipment identified in this table must comply with the latest edition of the referenced standard.

130.8 Work Within the Limited Approach Boundary or Arc Flash Boundary of Overhead Lines.

In most cases, overhead conductors are guarded by being elevated so that they are not subject to incidental contact. When working on an overhead conductor, employees are likely to be supported by an elevated or articulating platform. In some instances, employees might be supported by a scaffold or by a permanently installed platform or deck. When using such support methods, employees may have difficulty escaping from an arc flash or avoiding direct contact with energized conductors. This emphasizes the need for an emergency recovery plan.

(A) Uninsulated and Energized. Where work is performed in locations containing uninsulated energized overhead lines that are not guarded or isolated, precautions shall be taken to prevent employees from contacting such lines directly with any unguarded parts of their body or indirectly through conductive materials, tools, or equipment. Where the work to be performed is such that contact with uninsulated energized overhead lines is possible, the lines shall be de-energized and visibly grounded at the point of work or suitably guarded.

Electrical conductors that are not fully insulated for the circuit voltage are considered to have the same potential for shock and electrocution as conductors that are completely bare. Some overhead conductors have a covering as protection from environmental degradation, but the covering has no insulation rating.

Exhibit 130.16(a) illustrates the danger of operating equipment in close proximity of overhead power lines and the need to treat these lines as energized, unless they have been de-energized and visibly grounded. Exhibit 130.16(b) shows signage indicating that equipment booms should be kept a safe distance away from overhead power lines. See 130.8(F).

Employees should not work near overhead conductors unless they are protected from unintentional contact with the conductors. Personnel carrying conduits, pipes, ladders, and other long objects must exercise caution to avoid entering the space defined by the

(a)

(b)

EXHIBIT 130.16

(a) Equipment such as a crane, with the capability of reaching overhead power lines, presents a shock hazard to the equipment operator and to employees who are working in the vicinity of the equipment. (Courtesy of FEMA, photo by George Armstrong) (b) Signage shows universal symbology indicating prohibition of contact with overhead power lines and requirement to maintain safe distance.

limited approach boundary. When long objects are moved, employees should be assigned to each end of the object to maintain control of both ends. Any object not fully insulated for the circuit voltage is considered to be conductive. Unqualified persons must not approach an overhead line that is not in an electrically safe work condition. Qualified persons must observe and comply with the approach boundaries identified in Table 130.4(D)(a) for ac systems or Table 130.4(D)(b) for dc systems.

(B) Determination of Insulation Rating. A qualified person shall determine if the overhead electrical lines are insulated for the lines' operating voltage.

(C) De-energizing or Guarding. If the lines are to be de-energized, arrangements shall be made with the person or organization that operates or controls the lines to de-energize them and visibly ground them at the point of work. If arrangements are made to use protective measures, such as guarding, isolating, or insulation, these precautions shall prevent each employee from contacting such lines directly with any part of his or her body or indirectly through conductive materials, tools, or equipment.

The operation and maintenance of transmission and distribution lines is often the responsibility of a utility or other similar entity. The person responsible for operation and maintenance of the affected conductors must be consulted and directly involved in de-energizing and grounding the overhead conductors.

Suitable guards should be installed to prevent accidental contact with the overhead lines. However, the guards must be of sufficient strength to control the approach or any possible movement of the person or object, to eliminate the chance of unintentional contact. In most instances, line hose *is not satisfactory* to prevent unintentional contact.

Safety grounds, as shown in Exhibit 120.5, are required for overhead conductors, unless exceptional precautions are taken to avoid risk of injury. Safety grounds must be installed in a manner that provides an equipotential zone — having no (or unperceivable) voltage differences between exposed conductive surfaces that a worker might come in contact with while performing the task — for the work area. If there are other conductors in the immediate vicinity of the work area that could be contacted, they must be guarded from potential contact.

(D) Employer and Employee Responsibility. The employer and employee shall be responsible for ensuring that guards or protective measures are satisfactory for the conditions. Employees shall comply with established work methods and the use of protective equipment.

Employers are responsible for providing the electrical safety program, and employees are responsible for implementing the requirements of the program. Both employers and employees are responsible for ensuring that any installed guards are adequate for the conditions. The employer and employee must work together to make sure that effective procedures exist and that they are applied stringently and reviewed frequently.

(E) Approach Distances for Unqualified Persons. When unqualified persons are working on the ground or in an elevated position near overhead lines, the location shall be such that the employee and the longest conductive object the employee might contact do not come closer to any unguarded, energized overhead power line than the limited approach boundary in Table 130.4(D)(a), column 2 or Table 130.4(D)(b), column 2.

The approach distance for unqualified persons remains the same, regardless of the installation method for the conductor(s). The limited approach boundaries given in Table 130.4(D)(a) and Table 130.4(D)(b) define that distance. The limited approach distance depends on whether the distance between the conductor and the employee is under the employee's control. If the supporting platform can move, as would be the case for an articulating platform, column 2 of Table 130.4(D)(a) and Table 130.4(D)(b) applies.

If the conductor is supported on a messenger or similar support method, the conductor can move as the wind blows, and therefore the distance between the employee and the conductor is *not* under his or her control and column 2 still applies. If the overhead conductor is fixed into position, as is the case with solid bus conductors, and the employee is standing on a fixed platform or scaffold, the employee has control of the distance between himself or herself and the conductor, and column 3 applies.

> Informational Note: Objects that are not insulated for the voltage involved should be considered to be conductive.

Some conductors have a covering that is intended to serve as protection from the effects of the environment. Many electricians refer to such conductors as *weatherproof conductors*. This covering is *not* insulating material and generally has no established voltage rating. Weatherproof conductors must be considered to be uninsulated.

(F) Vehicular and Mechanical Equipment.

(1) Elevated Equipment. Where any vehicle or mechanical equipment structure will be elevated near energized overhead lines, it shall be operated so that the limited approach boundary distance of Table 130.4(D)(a), column 2 or Table 130.4(D)(b), column 2, is maintained. However, under any of the following conditions, the clearances shall be permitted to be reduced:

The limited approach boundary given in column 2 of Table 130.4(D)(a) and Table 130.4(D)(b) defines the closest dimension that any vehicle or mechanical equipment structure can be elevated to an exposed energized overhead conductor, unless conditions defined in this section permit closer approach. Distances can be difficult to estimate when standing on the ground or sitting in the seat of a crane or other mobile equipment.

(1) If the vehicle is in transit with its structure lowered, the limited approach boundary to overhead lines in Table 130.4(D)(a), column 2 or Table 130.4(D)(b), column 2, shall be permitted to be reduced by 1.83 m (6 ft). If insulated barriers, rated for the voltages involved, are installed and they are not part of an attachment to the vehicle, the clearance shall be permitted to be reduced to the design working dimensions of the insulating barrier.

Any elevating structure of a vehicle, such as a boom or dump truck bed, must be in the resting position. If such structures are not in a resting position, the limited approach boundary cannot be reduced.

(2) If the equipment is an aerial lift insulated for the voltage involved, and if the work is performed by a qualified person, the clearance (between the uninsulated portion of the aerial lift and the power line) shall be permitted to be reduced to the restricted approach boundary given in Table 130.4(D)(a), column 4 or Table 130.4(D)(b), column 4.

When qualified persons are supported by an aerial lifting device, such as a truck boom that is fully insulated from contact with earth, the minimum unprotected approach distance is defined as the restricted approach boundary given in column 4 of Table 130.4(D)(a) and Table 130.4(D)(b). The definition of a *qualified person* states that the employee has received training in the operation of the aerial lifting device in addition to all other required training.

(2) Equipment Contact. Employees standing on the ground shall not contact the vehicle or mechanical equipment or any of its attachments unless either of the following conditions apply:

(1) The employee is using protective equipment rated for the voltage.

(2) The equipment is located so that no uninsulated part of its structure (that portion of the structure that provides a conductive path to employees on the ground) can come closer to the line than permitted in 130.8(F)(1).

Although equipment contact with overhead conductors could be several feet away from the employee, they might provide the conductive path to earth when touching the equipment with unprotected hands or another body part. Handlines and tag lines sometimes serve as the point of contact for a person. This practice is dangerous, and these lines should not be used.

Employees who are outside the equipment that is in contact with an energized conductor have greater exposure to electrocution than those who are inside the equipment cab. Employees could be exposed to electrocution by step potential if they are standing on the ground near equipment that makes contact with an overhead line.

A barricade should be erected around the physical area to surround equipment that could contact an overhead line. Signs should be installed to warn people to stay out of the area. The barricaded area should not permit approach closer than the limited approach boundary.

(3) Equipment Grounding. If any vehicle or mechanical equipment capable of having parts of its structure elevated near energized overhead lines is intentionally grounded, employees working on the ground near the point of grounding shall not stand at the grounding location whenever there is a possibility of overhead line contact. Additional precautions, such as the use of barricades, dielectric overshoe footwear, or insulation, shall be taken to protect employees from hazardous ground potentials (step and touch potential).

Informational Note: Upon contact of the elevated structure with the energized lines, hazardous ground potentials can develop within a few feet or more outward from the grounded point.

Some safety programs require mobile equipment to be grounded with a temporary grounding conductor connected to an existing earth ground or a temporary ground rod. The grounding conductor expands the touch and step potential hazard to include the grounding conductor and the ground rod (or other earth ground connection point). If such a grounding conductor is installed, the employee must not be within the limited approach boundary of any portion of the temporary grounding circuit.

Barricades and warning signs should be in place to prevent any person from entering the space defined by the limited approach boundary.

130.9 Underground Electrical Lines and Equipment.

Before excavation starts where there exists a reasonable possibility of contacting electrical lines or equipment, the employer shall take the necessary steps to contact the appropriate owners or authorities to identify and mark the location of the electrical lines or equipment. When it has been determined that a reasonable possibility of contacting electrical lines or equipment exists, appropriate safe work practices and PPE shall be used during the excavation.

This section previously appeared as 110.5. Marking the location of underground conductors and equipment will help to minimize the possibility of accidental contact with buried electrical conductors during excavation. Safe work practices commensurate with the hazard must be implemented during the excavation. See Exhibit 130.17(a) for a dig locator.

All underground utilities, including gas, electricity, telephone, and cable TV companies, are members of 811. The call center notifies utility companies of excavation work near their underground installations and directs them to mark the approximate location of underground lines, pipes, and cables. See Exhibit 130.17(b).

(a)

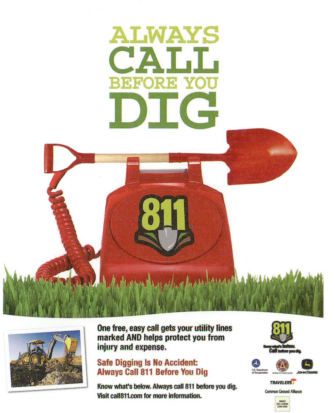

(b)

130.10 Cutting or Drilling.

Before cutting or drilling into equipment, floors, walls, or structural elements where a likelihood of contacting energized electrical lines or parts exists, the employer shall perform a risk assessment to:

(1)   Identify and mark the location of conductors, cables, raceways, or equipment
(2)   Create an electrically safe work condition
(3)   Identify safe work practices and PPE to be used

Safety-Related Maintenance Requirements

An electrical work environment consists of three interrelated components: installation, maintenance, and safe work practices. Safe work practices are most effective when the installation is code compliant and the equipment is maintained appropriately. The NFPA documents that address each aspect are *NFPA 70®*, *National Electrical Code®* (*NEC®*); NFPA 70B, *Recommended Practice for Electrical Equipment Maintenance*; and *NFPA 70E®*, *Standard for Electrical Safety in the Workplace®*. Exhibit 200.1 illustrates how these documents are interrelated. *NFPA 70E* considers equipment to be safe for operation if the equipment is installed according to *NFPA 70* and the manufacturer's instructions, and has been maintained in accordance with NFPA 70B in the absence of specific manufacturer's instructions. A deficiency in the installation or maintenance of a system has the potential to adversely impact electrical safety of employees and safe work practices.

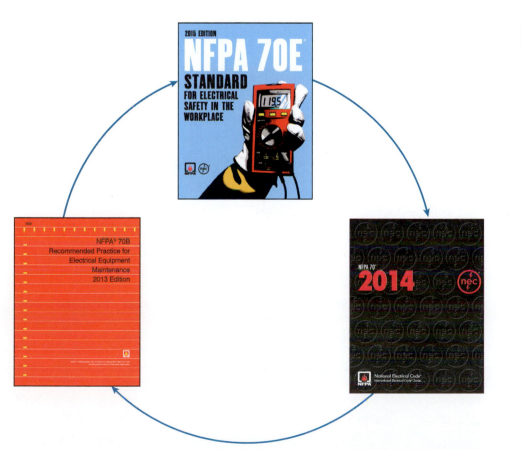

Properly maintained electrical equipment has proven reliable. Chapter 2 addresses maintenance of electrical equipment, which provides for the predictability and reliability necessary for safe operation. A companion document for *NFPA 70E* is NFPA 70B. The purpose of this recommended practice is to reduce hazards to life and property that can result from failure or malfunction of industrial-type electrical systems and equipment. It provides guidance on maintenance practices and on setting up a preventive maintenance program. NFPA 70B applies to preventive maintenance for electrical, electronic, and communication systems and equipment, and is not intended to duplicate or supersede instructions that manufacturers normally provide. *NFPA 70E* addresses the work practices that should be used during maintenance work.

Article 200 — Introduction

Summary of Article 200 Changes

200.1 Informational Note: Added IEEE 3007.2, *Recommended Practice for the Maintenance of Industrial and Commercial Power Systems*, for a more complete list of applicable safety standards.

A comprehensive electrical equipment maintenance program can increase the reliability of the electrical systems, which avoids electrical outages and malfunctions, and can decrease the exposure of employees to electrical hazards. Unsafe equipment is often the result of improper or inadequate maintenance. The risk of equipment failure is reduced when equipment is properly and adequately maintained. Inadequate maintenance can have a negative impact on personal safety. Section 130.5(3) requires that an arc flash risk assessment take into consideration the maintenance condition of overcurrent protective devices, because the condition can have an effect on the device's clearing time, thus increasing the incident energy.

200.1 Scope.

Chapter 2 addresses the requirements that follow.

(1) Chapter 2 covers practical safety-related maintenance requirements for electrical equipment and installations in workplaces as included in 90.2. These requirements identify only that maintenance directly associated with employee safety.

(2) Chapter 2 does not prescribe specific maintenance methods or testing procedures. It is left to the employer to choose from the various maintenance methods available to satisfy the requirements of Chapter 2.

Employers must determine a maintenance strategy and then implement the necessary components of that strategy. Some maintenance is necessary to support the implemented electrical safety program. For information on preventive maintenance programs, see NFPA 70B.

(3) For the purpose of Chapter 2, maintenance shall be defined as preserving or restoring the condition of electrical equipment and installations, or parts of either, for the safety of employees who work where exposed to electrical hazards. Repair or replacement of individual portions or parts of equipment shall be permitted without requiring modification or replacement of other portions or parts that are in a safe condition.

NFPA 70B provides information on commissioning and on an effective preventive maintenance program. Commissioning, or acceptance testing, verifies that the equipment functions as intended by the design specification. Acceptance testing generates baseline results which can help to identify equipment deterioration or a change in reliability or safety. Future trend analysis is useful in predicting when equipment failure or an out of tolerance condition will occur and can allow for convenient scheduling of outages.

All electrical equipment might have a predictable life cycle, and knowing the service life can be crucial in predicting the reliability and safe operation of the equipment. Routine maintenance and maintenance tests can be performed at regular intervals over the service life of equipment or when condition indicators warrant. Maintenance tests help identify changes in overcurrent protective device characteristics and potential failures before they occur. A shutdown can then be scheduled and repairs can be made before equipment damage and with minimum exposure to employees. An alternative method is utilizing Reliability-Centered Maintenance (RCM) techniques. See Chapter 30 of NFPA 70B for further information on RCM.

> Informational Note: Refer to NFPA 70B, *Recommended Practice for Electrical Equipment Maintenance*; ANSI/NETA MTS, *Standard for Maintenance Testing Specifications for Electrical Power Distribution Equipment and Systems*; and IEEE 3007.2, *IEEE Recommended Practice for the Maintenance of Industrial and Commercial Power Systems*, for guidance on maintenance frequency, methods, and tests.

Maintenance is often the most neglected component of a strategy to provide a safe work environment. NFPA 70B provides employees with solutions, techniques, and testing intervals for adequate maintenance to maximize the reliability of electrical equipment and systems. It describes electrical maintenance subjects and issues surrounding maintenance of electrical equipment.

General Maintenance Requirements   Article   205

Summary of Article 205 Changes

205.3: Replaced the phrase *risk of failure and the subsequent exposure of employees* with *risk associated with failure*. Added language to indicate that the *equipment owner or the owner's designated representative* is responsible for maintenance of the electrical equipment and documentation for consistency with other NFPA standards such as *NFPA 72®, National Fire Alarm and Signaling Code*.

205.7: Added new text regarding covers and doors.

205.13: Replaced the phrase *present a hazard to employees* with *expose employees to an electrical hazard* to provide clarity and consistency with other safety standards that address hazard, risk, and risk assessment.

205.14: Replaced the phrase *that to avoid strain and damage* with *preserve insulation integrity* to clarify the intent that insulation integrity be maintained regardless of the source of potential harm or physical damage.

205.14(1): Replaced the phrase *that present an electrical hazard to employees* with *that would expose employees to an electrical hazard* to provide clarity and consistency with other safety standards that address hazard, risk, and risk assessment. Damages to the items listed do not present a source of harm; rather, they expose the person to a source of harm.

205.14(3): Added a requirement to clarify that cord replacements and cord repairs to electrical equipment be performed by a knowledgeable qualified person and tested to ensure proper configuration.

205.15: Added a section to clarify that maintaining proper clearance of overhead conductors is critical to the prevention of unintentional contact.

205.1 Qualified Persons.

Employees who perform maintenance on electrical equipment and installations shall be qualified persons as required in Chapter 1 and shall be trained in, and familiar with, the specific maintenance procedures and tests required.

A qualified employee must be familiar with the operation, maintenance, and history of the equipment. Familiarity with equipment and tools directly affects an employee's ability to recognize and avoid electrical hazards. An employee who performs a task less frequently than once a year must receive supplemental training prior to performing the task in order to remain qualified.

205.2 Single-Line Diagram.

A single-line diagram, where provided for the electrical system, shall be maintained in a legible condition and shall be kept current.

Single-line diagrams are one of the best sources of information for locating the electrical hazards that might be encountered at a work site. Therefore, all qualified employees must have the ability to read and understand the single-line diagrams of the systems they work on.

Single-line diagrams are created for different purposes and may display different information. Some single-line diagrams are supplemented by equipment schedules that may or may not be included on the diagram. Some power sources, such as control power for a motor control center, may not be detailed on the single-line diagram. These sources may be detailed on a schematic or elementary diagram or in a panelboard schedule. Exhibit 205.1 shows a simple single-line diagram.

Legible, up-to-date single-line diagrams, along with any necessary supplemental documentation, enable an electrically safe work condition to be implemented. Maintaining these drawings provides valuable information, including the following:

- Sources of power to a specific piece of equipment
- The interrupting capacity of devices at each point in the system

• Possible paths of potential backfeed
• The correct rating for overcurrent devices

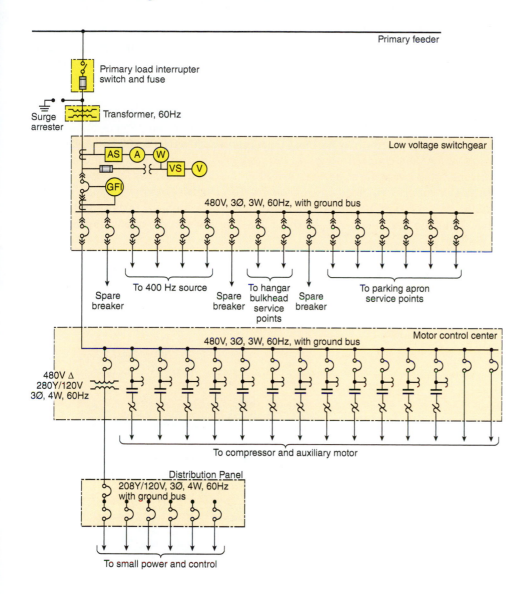

EXHIBIT 205.1

Simple single-line diagram.

205.3 General Maintenance Requirements.

Electrical equipment shall be maintained in accordance with manufacturers' instructions or industry consensus standards to reduce the risk associated with failure. The equipment owner or the owner's designated representative shall be responsible for maintenance of the electrical equipment and documentation.

Informational Note: Common industry practice is to apply test or calibration decals to equipment to indicate the test or calibration date and overall condition of equipment that has been tested and maintained in the field. These decals provide the employee immediate indication of last maintenance date and if the tested device or system was found acceptable on the date of test. This local information can assist the employee in the assessment of overall electrical equipment maintenance status.

Chapters 4, 5, and 6 of NFPA 70B contain recommendations for an effective electrical preventive maintenance (EPM) program. They have been prepared with the intent of providing a better understanding of benefits, both direct and intangible, that can be derived from a well-administered EPM program.

205.4 Overcurrent Protective Devices.

Overcurrent protective devices shall be maintained in accordance with the manufacturers' instructions or industry consensus standards. Maintenance, tests, and inspections shall be documented.

Following the maintenance schedule defined by the manufacturer or by a consensus standard reduces the risk of failure and the subsequent exposure of employees to electrical hazards such as shock, arc flash, or arc blast. Documents such as NFPA 70B and ANSI/NETA MTS, *Standard for Maintenance Testing Specification*, provide testing and maintenance instructions for some overcurrent devices. ANSI/NEMA AB 4, *Guidelines for Inspection and Preventive Maintenance of Molded Case Circuit Breakers Used in Commercial and Industrial Applications*, provides useful information on the type of maintenance, testing, and inspections that should be documented. See Exhibit 205.2 for an example of an adjustable-trip circuit breaker.

EXHIBIT 205.2

An adjustable-trip circuit breaker with a transparent, removable, and sealable cover. (Courtesy of Schneider Electric)

205.5 Spaces About Electrical Equipment.

All working space and clearances required by electrical codes and standards shall be maintained.

> Informational Note: For further information concerning spaces about electrical equipment, see Article 110, Parts II and III, of *NFPA 70, National Electrical Code*.

Adequate working space allows employees to perform tasks without jeopardizing worker safety. Sufficient clearance allows the use of tools and equipment while preventing inadvertent contact, which could result in an electrical incident and injury. Exhibit 205.3 illustrates the working space in front of a panelboard required by the *NEC*. Working spaces must be kept clear. Obstructions, even if temporary such as with stored equipment, restrict access to and egress from the working space.

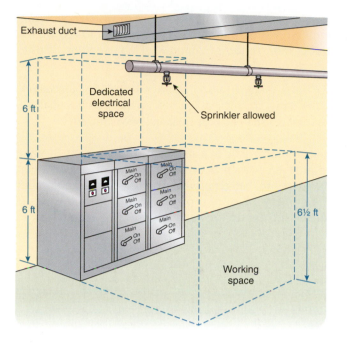

EXHIBIT 205.3

Working space in front of a panelboard.

205.6 Grounding and Bonding.

Equipment, raceway, cable tray, and enclosure bonding and grounding shall be maintained to ensure electrical continuity.

The electrical continuity achieved by grounding and bonding enables fault current to return to the source. During a short-circuit condition, an overcurrent device relies on an effective grounding path to operate as designed. The clearing time might be extended without effective grounding and bonding, thus increasing the amount of incident energy to which an employee could be exposed.

205.7 Guarding of Energized Conductors and Circuit Parts.

Enclosures shall be maintained to guard against accidental contact with energized conductors and circuit parts and other electrical hazards. Covers and doors shall be in place with all associated fasteners and latches secured.

Preventing access to energized conductors and circuits is a core concept for employee safety. Energized electrical conductors are required to be guarded against accidental contact and may be achieved by covering, shielding, enclosing, elevating, or otherwise preventing contact by unauthorized persons or objects. Access to exposed energized electrical components guarded by a locked fence or door can be restricted to only authorized and qualified personnel who have the key. Energized parts should not be exposed to workers unless safety measures are put in place.

205.8 Safety Equipment.

Locks, interlocks, and other safety equipment shall be maintained in proper working condition to accomplish the control purpose.

Locks and interlocks provide safety for employees by ensuring that only authorized and qualified persons have access to areas that contain exposed energized electrical conductors or circuit parts. An interlocking system of keys may be used to control the flow of electrical power through some systems and to control the sequence of switch operations. The system's keying mechanism must work smoothly and without incident to accomplish this sequence. Maintaining these locks and interlocks in good working condition helps to minimize exposure to electrical hazards.

205.9 Clear Spaces.

Access to working space and escape passages shall be kept clear and unobstructed.

Good housekeeping is an important characteristic of a safe work environment. Storage that blocks access or egress or prevents safe work practices must be avoided at all times. Maintaining adequate access is essential for an employee to operate the equipment in a safe and efficient manner. The primary intent of providing egress from the area is so that, in the event of an emergency such as an arc flash incident, the employee can escape.

205.10 Identification of Components.

Identification of components, where required, and safety-related instructions (operating or maintenance), if posted, shall be securely attached and maintained in legible condition.

It is crucial for electrical safety that identification on the single-line diagram is up-to-date and that it match the identification on the installed equipment. Up-to-date operating or maintenance instructions and necessary warnings are vital to ensure employee safety.

205.11 Warning Signs.

Warning signs, where required, shall be visible, securely attached, and maintained in legible condition.

Warning signs inform both qualified and unqualified employees of potential hazards that might be encountered. They must be clearly visible to qualified persons before examination, adjustment, servicing, or maintenance of the equipment. For example, according to 130.5(D), a warning label for a potential arc flash hazard must provide sufficient information to enable an employee to select appropriate personal protective equipment (PPE). The label must include the nominal system voltage, the arc flash boundary, and at least one of the following:

- Available incident energy and the corresponding working distance, or the arc flash PPE category in Table 130.7(C)(15)(A)(b) or Table 130.7(C)(15)(B) for the equipment, but not both
- Minimum arc rating of clothing
- Site-specific level of PPE

  Other warning signs may be required by installation standards and OSHA regulations.

205.12 Identification of Circuits.

Circuit or voltage identification shall be securely affixed and maintained in updated and legible condition.

Several sections of the *NEC* require that circuit identification be securely affixed to the equipment. *NEC* Section 110.22(A) requires that each disconnecting means indicate its purpose unless the purpose is obvious from the arrangement. Section 230.70(B) requires identification of the service disconnecting means, and where a structure is supplied by more than one service, 230.2(E) requires that each service disconnecting means location have a permanent plaque indicating the location of the other disconnecting means.

NEC Section 408.4 details the circuit identification information required for switchgear, switchboards and panelboards. The circuit identification is to be up-to-date, accurate, and legible. Mislabeled equipment endangers employees who might assume that they have de-energized the circuit feeding the equipment. However, circuit identification does not remove the employee's responsibility for verifying the absence of voltage when establishing an electrically safe work condition. Regardless of the presence of labels or warnings, the need to perform a risk assessment remains.

205.13 Single and Multiple Conductors and Cables.

Electrical cables and single and multiple conductors shall be maintained free of damage, shorts, and ground that would expose employees to an electrical hazard.

Cables and single and multiple conductors are often installed in cable trays. When the cable is installed in an open cable tray, cables and conductors should be protected from falling objects that could damage the cable. Temporary protection should be provided when working about cable trays so as not to damage the cable.

205.14 Flexible Cords and Cables.

Flexible cords and cables shall be maintained to preserve insulation integrity.

(1) Damaged Cords and Cables. Cords and cables shall not have worn, frayed, or damaged areas that would expose employees to an electrical hazard.

(2) Strain Relief. Strain relief of cords and cables shall be maintained to prevent pull from being transmitted directly to joints or terminals.

(3) Repair and Replacement. Cords and cord caps for portable electrical equipment shall be repaired and replaced by qualified personnel and checked for proper polarity, grounding, and continuity prior to returning to service.

The transient use of flexible cords and cables increases the possibility for cord and plug damage or interruption of the equipment grounding conductor. Before each use, extension cords must be inspected to ensure that there is no damage [see 110.4(B)(3)]. A damaged ground prong is a common problem with extension cords and cord caps for portable equipment. The ground prong provides the grounding path necessary to mitigate electrical shock or electrocution. Incorrect termination of flexible cords and cables at an enclosure is another common problem. Tension placed on the cable can allow conductors to be exposed and subject the employee to a hazard.

NEC Section 400.9 requires flexible cord to be used only in continuous lengths without splice or tap where initially installed. The repair of hard service cord and junior hard-service cord 14 AWG and larger is permitted if the conductors are spliced in accordance with NEC 110.14(B) and the completed splice retains the insulation, outer sheath properties, and usage characteristics of the cord being spliced. An in-line repair is not permitted if the cord is reused or reinstalled.

205.15 Overhead Line Clearances.

For overhead electric lines under the employer's control, grade elevation shall be maintained to preserve no less than the minimum designed vertical and horizontal clearances necessary to minimize risk of unintentional contact.

Maintaining proper clearance of overhead lines is critical to the prevention of unintentional contact. Unintentional contact with overhead lines is one of the leading causes of occupational electrical fatalities in the United States. Examples of equipment that can contact power lines include the following: metal ladders, scaffolds, long-handled tools such as cement finishing floats and aluminum paint rollers, cranes, backhoes, concrete pumpers, and raised dump truck beds.

Article 210 Substations, Switchgear Assemblies, Switchboards, Panelboards, Motor Control Centers, and Disconnect Switches

Summary of Article 210 Changes

210.1: Replaced the phrase *create an electrical hazard* with *expose employees to an electrical hazard* to provide clarity and consistency with other safety standards that address hazard. (Material in enclosures does not create a source of injury or damage to health; rather, it exposes the person to a source of harm.)

210.5 Informational Note: Revised to provide clarity and consistency with other safety standards that address hazard, risk, and risk assessment.

210.1 Enclosures.

Enclosures shall be kept free of material that would expose employees to an electrical hazard.

Housekeeping is a critical action that must be performed before a work task is completed. Materials or tools left in enclosures are a common cause of a fault and can initiate an arc flash event. Employees must remove all extraneous materials and all tools from and around enclosures for electrical safety.

210.2 Area Enclosures.

Fences, physical protection, enclosures, or other protective means, where required to guard against unauthorized access or accidental contact with exposed energized conductors and circuit parts, shall be maintained.

Fences and other enclosures should be inspected regularly to ensure that they continue to guard against entry of unauthorized personnel or animals. Gates and doors, especially if equipped with panic hardware, should be checked regularly for security and proper operation. Any defect or damage must be repaired promptly and sufficiently to afford, at minimum, equivalent protection to the initial installation.

210.3 Conductors.

Current-carrying conductors (buses, switches, disconnects, joints, and terminations) and bracing shall be maintained to perform as follows:

(1) Conduct rated current without overheating

Discoloration of conductors or terminals is evidence of overheating. Infrared thermography performed while the equipment is operating is one method of investigating overheating. Thermography may be considered a hazardous task depending upon how it is performed. The use of properly installed infrared windows in enclosures is one way to lower the risk associated with infrared scanning. If evidence of overheating is found, the equipment should be de-energized and the problem investigated and repaired in accordance with manufacturer's specifications.

(2) Withstand available fault current

Short circuits or fault currents present a significant amount of destructive energy that can cause serious damage to electrical equipment and create the potential for serious injury to personnel. The short-circuit current rating of electrical equipment is the amount of current that it can carry safely for a specific period of time before it is damaged. For example, a bus duct may have a short-circuit current rating of 22,000 rms symmetrical amperes for three cycles. It might be damaged if 30,000 amperes were to flow through the bus for three cycles or if 22,000 amperes were to flow for six cycles.

210.4 Insulation Integrity.

Insulation integrity shall be maintained to support the voltage impressed.

Temperature extremes, chemical contamination, operating conditions, and aging are common causes that can degrade insulation performance and jeopardize the safety of personnel. Periodic insulation testing performed on a regular basis, along with the acquired test results, can be used to indicate if the insulation is deteriorating over time. If the insulation resistance falls below an accepted value or is declining rapidly over a period of time, corrective measures can be taken to prevent damage to equipment and injury to personnel.

210.5 Protective Devices.

Protective devices shall be maintained to adequately withstand or interrupt available fault current.

> Informational Note: Improper or inadequate maintenance can result in increased opening time of the overcurrent protective device, thus increasing the incident energy.

Protective devices are designed to operate within a prescribed range and to disconnect the power to equipment in a timely manner in order to minimize damage to equipment and injury to personnel. If the amount of available fault current increases for any reason — due to a change in upstream components, for example — each protective device must be analyzed to determine if it is adequate for interrupting the new fault current.

When a protective device fails to operate as intended, employees can be exposed to the incident energy generated in an arcing fault. If the designed clearing time of the protective device is delayed, an incident energy value greater than anticipated can occur, rendering the employee's selected PPE inadequate.

See Article 225 for further information regarding the maintenance of fuses and circuit breakers.

Article 215   Premises Wiring

215.1 Covers for Wiring System Components.

Covers for wiring system components shall be in place with all associated hardware, and there shall be no unprotected openings.

In order to protect employees from contact with energized electrical components, all covers and doors must be closed and latched using all fasteners provided with the equipment. All unused openings other than those intended for the operation of equipment, or those permitted as part of the design for listed equipment, must be closed to afford protection substantially equivalent to the wall of the equipment.

215.2 Open Wiring Protection.

Open wiring protection, such as location or barriers, shall be maintained to prevent accidental contact.

Open wiring as used in this section includes wiring that is not enclosed in a cable assembly (outer jacket), cable tray, or raceway. The intended function of protection by barriers or equipment location must continue to be effective. A damaged or moved barrier will not provide the protection intended. The protection provided by elevating equipment may be breached if a new means of access, such as a ladder, is installed. See Exhibit 215.1.

215.3 Raceways and Cable Trays.

Raceways and cable trays shall be maintained to provide physical protection and support for conductors.

Periodic inspection of raceway and cable tray systems will ensure that the systems will function as intended, which is not limited to only providing physical protection and support for conductors. Metal raceway and metal cable tray systems are recognized by the *NEC* as equipment grounding conductors. And when these systems are used as an equipment grounding conductor, electrical continuity must be maintained to ensure they have the capacity to conduct safely any fault current likely to be imposed and have sufficiently low impedance to limit the voltage to ground to cause operation of the circuit protective device.

EXHIBIT 215.1

Protection of open wiring by location in a battery room accessible only to qualified persons. This method of guarding exposed live parts is recognized by 110.27(A)(1) of the NEC. (Courtesy of International Association of Electrical Inspectors)

Controller Equipment

Article | 220

220.1 Scope.

This article shall apply to controllers, including electrical equipment that governs the starting, stopping, direction of motion, acceleration, speed, and protection of rotating equipment and other power utilization apparatus in the workplace.

A controller can be a remote-controlled magnetic contactor, switch, circuit breaker, or device that normally is used to start and stop motors and other apparatus. Stop-and-start stations and similar control circuit components that do not open the power conductors to the motor are not considered to be controllers.

220.2 Protection and Control Circuitry.

Protection and control circuitry used to guard against accidental contact with energized conductors and circuit parts and to prevent other electrical or mechanical hazards shall be maintained.

Some controller equipment is designed with removable protective components that are used to prevent or minimize exposure to an electrical hazard. If these protective components are removed for repairs or maintenance to the equipment, they must be reinstalled after the task is complete to ensure the continued integrity of the installed components.

Fuses and Circuit Breakers

Article | 225

Summary of Article 225 Changes

225.1: Added a new last sentence: *Non-current limiting fuses shall not be modified to allow their insertion into current-limiting fuseholders.*

Overcurrent devices play an important role in electrical safety. Not only do they protect conductors and equipment, but they also protect employees. To do this, fuses and circuit breakers must operate within their published time–current characteristic curves safely and correctly. Section 130.5(3) requires that the condition of overcurrent protective devices be taken into consideration for an arc flash risk assessment, because the condition can have an effect on the device's clearing time. Therefore, adequate maintenance is essential to maintaining a safe work environment.

225.1 Fuses.

Fuses shall be maintained free of breaks or cracks in fuse cases, ferrules, and insulators. Fuse clips shall be maintained to provide adequate contact with fuses. Fuseholders for current-limiting fuses shall not be modified to allow the insertion of fuses that are not current-limiting. Non-current limiting fuses shall not be modified to allow their insertion into current-limiting fuseholders.

Discoloration of fuse terminals and fuse clips could be due to heat from poor contact or corrosion. Fuseholders with rejection features need to be maintained so that they will only accept current-limiting fuses. A fuseholder should never be altered or forced to accept a fuse for which it is not designed. Any damaged fuse should be promptly replaced with an identical fuse. Exhibit 225.1 shows examples of fuses that include a rejection feature to prohibit the installation of non-current-limiting fuses.

Different types of fuses are used throughout an electrical system, and fuses from different manufacturers should not be mixed in the same circuit. Fuses differ by performance,

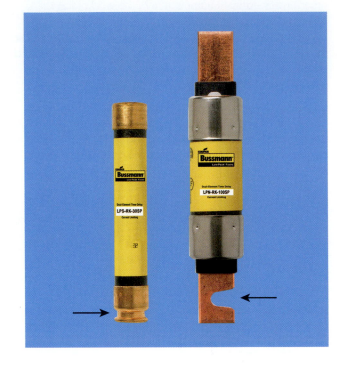

EXHIBIT 225.1

Fuses with a rejection feature to prohibit the installation of non-current-limiting fuses. (Courtesy of Cooper Bussman, a division of Cooper Industries, Inc.)

characteristics, and physical size. Although many fuses might have the same rating, their operating characteristics may differ, making coordination unlikely. Replacement fuses must conform to all requirements detailed in the electrical hazards risk assessment. For further information on the electrical maintenance of fuses, see Chapter 18 of NFPA 70B.

225.2 Molded-Case Circuit Breakers.

Molded-case circuit breakers shall be maintained free of cracks in cases and cracked or broken operating handles.

Although molded-case circuit breakers can be in service for years and may never be called upon to perform their overload– or short-circuit–tripping functions, they are not "maintenance-free" devices. They require both mechanical and electrical maintenance. Mechanical maintenance consists of inspection and adjustment as needed of mechanical mounting and electrical connections, and manual operation of the circuit breaker, which will help keep the contacts clean and will help the lubrication perform properly. Electrical maintenance verifies that the circuit breaker will trip at its desired set point.

Excessive heat in a circuit breaker can cause nuisance tripping and an eventual failure. Molded-case circuit breakers should be kept free of external contamination so that internal heat can be dissipated normally. A clean circuit breaker enclosure also reduces the potential for arcing between energized conductors and between energized conductors and ground. Loose connections are a common cause of excessive heat, and maintenance should involve checking for loose connections or evidence of overheating. All connections should be maintained in accordance with manufacturers' instructions.

The structural strength of the case is important in withstanding the stresses imposed during fault current operation. Therefore, an inspection should be made for cracks in the case, and replacement made if necessary.

Manual operation of the circuit breaker helps keep the contacts clean, but none of the mechanical linkages in the tripping mechanisms are moved with this exercise. Still, manual operation of the circuit breaker helps in assuring that the circuit breaker will operate properly. Some circuit breakers have push-to-trip buttons that should be operated periodically to exercise the tripping mechanical linkages. See Chapter 17 of NFPA 70B and ANSI/NEMA AB 4, *Guidelines for Inspection and Preventive Maintenance of Molded Case Circuit Breakers Used in Commercial and Industrial Applications,* for more information on electrical maintenance of molded-case circuit breakers.

225.3 Circuit Breaker Testing After Electrical Faults.

Circuit breakers that interrupt faults approaching their interrupting ratings shall be inspected and tested in accordance with the manufacturer's instructions.

A high-level fault current can cause damage even when catastrophic failure does not occur. Testing of the device will ensure that the circuit breaker is not damaged and that it will operate at its set point if called upon again. Circuit breakers that encounter high short-circuit currents should receive a thorough inspection and be replaced as necessary.

The result of a circuit breaker not operating within its designed parameters can be disastrous. For example, a circuit breaker could explode if the case is cracked or a higher current than rated is induced through a fault condition. The incident energy may be increased if the circuit breaker does not trip within its set clearing time. For example, an employee 18 inches from a 20-kA short-circuit and 5-cycle tripping time has a potential incident energy exposure of 6.44 cal/cm^2. If the tripping time is increased to 30 cycles, due to improper maintenance or the circuit breaker being out of calibration, the incident energy is increased to 38.64 cal/cm^2.

Circuit breakers should have an initial acceptance test and subsequent maintenance testing at recommended intervals. Following the maintenance schedule defined by the manufacturer or by a consensus standard reduces the risk of failure and the subsequent exposure of employees to electrical hazards such as shock, arc flash, or arc blast. NFPA 70B, ANSI/NETA MTS, and ANSI/NEMA AB 4 are documents that can assist a company in understanding the specific tests and testing intervals required to ensure reliability and safety.

Rotating Equipment   Article   230

230.1 Terminal Boxes.

Terminal chambers, enclosures, and terminal boxes shall be maintained to guard against accidental contact with energized conductors and circuit parts and other electrical hazards.

Vibration and movement of a motor terminal box could exert pressure on the conductors that are terminated or spliced within it. The terminal box must be securely mounted in place by the complete set of hardware supplied by the manufacturer.

230.2 Guards, Barriers, and Access Plates.

Guards, barriers, and access plates shall be maintained to prevent employees from contacting moving or energized parts.

Inspection and maintenance of rotating equipment and motor guards are necessary to prevent an employee from contacting or becoming entangled in the moving part. Should any guard, barrier, or access plate be removed for repairs or maintenance of the rotating equipment, it must be properly restored to its original integrity.

Article 235 — Hazardous (Classified) Locations

Confined spaces, toxic chemicals, and radiation exposure are often associated with the term *hazardous location*. While each of these may qualify as a hazardous location with the presence of the right type of material, they are not necessarily hazardous locations as defined by the *NEC*. A flammable or combustible concentration of a material must be available in order for a location to be considered hazardous within the scope of the *NEC*.

Employees should not carry flashlights, radios, cell phones, computers, multimeters, or other devices into a hazardous location unless the devices have been evaluated for use in the specific hazardous location.

235.1 Scope.

This article covers maintenance requirements in those areas identified as hazardous (classified) locations.

> Informational Note No. 1: These locations need special types of equipment and installation to ensure safe performance under conditions of proper use and maintenance. It is important that inspection authorities and users exercise more than ordinary care with regard to installation and maintenance. The maintenance for specific equipment and materials is covered elsewhere in Chapter 2 and is applicable to hazardous (classified) locations. Other maintenance will ensure that the form of construction and of installation that makes the equipment and materials suitable for the particular location are not compromised.

> Informational Note No. 2: The maintenance needed for specific hazardous (classified) locations depends on the classification of the specific location. The design principles and equipment characteristics, for example, use of positive pressure ventilation, explosionproof, nonincendive, intrinsically safe, and purged and pressurized equipment, that were applied in the installation to meet the requirements of the area classification must also be known. With this information, the employer and the inspection authority are able to determine whether the installation as maintained has retained the condition necessary for a safe workplace.

Hazardous locations are required by the *NEC* to be documented. This document often shows the source of the material, process parameters (such as temperature, flow, pressure), hazardous location boundaries, and any other pertinent information. Personnel responsible for the design, installation, inspection, operation, and maintenance of electrical equipment are required to have access to this document.

Maintenance personnel must be trained to understand the explosive nature of the material, the type of protection employed, and how equipment maintenance is important to a safe environment. There are 15 different types of protection recognized by the *NEC*, and each prevents ignition of the atmosphere in a different manner. Misunderstanding the protection technique, or applying inappropriate maintenance methods, can be catastrophic.

Troubleshooting equipment in a hazardous location presents a special problem. Most equipment cannot be opened while energized in the presence of explosive or combustible material. Most portable troubleshooting instruments are powered by a battery; however, "battery operated" does not equate to being safe for use in a hazardous location. A spark is likely to occur when the testing device contacts a conductor, and an explosion is possible if an explosive atmosphere exists. Before using any of these devices in a hazardous area, it should be determined that an explosive atmosphere does not exist.

235.2 Maintenance Requirements for Hazardous (Classified) Locations.

Equipment and installations in these locations shall be maintained such that the following criteria are met:

(1) No energized parts are exposed.

Exception to (1): Intrinsically safe and nonincendive circuits.

(2) There are no breaks in conduit systems, fittings, or enclosures from damage, corrosion, or other causes.

(3) All bonding jumpers are securely fastened and intact.

(4) All fittings, boxes, and enclosures with bolted covers have all bolts installed and bolted tight.

(5) All threaded conduit are wrenchtight and enclosure covers are tightened in accordance with the manufacturer's instructions.

(6) There are no open entries into fittings, boxes, or enclosures that would compromise the protection characteristics.

(7) All close-up plugs, breathers, seals, and drains are securely in place.

(8) Marking of luminaires (lighting fixtures) for maximum lamp wattage and temperature rating is legible and not exceeded.

(9) Required markings are secure and legible.

Equipment maintenance in hazardous locations should be performed only by personnel trained to maintain the special electrical equipment. Workers should be trained to identify and eliminate ignition sources, such as high surface temperatures, stored electrical energy, and the buildup of static charges, and to identify the need for special tools, equipment, and tests. These individuals should be familiar with the requirements for the electrical installation of the equipment and protection technique employed. For example, they should understand that joint compound or tape may weaken an explosionproof fitting during an ignition or may interrupt the required ground path.

Maintenance personnel should be trained to look for cracked viewing windows, missing fasteners, and damaged threads that may affect the integrity of the protection system. All bolts, screws, fittings, and covers must be properly installed. Every missing or damaged fastener must be replaced with those specified by the manufacturer to provide sufficient strength to withstand an internal ignition.

After equipment maintenance is performed, the integrity of the protective scheme that prevents an explosion must be restored. Re-establishing the required air flow for a purged system or sealing a cable within a conduit fitting of an explosionproof system are two examples of restoring the protective scheme.

Batteries and Battery Rooms   Article 240

Article 480 of the NEC applies to installations of stationary storage batteries. The standards that follow are also referenced for the installation of stationary batteries:

- IEEE 484, *Recommended Practice for Installation Design and Installation of Vented Lead-Acid Batteries for Stationary Applications*
- IEEE 485, *Recommended Practice for Sizing Vented Lead-Acid Storage Batteries for Stationary Applications*
- IEEE 1187, *Recommended Practice for Installation Design, and Installation of Valve-Regulated Lead-Acid Batteries for Stationary Applications*
- IEEE 1375, *IEEE Guide for the Protection of Stationary Battery Systems*
- IEEE 1578, *Recommended Practice for Stationary Battery Spill Containment and Management*
- IEEE 1635/ASHRAE 21, *Guide for the Ventilation and Thermal Management of Stationary Battery Installations*

240.1 Ventilation.

Ventilation systems, forced or natural, shall be maintained to prevent buildup of explosive mixtures. This maintenance shall include a functional test of any associated detection and alarm systems.

Depending on the battery construction and chemistry, ventilation of the battery room may not be required. A ventilation system is designed to provide for sufficient diffusion and ventilation of gases to prevent the accumulation of an explosive mixture. Mechanical ventilation is not mandated and ventilation may be achieved by other means. Maintenance of ventilation systems not only includes any electrical system but also maintenance of the associated mechanical systems such as duct work, screens, louvers, and exhaust ports. Where necessary, NFPA 1, *Fire Code,* requires ventilation in accordance with the mechanical code, and to either limit the maximum concentration of hydrogen to 1.0 percent of the total volume of the room or provide ventilation at a rate of not less than 1 ft³/min/ft² (5.1 L/sec/m²) of floor area.

240.2 Eye and Body Wash Apparatus.

Eye and body wash apparatus shall be maintained in operable condition.

Proper maintenance of eye and body wash apparatus ensures that they supply clean, potable water, and that they are in proper working order. A maintenance program should define guidelines for inspection, testing, and maintenance that includes procedures for flushing and flow rate testing. Exhibit 240.1 shows an example of an eye and body wash apparatus.

EXHIBIT 240.1

Eye and body wash station. (Courtesy of Haws Co.)

Article **245** # Portable Electric Tools and Equipment

Fixed equipment is typically included in a maintenance program, but portable tools are commonly omitted. The intermittent use of portable tools by many users for a multitude of tasks in various locations often subjects the tools to damage. Electrical shock and electrocution from portable tool use is often the result of improper handling or storage. A facility's electrical safety program must include the maintenance and inspection of portable tools and equipment.

245.1 Maintenance Requirements for Portable Electric Tools and Equipment.

Attachment plugs, receptacles, cover plates, and cord connectors shall be maintained such that the following criteria are met:

(1) There are no breaks, damage, or cracks exposing energized conductors and circuit parts.
(2) There are no missing cover plates.
(3) Terminations have no stray strands or loose terminals.
(4) There are no missing, loose, altered, or damaged blades, pins, or contacts.
(5) Polarity is correct.

A visual inspection should be conducted both when a tool is issued and when the tool is returned to the storage area after each use. Employees should be trained to recognize visible defects such as cut, frayed, spliced, or broken cords; cracked or broken attachment plugs; and missing or deformed grounding prongs. Damaged housings, broken switches, and missing parts should also be detected during a visual inspection. Any defect should be reported immediately and the tool removed from service and tagged "Do Not Use" until it is repaired.

Periodic electrical testing of portable electric tools can uncover operating defects. Nonfunctioning and malfunctioning equipment should be returned for repair before continued use. Immediate correction of a defect ensures safe operation, prevents breakdown, and limits more costly repairs.

Employees should be instructed to report all shocks immediately, no matter how minor, and to cease using the tool. The tool must be immediately removed from service, tagged "Do Not Use," examined, and repaired before further use. Tools that trip GFCI devices must also be removed from service until the cause has been determined and corrected. Also, a record of the GFCI tripping should be given to the next work shift.

Personal Safety and Protective Equipment   Article | 250

Summary of Article 250 Changes

250.1: Replaced item (4) *voltage test indicators* with *test instruments* and item (10) *safety grounding equipment* with *temporary protective grounding equipment*.

250.2(B): Added the phrase *that is used as primary protection from shock hazards and requires an insulation system to ensure protection of personnel* to the first sentence. Deleted the reference to ASTM standards.

250.4: Added a new requirement for the maintenance of test instruments used for the verification of the absence or presence of voltage.

Since PPE is an employee's final chance to avoid injury in the event of an incident, employees have a vested interest in maintaining PPE. The condition of the PPE has a direct impact on the employee's well-being. Therefore, employees should take an active role in inspecting and maintaining this special equipment. See 130.7 for additional information and requirements for the selection of PPE. For guidance on the selection of PPE, see Annex H.

250.1 Maintenance Requirements for Personal Safety and Protective Equipment.

Personal safety and protective equipment such as the following shall be maintained in a safe working condition:

(1) Grounding equipment
(2) Hot sticks
(3) Rubber gloves, sleeves, and leather protectors
(4) Test instruments
(5) Blanket and similar insulating equipment
(6) Insulating mats and similar insulating equipment
(7) Protective barriers
(8) External circuit breaker rack-out devices
(9) Portable lighting units
(10) Temporary protective grounding equipment
(11) Dielectric footwear
(12) Protective clothing
(13) Bypass jumpers
(14) Insulated and insulating hand tools

This is not an all-inclusive list of PPE that may be used by the worker. To ensure reliability, all equipment must be maintained in accordance with manufacturers' instructions or listings.

250.2 Inspection and Testing of Protective Equipment and Protective Tools.

(A) Visual. Safety and protective equipment and protective tools shall be visually inspected for damage and defects before initial use and at intervals thereafter, as service conditions require, but in no case shall the interval exceed 1 year, unless specified otherwise by the respective ASTM standards.

Although an inspection of PPE may be conducted at regular intervals, the employee should visually inspect each component immediately before use to verify that no visual defects exist in the equipment. The employee is the last one to inspect the equipment before it may be called upon to prevent an injury. See Table 130.7(C)(14) for specific ASTM standards which describe what aspects of the equipment should be included in the visual inspection.

In some instances, such as rubber insulating equipment, the PPE should have a date stamp or other means of identification that indicates when the equipment must be retested. The visual inspection must verify that the equipment is not past the date retesting is required. See Table 130.7(C)(7)(c) for test intervals and for specific ASTM standards for rubber insulating equipment.

(B) Testing. The insulation of protective equipment and protective tools, such as items specified in 250.1(1) through 250.1 (14), that is used as primary protection from shock hazards and requires an insulation system to ensure protection of personnel, shall be verified by the appropriate test and visual inspection to ascertain that insulating capability has been retained before initial use, and at intervals thereafter, as service conditions and applicable standards and instructions require, but in no case shall the interval exceed 3 years.

See Table 130.7(C)(14) for ASTM standards that describe testing requirements.

250.3 Safety Grounding Equipment.

Temporary protective grounding equipment (TPGE), safety grounds, and *ground sets* are terms used to refer to personal protective grounding equipment. The preferred terminology is *temporary protective grounding equipment.* Temporary protective grounding equipment is normally constructed with insulated conductors terminated in devices intended for connection to a bare conductor or part. See 120.1(6) and 120.3 for further information regarding the use of this equipment.

TPGE should be assigned an identifying mark for record keeping. The identifying mark can be recorded when the equipment is installed on a circuit. After the task has been performed, the equipment can be removed and the identifying mark logged. This will confirm that all temporary protective grounding equipment has been removed prior to re-energizing the circuit.

(A) Visual. Personal protective ground cable sets shall be inspected for cuts in the protective sheath and damage to the conductors. Clamps and connector strain relief devices shall be checked for tightness. These inspections shall be made at intervals thereafter as service conditions require, but in no case shall the interval exceed 1 year.

TPGE should be visually inspected before each use.

(B) Testing. Prior to being returned to service, temporary protective grounding equipment that has been repaired or modified shall be tested.

> Informational Note: Guidance for inspecting and testing safety grounds is provided in ASTM F2249, *Standard Specification for In-Service Test Methods for Temporary Grounding Jumper Assemblies Used on De-Energized Electric Power Lines and Equipment.*

TPGE must be capable of conducting any available fault current long enough for the overcurrent protection to clear the fault. A destructive test is normally performed when a manufacturer determines the rating of specific devices. However, destructive testing is not an option for equipment that will be used again. For maintenance testing of TPGE, see the document referenced in the informational note above.

(C) Grounding and Testing Devices. Grounding and testing devices shall be stored in a clean and dry area. Grounding and testing devices shall be properly inspected and tested before each use.

> Informational Note: Guidance for testing of grounding and testing devices is provided in Section 9.5 of IEEE C37.20.6, *Standard for 4.76 kV to 38 kV-Rated Ground and Test Devices Used in Enclosures.*

Grounding and testing devices are designed to be inserted (racked) into a compartment from which a circuit breaker or disconnect has been removed. These devices can be inserted only into specific spaces. See Exhibit 250.1 for an example of a manual grounding and test device. Grounding and testing devices must not only be visually inspected for defects but also tested before each use. The IEEE standard referenced in the informational note above provides information on integrity tests for grounding and testing devices.

250.4 Test Instruments.

Test instruments and associated test leads used to verify the absence or presence of voltage shall be maintained to assure functional integrity. The maintenance program shall include functional verification as described in 110.4(A)(5).

EXHIBIT 250.1

Manual ground and test device. (Courtesy of Schneider Electric)

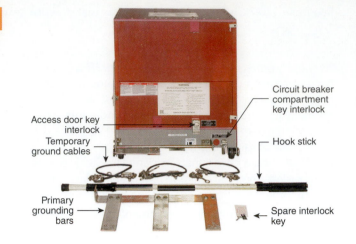

Access door key interlock

Temporary ground cables

Primary grounding bars

Circuit breaker compartment key interlock

Hook stick

Spare interlock key

Many consider voltage measuring test instruments to be as important as their PPE. Test instruments utilized in the verification of the absence or presence of voltage are critical to worker safety. The proper operation of the test instrument must be verified on a known voltage source before testing for absence of voltage. After the absence of voltage test has been conducted, the proper operation of the test instrument must again be verified on a known voltage source to ensure that a failure did not occur during the testing for absence of voltage.

Safety Requirements for Special Equipment

Some facilities use electrical energy in unique ways that differ from most general industries. In some cases, the electrical energy is an integral part of the manufacturing process. In others, the electrical energy is converted to a form that presents unique hazards. When electrical energy is used as a process variable, the general safe work practices defined in Chapter 1 can become unsafe or produce unsafe conditions. Chapter 3 modifies the requirements of Chapter 1 as necessary for use in special situations.

Some workplaces require equipment that is unique. For example, research and development facilities frequently use equipment that expose employees to unique hazards. General safe work practices might not mitigate that exposure adequately. Chapter 3 permits an employer to comply with appropriate requirements from Chapter 1 by amending requirements that are not appropriate for the specific conditions. Chapter 3 supplements or modifies the safety-related work practices in Chapter 1 with safety requirements for special equipment.

Introduction

Article **300**

300.1 Scope.

Chapter 3 covers special electrical equipment in the workplace and modifies the general requirements of Chapter 1.

Chapter 3 covers additional safety-related work practices that are necessary for the practical safeguarding of employees relative to the electrical hazards associated with special equipment and processes that have not been excluded by 90.2(B).

300.2 Responsibility.

The employer shall provide safety-related work practices and employee training. The employee shall follow those work practices.

Employers are assigned the same responsibility for Chapter 3 requirements as for Chapters 1 and 2. The employer must define the electrical safety program, and employees must implement the requirements defined in the program. An electrical safety program is most effective when employers and employees work together to accomplish both needs.

300.3 Organization.

Chapter 3 of this standard is divided into articles. Article 300 applies generally. Article 310 applies to electrolytic cells. Article 320 applies to batteries and battery rooms. Article 330 applies to lasers. Article 340 applies to power electronic equipment. Article 350 applies to research and development (R&D) laboratories.

Each article in Chapter 3 addresses a single unique equipment type or work area, and each article stands alone. Requirements defined in one article apply only to that special equipment type or work area and amend requirements of Chapter 1 for only that purpose.

Article 310 — Safety-Related Work Practices for Electrolytic Cells

Summary of Article 310 Changes

310.2 Battery Effect Informational Note: Replaced the phrase *thus, a hazardous voltage* with *a shock hazard* to provide clarity and consistency with other safety standards that address hazard, risk, and risk assessment.

310.3(A): Replaced the phrase *who are exposed to the risk of electrical hazard* with *exposed to electrical hazards* to provide clarity and promote the consistent use of terminology associated with hazard and risk.

310.3(B): Added the term *electrical* before *hazards* to provide clarity and consistency, since the scope of the standard pertains to electrical hazards.

310.4(A)(1): Restructured the section into a list format for clarity and consistency.

310.4(B)(1): Replaced the term *recognize* with *identify* for clarity and consistency.

310.4(B)(2): Revised to explain that the role of the designated qualified person in charge includes notifying unqualified person(s) of hazards.

310.5(A): Replaced the phrase *hazardous electrical condition* with *an electrical hazard* to provide clarity and promote the consistent use of terminology associated with hazard and risk. Also, updated the references to tables found in Section 130.

310.5(C)(1): Replaced the terms *risk* with *likelihood*, *eliminate* with *reduce*, *possibility* with *likelihood*, *hazard* with *incident*, and *PPE* with *personal protective equipment*. Replaced the phrase *Hazard Analysis* with *Procedure Risk Assessment* in the title for further clarification.

310.5(C)(2): Revised to provide clarity and consistency with definitions of hazard and risk.

310.5(C)(3): Replaced the phrase *hazard risk analysis* with *risk assessment* for consistency with definitions of hazard, risk, and risk assessment. Replaced the phrase *possibility of a hazardous* with *risk associated with* to indicate that the reference is to minimizing the risk (i.e., combination of likelihood of occurrence of harm and the severity of harm) associated with arc flash.

310.5(C)(4): Replaced the phrase *possibility* to *likelihood of occurrence* for consistency with definition of *risk*.

310.5(D)(2): Replaced the phrases *hazardous electrical conditions* with *electrical hazards*, and replaced *shoes, boots, or overshoes* with *footwear* to provide consistent use of terminology with other referenced standards that address safety footwear, such as the ASTM family of standards. Updated the references to tables found in Section 130.

310.5(D)(3): Replaced the phrase *a hazardous electrical condition* with *an electrical hazard* to clarify and promote the consistent use of terminology associated with hazard and risk.

310.5(D)(5): Replaced the phrase *a hazardous electrical condition* with *an electrical hazard* to clarify and promote the consistent use of terminology associated with hazard and risk.

Article 310 identifies the supplementary or replacement safe work practices that employees should use in electrolytic cell line working zones and the special hazards of working with these ungrounded direct current (dc) systems. A cell line is a series of individual cells that are connected together electrically. Generally, the process requires a significant amount of direct current and is ungrounded. See Exhibit 310.1.

EXHIBIT 310.1

An electrolytic cell line. (Photo by David Pace and Michael Petry, Courtesy of Olin Corporation, McIntosh, AL)

Working on an electrolytic cell line is always considered energized electrical work. Each individual cell of an electrolytic cell line is a battery and cannot be de-energized without removing the electrolyte in the vessel. Therefore, establishing an electrically safe work condition is not by itself a viable method for avoiding injury.

310.1 Scope.

The requirements of this article shall apply to the electrical safety-related work practices used in the types of electrolytic cell areas.

Informational Note No. 1: See Informative Annex L for a typical application of safeguards in the cell line working zone.

Informational Note No. 2: For further information about electrolytic cells, see *NFPA 70, National Electrical Code*, Article 668.

310.2 Definitions.

For the purposes of this article, the definitions that follow shall apply.

Battery Effect. A voltage that exists on the cell line after the power supply is disconnected.

> Informational Note: Electrolytic cells can exhibit characteristics similar to an electrical storage battery and a shock hazard could exist after the power supply is disconnected from the cell line.

Safeguarding. Safeguards for personnel include the consistent administrative enforcement of safe work practices. Safeguards include training in safe work practices, cell line design, safety equipment, PPE, operating procedures, and work checklists.

310.3 Safety Training.

(A) General. The training requirements of this chapter shall apply to employees exposed to electrical hazards in the cell line working zone defined in 110.2 and shall supplement or modify the requirements of 120.1, 130.2, 130.3, and 130.8.

(B) Training Requirements. Employees shall be trained to understand the specific electrical hazards associated with electrical energy on the cell line. Employees shall be trained in safety-related work practices and procedural requirements to provide protection from the electrical hazards associated with their respective job or task assignment.

Employees who work in the vicinity of the cells and interconnecting bus must be trained to understand the hazards associated with an unintentional grounded condition of either an individual cell or the interconnecting bus. Employees must also understand that the significant magnetic field generated by current flowing in the interconnecting bus might interfere with certain medical devices.

310.4 Employee Training.

(A) Qualified Persons.

(1) Training. Qualified persons shall be trained and knowledgeable in the operation of cell line working zone equipment and specific work methods and shall be trained to avoid the electrical hazards that are present. Such persons shall be familiar with the proper use of precautionary techniques and PPE. Training for a qualified person shall include the following:

(1) Skills and techniques to avoid a shock hazard:
   a. Between energized surfaces, which might include temporarily insulating or guarding parts to permit the employee to work on energized parts
   b. Between energized surfaces and grounded equipment, other grounded objects, or the earth itself, that might include temporarily insulating or guarding parts to permit the employee to work on energized parts
(2) Method of determining the cell line working zone area boundaries

Employees who work within the area of the dc bus must be trained to understand the unique hazards associated with ungrounded dc voltage. Normally the dc voltage is ungrounded; therefore hand tools that might contact the dc bus work must not be grounded.

(2) Qualified Persons. Qualified persons shall be permitted to work within the cell line working zone.

(B) Unqualified Persons.

(1) Training. Unqualified persons shall be trained to identify electrical hazards to which they could be exposed and the proper methods of avoiding the hazards.

(2) In Cell Line Working Zone. When there is a need for an unqualified person to enter the cell line working zone to perform a specific task, that person shall be advised of the electrical hazards by the designated qualified person in charge to ensure that the unqualified person is safeguarded.

310.5 Safeguarding of Employees in the Cell Line Working Zone.

(A) General. Operation and maintenance of electrolytic cell lines might require contact by employees with exposed energized surfaces such as buses, electrolytic cells, and their attachments. The approach distances referred to in Table 130.4(D)(a) and Table 130.4(D)(b) shall not apply to work performed by qualified persons in the cell line working zone. Safeguards such as safety-related work practices and other safeguards shall be used to protect employees from injury while working in the cell line working zone. These safeguards shall be consistent with the nature and extent of the related electrical hazards. Safeguards might be different for energized cell lines and de-energized cell lines. Hazardous battery effect voltages shall be dissipated to consider a cell line de-energized.

> Informational Note No. 1: Exposed energized surfaces might not present an electrical hazard. Electrical hazards are related to current flow through the body, causing shock and arc flash burns and arc blasts. Shock is a function of many factors, including resistance through the body and the skin, return paths, paths in parallel with the body, and system voltages. Arc flash burns and arc blasts are a function of the current available at the point involved and the time of arc exposure.

> Informational Note No. 2: A cell line or group of cell lines operated as a unit for the production of a particular metal, gas, or chemical compound might differ from other cell lines producing the same product because of variations in the particular raw materials used, output capacity, use of proprietary methods or process practices, or other modifying factors. Detailed standard electrical safety-related work practice requirements could become overly restrictive without accomplishing the stated purpose of Chapter 1.

Employers must institute an electrical safety program that addresses the issues identified in Chapter 1. However, the work practices can be modified as necessary to recognize the different types of exposure to electrical hazards. For instance, because each cell acts like a battery, the employer must define actions that are necessary if an employee needs to contact the dc bus structure. Those procedures must be consistent with the risk associated with the work task. See Informative Annex L for a typical application of safeguards in the cell line working zone.

(B) Signs. Permanent signs shall clearly designate electrolytic cell areas.

(C) Electrical Arc Flash Hazard Analysis. The requirements of 130.5, Arc Flash Risk Assessment, shall not apply to electrolytic cell line work zones.

Hazard identification and risk assessment should take into account the unique characteristics of dc systems. Table 130.4(D)(b) provides shock approach boundaries for dc systems.

Table 130.7(C)(15)(A)(a) indicates when arc flash protective equipment is necessary for the identified tasks for dc systems. Table 130.7(C)(15)(B) provides the PPE category and arc flash boundary for the identified equipment. These tables may be used in lieu of conducting an incident energy calculation.

Annex D, Section D.5, provides information on a method for calculating direct current incident energy. The basis for the calculation is that the maximum power possible in a dc arc will occur when the arcing voltage is one-half of the system voltage. Annex H, Table H.3(b), provides suggestions on selection of PPE when using the calculation method.

(1) Arc Flash Risk Assessment. Each task performed in the electrolytic cell line working zone shall be analyzed for the likelihood of arc flash injury. If there is a likelihood of personal injury, appropriate measures shall be taken to protect persons exposed to the arc flash hazards, including one or more of the following:

    (1) Providing appropriate PPE *[see 310.5(D)(2)]* to prevent injury from the arc flash hazard

    (2) Altering work procedures to reduce the likelihood of occurrence of an arc flash incident

    (3) Scheduling the task so that work can be performed when the cell line is de-energized

(2) Routine Tasks. Arc flash risk assessment shall be done for all routine tasks performed in the cell line work zone. The results of the arc flash risk assessment shall be used in training employees in job procedures that minimize the possibility of arc flash hazards. The training shall be included in the requirements of 310.3.

(3) Nonroutine Tasks. Before a nonroutine task is performed in the cell line working zone, an arc flash risk assessment shall be done. If an arc flash hazard is a possibility during nonroutine work, appropriate instructions shall be given to employees involved on how to minimize the risk associated with arc flash.

(4) Arc Flash Hazards. If the likelihood of occurrence of an arc flash hazard exists for either routine or nonroutine tasks, employees shall use appropriate safeguards.

(D) Safeguards. Safeguards shall include one or a combination of the following means.

(1) Insulation. Insulation shall be suitable for the specific conditions, and its components shall be permitted to include glass, porcelain, epoxy coating, rubber, fiberglass, and plastic and, when dry, such materials as concrete, tile, brick, and wood. Insulation shall be permitted to be applied to energized or grounded surfaces.

(2) Personal Protective Equipment (PPE). PPE shall provide protection from electrical hazards. PPE shall include one or more of the following, as determined by authorized management:

    (1) Footwear for wet service

    (2) Gloves for wet service

    (3) Sleeves for wet service

    (4) Footwear for dry service

    (5) Gloves for dry service

    (6) Sleeves for dry service

    (7) Electrically insulated head protection

    (8) Protective clothing

    (9) Eye protection with nonconductive frames

   (10) Face shield (polycarbonate or similar nonmelting type)

         a. Standards for PPE. Personal and other protective equipment shall be appropriate for conditions, as determined by authorized management, and shall not be required to meet the equipment standards in 130.7(C)(14) through 130.7(F) and in Table 130.7(C)(14) and Table 130.7(F).

         b. Testing of PPE. PPE shall be verified with regularity and by methods that are consistent with the exposure of employees to electrical hazards.

(3) Barriers. Barriers shall be devices that prevent contact with energized or grounded surfaces that could present an electrical hazard.

(4) Voltage Equalization. Voltage equalization shall be permitted by bonding a conductive surface to an electrically energized surface, either directly or through a resistance, so that there is insufficient voltage to create an electrical hazard.

(5) Isolation. Isolation shall be the placement of equipment or items in locations such that employees are unable to simultaneously contact exposed conductive surfaces that could present an electrical hazard.

(6) Safe Work Practices. Employees shall be trained in safe work practices. The training shall include why the work practices in a cell line working zone are different from similar work situations in other areas of the plant. Employees shall comply with established safe work practices and the safe use of protective equipment.

    (a) Attitude Awareness. Safe work practice training shall include attitude awareness instruction. Simultaneous contact with energized parts and ground can cause serious electrical shock. Of special importance is the need to be aware of body position where contact may be made with energized parts of the electrolytic cell line and grounded surfaces.

    (b) Bypassing of Safety Equipment. Safe work practice training shall include techniques to prevent bypassing the protection of safety equipment. Clothing may bypass protective equipment if the clothing is wet. Trouser legs should be kept at appropriate length, and shirt sleeves should be a good fit so as not to drape while reaching. Jewelry and other metal accessories that may bypass protective equipment shall not be worn while working in the cell line working zone.

(7) Tools. Tools and other devices used in the energized cell line work zone shall be selected to prevent bridging between surfaces at hazardous potential difference.

    Informational Note: Tools and other devices of magnetic material could be difficult to handle in the area of energized cells due to their strong dc magnetic fields.

(8) Portable Cutout-Type Switches. Portable cell cutout switches that are connected shall be considered as energized and as an extension of the cell line working zone. Appropriate procedures shall be used to ensure proper cutout switch connection and operation.

(9) Cranes and Hoists. Cranes and hoists shall meet the requirements of 668.32 of *NFPA 70, National Electrical Code.* Insulation required for safeguarding employees, such as insulated crane hooks, shall be periodically tested.

(10) Attachments. Attachments that extend the cell line electrical hazards beyond the cell line working zone shall use one or more of the following:

(1) Temporary or permanent extension of the cell line working zone

(2) Barriers

(3) Insulating breaks

(4) Isolation

(11) Pacemakers and Metallic Implants. Employees with implanted pacemakers, ferromagnetic medical devices, or other electronic devices vital to life shall not be permitted in cell areas unless written permission is obtained from the employee's physician.

> Informational Note: The American Conference of Government Industrial Hygienists (ACGIH) recommends that persons with implanted pacemakers should not be exposed to magnetic flux densities above 10 gauss.

Employers must take steps to ensure that employees who wear pacemakers and similar medical devices are not exposed to the magnetic fields that normally exist in the cell area.

(12) Testing. Equipment safeguards for employee protection shall be tested to ensure they are in a safe working condition.

310.6 Portable Tools and Equipment.

> Informational Note: The order of preference for the energy source for portable hand-held equipment is considered to be as follows:

(1) Battery power

(2) Pneumatic power

(3) Portable generator

(4) Nongrounded-type receptacle connected to an ungrounded source

Grounded portable tools and equipment must not be used in the area containing the cells or interconnecting bus. Although in normal settings an equipment grounding conductor decreases exposure to an electrical hazard, in cell areas any conductor that is grounded increases exposure to an electrical hazard. All equipment and tools, including pneumatic tools, must be free from any grounding circuit. Pneumatic tools must be fitted with nonconductive hoses.

(A) Portable Electrical Equipment. The grounding requirements of 110.4(B)(2) shall not be permitted within an energized cell line working zone. Portable electrical equipment shall meet the requirements of 668.20 of *NFPA 70, National Electrical Code.* Power supplies for portable electric equipment shall meet the requirements of 668.21 of *NFPA 70.*

(B) Auxiliary Nonelectric Connections. Auxiliary nonelectric connections such as air, water, and gas hoses shall meet the requirements of 668.31 of *NFPA 70, National Electrical Code.* Pneumatic-powered tools and equipment shall be supplied with nonconductive air hoses in the cell line working zone.

(C) Welding Machines. Welding machine frames shall be considered at cell potential when within the cell line working zone. Safety-related work practices shall require that the cell line not be grounded through the welding machine or its power supply. Welding machines located outside the cell line working zone shall be barricaded to prevent employees from touching the welding machine and ground simultaneously where the welding cables are in the cell line working zone.

(D) Portable Test Equipment. Test equipment in the cell line working zone shall be suitable for use in areas of large magnetic fields and orientation.

> Informational Note: Test equipment that is not suitable for use in such magnetic fields could result in an incorrect response. When such test equipment is removed from the cell line working zone, its performance might return to normal, giving the false impression that the results were correct.

Safety Requirements Related to Batteries and Battery Rooms

Article 320

Summary of Article 320 Changes

320.2, Prospective Short-Circuit Current: Changed the name of the term from *Prospective Fault Current* to *Prospective Short-Circuit Current* to promote consistent use of terminology throughout the standard. Replaced the phrase *that can occur* with *that could theoretically occur*.

320.3(A)(1): Added a new requirement for a risk assessment associated with battery work.

320.3(A)(4): Revised previous 320.3(A)(3) to simplify the requirement for annual testing of battery alarm functionality.

320.3(A)(5): Revised former 320.3(A)(4)(1) by replacing the phrase *and thermal hazards* with *hazard*. Added the phrase *and thermal hazards* to the end of the sentence.

320.3(B)(1): Changed title from *Batteries with Liquid Electrolyte* to *Battery Activities That Include Handling of Liquid Electrolyte*. Provided requirements for capacity of eye wash facilities by introducing the concept that the duration of the flushing should be specified by the electrolyte or battery manufacturer.

320.3(B)(2): Changed the title from *Batteries with Solid or Immobilized Electrolyte* to *Activities That Do Not Include Handling of Electrolyte*.

320.3(C): Deleted former subdivision (1) for being redundant with 320.3(A).

320.3(C)(2)(c): Replaced the phrase *hazard identification and risk assessment* with *risk assessment*.

320.3(D): Combined former sections 320.3(D) and 320.3(E). Added requirement for cell flame arresters to be replaced when necessary.

Article 320 identifies work practices associated with installation and maintenance of batteries containing many cells, such as those used with uninterruptible power supplies (UPS), telecommunications systems, and unit substation dc power supplies.

Working with batteries can expose an employee to both potential shock and arc flash hazards. A person's body might react to contact with dc voltage differently than from contact with ac voltage (see Commentary Table 340.1). Section 340.5 provides some information on the hazardous effects of electricity on the human body. Table 130.4(D)(b) gives the dc shock boundaries. Although shock approach boundaries are not specified for 100 volts or less, 130.2(A)(3) requires an energized work permit for nominal dc voltages 50 volts or greater, because there is a potential arc flash hazard risk. It is expected that a vast majority of dc systems encountered will be below 1000 V dc.

Batteries can also expose an employee to hazards associated with the chemical electrolyte used in the battery. Battery charging can sometimes generate flammable gases, so it is important for the employee to avoid anything that could cause open flame or sparks. The employee must consider exposure to these hazards when selecting work practices and PPE.

320.1 Scope.

This article covers electrical safety requirements for the practical safeguarding of employees while working with exposed stationary storage batteries that exceed 50 volts, nominal.

Informational Note: For additional information on best practices for safely working on stationary batteries, see the following documents:

(1) NFPA 1, *Fire Code*, Chapter 52, Stationary Storage Battery Systems, 2009
(2) *NFPA 70, National Electrical Code*, Article 480, Storage Batteries, 2011
(3) IEEE 450, *IEEE Recommended Practice for Maintenance, Testing, and Replacement of Vented Lead-Acid Batteries for Stationary Applications*, 2010
(4) IEEE 937, *Recommended Practice for Installation and Maintenance of Lead-Acid Batteries for Photovoltaic Systems*, 2007
(5) IEEE 1106, *IEEE Recommended Practice for Installation, Maintenance, Testing, and Replacement of Vented Nickel-Cadmium Batteries for Stationary Applications*, 2005 (R 2011)
(6) IEEE 1184, *IEEE Guide for Batteries for Uninterruptible Power Supply Systems*, 2006
(7) IEEE 1188, *IEEE Recommended Practice for Maintenance, Testing, and Replacement of Valve-Regulated Lead-Acid (VRLA) Batteries for Stationary Applications*, 2005 (R 2010)
(8) IEEE 1657, *Recommended Practice for Personnel Qualifications for Installation and Maintenance of Stationary Batteries*, 2009
(9) OSHA 29 CFR 1910.305(j)(7), "Storage batteries"
(10) OSHA 29 CFR 1926.441, "Batteries and battery charging"
(11) DHHS (NIOSH) Publication No. 94-110, *Applications Manual for the Revised NIOSH Lifting Equation*, 1994

320.2 Definitions.

For the purposes of this article definitions that follow shall apply.

Authorized Personnel. The person in charge of the premises, or other persons appointed or selected by the person in charge of the premises who performs certain duties associated with stationary storage batteries.

Battery. A system consisting of two or more electrochemical cells connected in series or parallel and capable of storing electrical energy received and that can give it back by reconversion.

Battery Room. A room specifically intended for the installation of batteries that have no other protective enclosure.

Cell. The basic electrochemical unit, characterized by an anode and a cathode used to receive, store, and deliver electrical energy.

Electrolyte. A solid, liquid, or aqueous immobilized liquid medium that provides the ion transport mechanism between the positive and negative electrodes of a cell.

Nominal Voltage. The value assigned to a cell or battery of a given voltage class for the purpose of convenient designation; the operating voltage of the cell or system may vary above or below this value.

Pilot Cell. One or more cells chosen to represent the operating parameters of the entire battery (sometimes called "temperature reference" cell).

Prospective Short-Circuit Current. The highest level of fault current that could theoretically occur at a point on a circuit. This is the fault current that can flow in the event of a zero impedance short circuit and if no protection devices operate.

Valve-Regulated Lead Acid (VRLA) Cell. A lead-acid cell that is sealed with the exception of a valve that opens to the atmosphere when the internal pressure in the cell exceeds atmospheric pressure by a pre-selected amount, and that provides a means for recombination of internally generated oxygen and the suppression of hydrogen gas evolution to limit water consumption.

Vented Cell. A type of cell in which the products of electrolysis and evaporation are allowed to escape freely into the atmosphere as they are generated. (Also called "flooded cell.")

320.3 Safety Procedures.

(A) General Safety Hazards.

(1) Battery Risk Assessment. Prior to any work on a battery system, a risk assessment shall be performed to identify the chemical, electrical shock, and arc flash hazards and assess the risks associated with the type of tasks to be performed.

Batteries are sources of energy. Therefore, isolating the source of voltage from a cell is not possible. Working on a battery system is always considered energized electrical work.

Risk associated with batteries can be mitigated starting with the system design. For example, a battery system could be designed to allow the battery to be partitioned into low voltage segments before work is conducted on it. Other system design mitigation methods include widely separating the positive and negative conductors and installing insulated covers on battery inter-cell connector bus bars or terminals.

Table 130.7(C)(15)(A)(a) indicates when arc flash protective equipment is necessary for the identified tasks for dc systems. Table 130.7(C)(15)(B) provides the PPE category and arc flash boundary for the identified equipment. These tables may be used in lieu of conducting an incident energy calculation.

Annex D, Section D.5, provides information on a method for calculating direct current incident energy. The basis for the calculation is that the maximum power possible in a dc arc will occur when the arcing voltage is one-half of the system voltage. This calculation is conservative. Annex H, Table H.3(b), provides suggestions on selection of PPE when using the calculation method.

Exhibit 320.1 illustrates how a risk assessment of a battery system for the various types of hazards (shock, chemical, arc flash, and thermal) might be conducted. It also illustrates how the likely exposure (risk) depends on the type of task being performed. The same process could be applied to other types of systems. This figure is for illustrative purposes and may not apply to a specific installation or electrical safety program.

(2) Battery Room or Enclosure Requirements.

(a) Personnel Access to Energized Batteries. Each battery room or battery enclosure shall be accessible only to authorized personnel.

(b) Illumination. Employees shall not enter spaces containing batteries unless illumination is provided that enables the employees to perform the work safely.

> Informational Note: Battery terminals are normally exposed and pose possible shock hazard. Batteries are also installed in steps or tiers that can cause obstructions.

(3) Apparel. Personnel shall not wear electrically conductive objects such as jewelry while working on a battery system.

(4) Abnormal Battery Conditions. Instrumentation that provides alarms for early warning of abnormal conditions of battery operation, if present, shall be tested annually.

> Informational Note: Battery monitoring systems typically include alarms for such conditions as overvoltage, undervoltage, overcurrent, ground fault, and overtemperature. The type of conditions monitored will vary depending upon the battery technology. One source of guidance on monitoring battery systems is IEEE 1491, *Guide for the Selection and Use of Battery Monitoring Equipment in Stationary Applications.*

EXHIBIT 320.1

Risk assessment of a battery system.

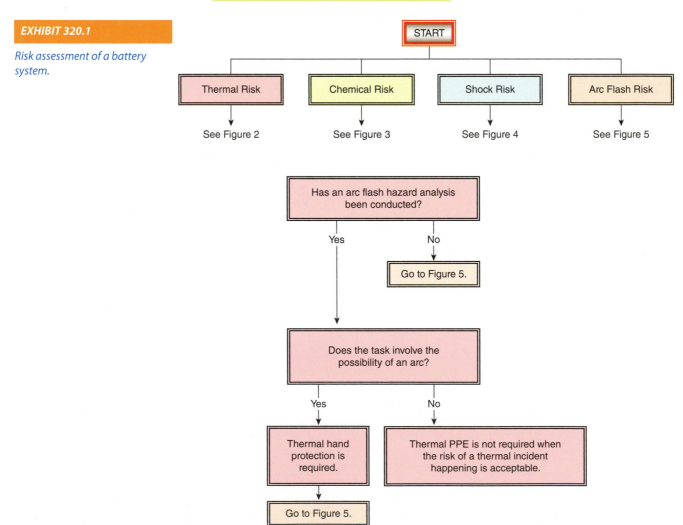

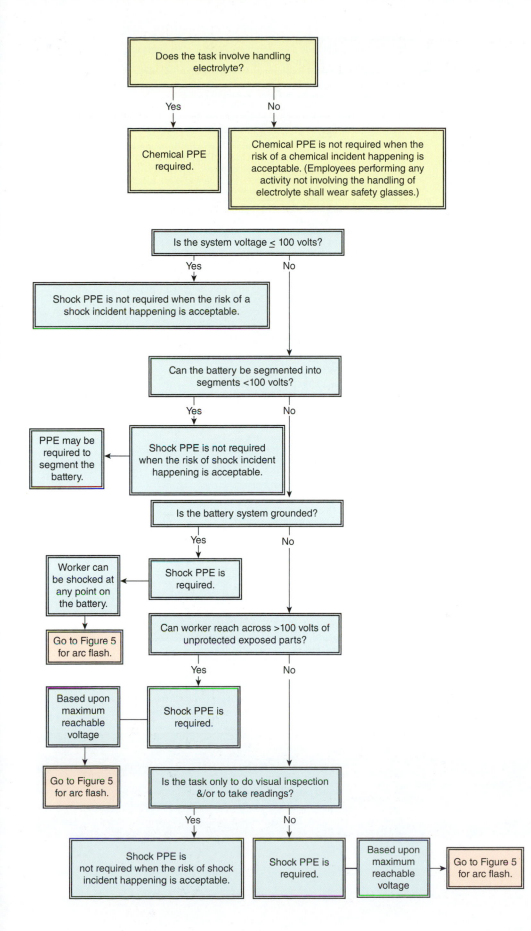

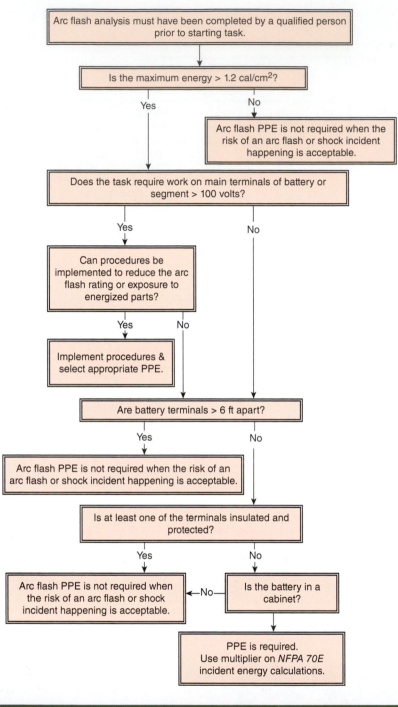

Note

Three simple messages are embedded in all these figures:

1. The hazard always exists unless it can be engineered out of the design.
2. Risk increases in proportion to one's exposure to the hazard. The risk is zero if one is not exposed to the hazard.
3. The level of PPE should be proportionate to the degree of risk.

The figures shown here are somewhat simplistic and are meant only to be models for evaluating risk.

(5) Warning Signs. The following warning signs or labels shall be posted in appropriate locations:

(1) Electrical hazard warnings indicating the shock hazard due to the battery voltage and the arc flash hazard due to the prospective short-circuit current, and the thermal hazard.

> **Informational Note No. 1:** Because internal resistance, prospective short-circuit current, or both are not always provided on battery container labels or data sheets, and because many variables can be introduced into a battery layout, the battery manufacturer should be consulted for accurate data. Variables can include, but are not limited to, the following:
>
> a. Series connections
> b. Parallel connections
> c. Charging methodology
> d. Temperature
> e. Charge status
> f. Dc distribution cable size and length
>
> **Informational Note No. 2:** See 130.5(D) for requirements for equipment labeling.

(2) Chemical hazard warnings, applicable to the worst case when multiple battery types are installed in the same space, indicating the following:

    a. Potential presence of explosive gas (when applicable to the battery type)
    b. Prohibition of open flame and smoking
    c. Danger of chemical burns from the electrolyte (when applicable to the battery type)

(3) Notice for personnel to use and wear protective equipment and apparel appropriate to the hazard for the battery

(4) Notice prohibiting access to unauthorized personnel

(B) Electrolyte Hazards.

Batteries are somewhat unique in that they present chemical hazards as well as electrical hazards. Electrolyte (chemical) hazards vary depending on the type of battery, so the risk is product-specific as well as activity-specific. For example, nickel-cadmium (NiCd) and vented lead-acid (VLA) batteries allow access to liquid electrolyte, thereby potentially exposing a worker to a chemical hazard when performing certain tasks. By contrast, valve-regulated lead-acid (VRLA) and certain lithium batteries are designed with solid or immobilized electrolyte so that a worker can only be exposed to electrolyte under failure conditions.

(1) Battery Activities That Include Handling of Liquid Electrolyte. The following protective equipment shall be available to employees performing any type of service on a battery with liquid electrolyte:

    (1) Goggles and face shield appropriate for the electrical hazard and the chemical hazard
    (2) Gloves and aprons appropriate for the chemical hazards
    (3) Portable or stationary eye wash facilities within the work area that are capable of drenching or flushing of the eyes and body for the duration necessary to the hazard

Informational Note: Guidelines for the use and maintenance of eye wash facilities for vented batteries in nontelecom environments can be found in ANSI/ISEA Z358.1, *American National Standard for Emergency Eye Wash and Shower Equipment.*

The requirements of 320.3(B)(1) only apply if electrolyte is being handled, which is possible only with batteries utilizing free-flowing liquid electrolyte. Activities in which "electrolyte is being handled" would include acid adjustment, removal of excess electrolyte, or clean-up of an electrolyte leak or spill. Most battery maintenance activities do not involve handling of electrolyte, so the requirements of 320.3(B)(2) would apply. Exhibit 320.2 illustrates a type of eye wash station that is located in a battery room to mitigate injury to employees.

(2) Activities That Do Not Include Handling of Electrolyte. Employees performing any activity not involving the handling of electrolyte shall wear safety glasses.

•

Informational Note: Battery maintenance activities usually do not involve handling electrolyte. Batteries with solid electrolyte (such as most lithium batteries) or immobilized electrolyte (such as valve-regulated lead acid batteries) present little or no electrolyte hazard. Most modern density meters expose a worker to a quantity of electrolyte too minute to be considered hazardous, if at all. Such work would not be considered handling electrolyte. However, if specific gravity readings are taken using a bulb hydrometer, the risk of exposure is higher — this could be considered to be handling electrolyte, and the requirements of 320.3(B)(1) would apply.

(C) Testing, Maintenance, and Operation.

•

(1) Direct-Current Ground-Fault Detection. Ground-fault detection shall be based on the type of dc grounding systems utilized.

Informational Note: Not all battery systems have dc ground-fault detection systems. For personnel safety reasons, it is important to understand the grounding methodology being used and to determine the appropriate manner of detecting ground faults. If an unintended ground develops within the system (e.g., dirt and acid touching the battery rack), it can create a short circuit that could cause a fire. Commonly used dc grounding systems include, but are not limited to, the following:

(1) Type 1. An ungrounded dc system, in which neither pole of the battery is connected to ground. If an unintentional ground occurs at any place in the battery, an increased potential would exist, allowing fault current to flow between the opposite end of the battery and the ground. An ungrounded dc system is typically equipped with an alarm to indicate the presence of a ground fault.

(2) Type 2. A solidly grounded dc system, in which either the most positive or most negative pole of the battery is connected directly to ground. If an unintentional ground occurs, it introduces a path through which fault current can flow. A ground detection system is not typically used on this type of grounded system.

(3) Type 3. A resistance grounded dc system, which is a variation of a Type 1 system, in which the battery is connected to ground through a resistance. Detection of a change in the resistance typically enables activation of a ground-fault alarm. Introducing an unintentional ground at

EXHIBIT 320.2

Eye wash station located in a battery room. (Courtesy of Honeywell Safety Products)

one point of the battery could be detected and alarmed. A second unintentional ground at a different point in the battery would create a path for short-circuit current to flow.

(4) Type 4. A solidly grounded dc system, either at the center point or at another point to suit the load system. If an unintentional ground occurs on either polarity, it introduces a path through which short circuit current can flow. A ground detection system is not typically used on this type of grounded system.

The worker must know the type of grounding employed in order to understand the type of dc ground-fault detection required. The informational note describes four different battery grounding types. Some types will normally have ground-fault detection, whereas in others a ground fault detector would be useless or could even create a safety hazard.

(2) Tools and Equipment.

(a) Tools and equipment for work on batteries shall be equipped with handles listed as insulated for the maximum working voltage.

(b) Battery terminals and all electrical conductors shall be kept clear of unintended contact with tools, test equipment, liquid containers, and other foreign objects.

(c) Nonsparking tools shall be required when the risk assessment required by 110.1(F) justifies their use.

(D) Cell Flame Arresters and Cell Ventilation. When present, battery cell ventilation openings shall be unobstructed. Cell flame arresters shall be inspected for proper installation and unobstructed ventilation and shall be replaced when necessary in accordance with the manufacturer's instructions.

Article 330 Safety-Related Work Practices for Use of Lasers

Summary of Article 330 Changes

330.2, Fail Safe: Replaced the phrase *increase the hazard* with *create additional hazards or increased risk* to clarify and promote the consistent use of terminology associated with hazard and risk.

330.3(B)(2)(b): Replaced the phrase *hazard control procedures* with *risk assessment, and risk control procedures* to provide clarity and consistency with the definitions of hazard, risk, and risk assessment and with risk management principles. (Hazards are identified, whereas risk is assessed and controlled.)

330.3(C): Replaced the phrase *available and in possession of the operator at all times* with *readily available* to clarify that the operator need only provide proof of qualification while outside the zone where the risk of hazard could exist.

330.1 Scope.

The requirements of this article shall apply to the use of lasers in the laboratory and the workshop.

Article 330 is limited to tasks performed in the laboratory or in the workshop using a laser. It does not cover the general application and use of lasers in the workplace. Employees who work with lasers might be exposed to hazards associated with the laser output in addition to the electrical hazards associated with the equipment.

330.2 Definitions.

For the purposes of this article, the definitions that follow shall apply.

Fail Safe. The design consideration in which failure of a component does not create additional hazards or increased risk. In the failure mode, the system is rendered inoperative or nonhazardous.

Fail-Safe Safety Interlock. An interlock that in the failure mode does not defeat the purpose of the interlock; for example, an interlock that is positively driven into the off position as soon as a hinged cover begins to open, or before a detachable cover is removed, and that is positively held in the off position until the hinged cover is closed or the detachable cover is locked in the closed position.

Laser. Any device that can be made to produce or amplify electromagnetic radiation in the wavelength range from 100 nm to 1 mm primarily by the process of controlled stimulated emission.

Laser Energy Source. Any device intended for use in conjunction with a laser to supply energy for the excitation of electrons, ions, or molecules. General energy sources, such as electrical supply services or batteries, shall not be considered to constitute laser energy sources.

Laser Product. Any product or assembly of components that constitutes, incorporates, or is intended to incorporate a laser or laser system.

Laser Radiation. All electromagnetic radiation emitted by a laser product between 100 nm and 1 mm that is produced as a result of a controlled stimulated emission.

Laser System. A laser in combination with an appropriate laser energy source with or without additional incorporated components.

There are two classification systems in use, the FDA laser regulations and the international standard IEC 60825, *Safety of Laser Products*. At this time the FDA and IEC 60825 laser classification systems are not fully harmonized. Laser systems are classified by output power and wavelength. The classifications described by physical hazard are as follows:

- *Class 1 laser system* — Incapable of producing damaging radiation levels during operation. Class 1 systems may contain laser sources of a higher class but are intrinsically safe when used as intended. Examples are found in DVD players and laser printers.
- *Class 1M laser system* — Incapable of producing hazardous exposure conditions during normal operation unless the beam is viewed with an optical instrument.
- *Class 2 laser system (Class 2 and 2M)* — Emits in the visible portion of the spectrum (0.4 to 0.7 µm) and eye protection is normally afforded by the aversion response (blinking reflex). Examples of Class 2 systems are found in laser pointers and some point-of-sale scanners.
- *Class 3 laser system (Class 3R and Class 3B)* — Hazardous if viewed directly but is considered safe if handled carefully and with restricted beam viewing. Examples of Class 3 lasers are found inside DVD writers.
- *Class 4 laser system* — Can burn the skin or cause permanent eye damage as a result of direct, diffuse, or indirect beam viewing. May ignite combustible materials. Class 4 lasers are typically limited to industrial, scientific, military, and medical use.

330.3 Safety Training.

(A) Personnel to Be Trained. Employers shall provide training for all operator and maintenance personnel.

Federal regulations recommend assignment of a Laser Safety Officer and inclusion of a standard operating procedure for facilities that use Class 3B and 4 lasers.

(B) Scope of Training. The training shall include, but is not limited to, the following:

(1) Familiarization with laser principles of operation, laser types, and laser emissions

Understanding the laser class is essential to addressing safety issues. Although any class of laser may be found in the laboratory or workshop, many are considered to be not hazardous under normal operation or viewing. See the commentary following the definition of *Laser System* in 330.2 above.

(2) Laser safety, including the following:

    a. System operating procedures

    b. Risk assessment and risk control procedures

    c. Need for personnel protection

    d. Accident reporting procedures

    e. Biological effects of the laser upon the eye and the skin

    f. Electrical and other hazards associated with the laser equipment, including the following:

        i. High voltages (>1 kV) and stored energy in the capacitor banks

        ii. Circuit components, such as electron tubes, with anode voltages greater than 5 kV emitting X-rays

iii. Capacitor bank explosions

iv. Production of ionizing radiation

v. Poisoning from the solvent or dye switching liquids or laser media

vi. High sound intensity levels from pulsed lasers

Great care is needed when working with the various types of laser systems, and electrical safety training is paramount. Lasers that are electrically powered can be hazardous due to their radiation as well as their power sources. High energy density batteries, generators, ultracapacitors, and associated power distribution panels present potential shock, arc flash, and fire hazards. Some source batteries are megajoule systems. A typical welding laser will have a prime power source of 50 to 100 kVA, with larger systems approaching 500 kVA. Large systems are often water cooled, so wet environments might be present as well. These units can be mobile units powered by diesel generators.

(C) Proof of Qualification. Proof of qualification of the laser equipment operator shall be readily available.

330.4 Safeguarding of Employees in the Laser Operating Area.

The ANSI Z136 series addresses hazards and associated regulatory information in detail. The hazards are also addressed by the IEC 60825 series. The ANSI series contains nine ANSI standards for lasers. Those following documents are applicable to facilities addressed by Article 330:

- ANSI Z136.1-2014, *American National Standard for Safe Use of Lasers*
- ANSI Z136.4-2010, *Recommended Practice for Laser Safety Measurements for Hazard Evaluation*
- ANSI Z136.5-2000, *American National Standard for Safe Use of Lasers in Educational Institutions*
- ANSI Z136.6-2005, *American National Standard for Safe Use of Lasers Outdoors*
- ANSI Z136.8-2012, *American National Standard for Safe Use of Lasers in Research, Development, or Testing*

(A) Eye Protection. Employees shall be provided with eye protection as required by federal regulation.

Typically, only Class 3B and Class 4 lasers require protective eyewear for operation, such as that shown in Exhibit 330.1. Other laser classes may be hazardous if viewed with optical instruments or during repair.

(B) Warning Signs. Warning signs shall be posted at the entrances to areas or protective enclosures containing laser products.

Warning signs are specified by OSHA and are currently required to comply with ANSI Z535, *Series of Standards for Safety Signs and Tags*.

(C) Master Control. High-power laser equipment shall include a key-operated master control.

(D) High-Power Radiation Emission Warning. High-power laser equipment shall include a fail-safe laser radiation emission audible and visible warning when it is switched on or if the capacitor banks are charged.

(E) Beam Shutters or Caps. Beam shutters or caps shall be used, or the laser switched off, when laser transmission is not required. The laser shall be switched off when unattended for 30 minutes or more.

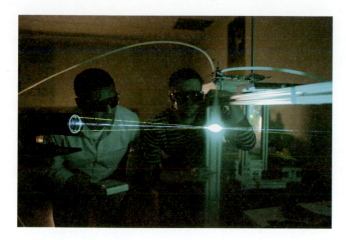

(F) Aiming. Laser beams shall not be aimed at employees.

Federal regulations recommend standard operating and alignment procedures for facilities using Class 3B and Class 4 lasers.

(G) Label. Laser equipment shall bear a label indicating its maximum output.

(H) Personal Protective Equipment (PPE). PPE shall be provided for users and operators of high-power laser equipment.

330.5 Employee Responsibility.

Employees shall be responsible for the following:

(1) Obtaining authorization for laser use
(2) Obtaining authorization for being in a laser operating area
(3) Observing safety rules
(4) Reporting laser equipment failures and accidents to the employer

Safety-Related Work Practices: Power Electronic Equipment

Article | 340

Summary of Article 340 Changes

340.5: Deleted the term *Hazardous* from the title. Replaced *10 mA* with *40 mA*, *may* with *can*, and *microfarad* with *μF*.

340.5(2): Replaced the term *result* with *effects* in the title to help clarify the main title from the subsection titles and to provide consistency with the rest of the standard and risk assessment principles.

340.7(A): Replaced *nature* with *identification*; *hazard* with *hazards*; *minimize the hazard* with *reduce the risk associated with the hazards*; and *hazardous incident* with *incident that resulted in, or could have resulted in, injury or damage to health*. These changes clarify and promote the consistent use of terminology associated with hazard and risk.

340.7(B): Replaced *is* with *shall be*; *hazardous components location* with *location of components that present an electrical hazard*; and *hazardous incident* with *incident that resulted in, or could have resulted in, injury or damage to health*. These changes clarify and promote the consistent use of terminology associated with hazard and risk.

Chapter 1 applies to electrical equipment that operates at frequencies normally supplied for consumer use. The reaction of the human body to current flow changes as the frequency increases or decreases. When a frequency reaches the microwave band, joule heating can result in internal burns.

Employees who work on or with equipment within the scope of Article 340 must be qualified to perform tasks using the specific electronic equipment, and they should be trained to understand the unique hazards associated with the specific equipment and how to avoid exposure to those hazards.

340.1 Scope.

This article shall apply to safety-related work practices around power electronic equipment, including the following:

(1) Electric arc welding equipment
(2) High-power radio, radar, and television transmitting towers and antennas
(3) Industrial dielectric and radio frequency (RF) induction heaters
(4) Shortwave or RF diathermy devices
(5) Process equipment that includes rectifiers and inverters such as the following:

    a. Motor drives
    b. Uninterruptible power supply systems
    c. Lighting controllers

Article 340 includes some common types of equipment that are found on construction sites or at industrial or commercial facilities. It also includes some types of equipment that are not as common. This list is not all inclusive. For example, photovoltaic, fuel cell, and wind systems used for the generation of electricity generally incorporate power electronic devices (power converters).

340.2 Definition.

For the purposes of this article, the definition that follows shall apply.

Radiation Worker. A person who is required to work in electromagnetic fields, the radiation levels of which exceed those specified for nonoccupational exposure.

340.3 Application.

The purpose of this article is to provide guidance for safety personnel in preparing specific safety-related work practices within their industry.

340.4 Reference Standards.

The following are reference standards for use in the preparation of specific guidance to employees as follows:

(1) International Electrotechnical Commission IEC 60479, *Effects of Current Passing Through the Body*:

   a.   60479-1 Part 1: General aspects

   b.   60479-1-1 Chapter 1: Electrical impedance of the human body

   c.   60479-1-2 Chapter 2: Effects of ac in the range of 15 Hz to 100 Hz

   d.   60479-2 Part 2: Special aspects

   e.   60479-2-4 Chapter 4: Effects of ac with frequencies above 100 Hz

   f.   60479-2-5 Chapter 5: Effects of special waveforms of current

   g.   60479-2-6 Chapter 6: Effects of unidirectional single impulse currents of short duration

(2) International Commission on Radiological Protection (ICRP) Publication 33, *Protection Against Ionizing Radiation from External Sources Used in Medicine*

340.5 Effects of Electricity on the Human Body.

The employer and employees shall be aware of the following hazards associated with power electronic equipment.

(1) Effects of Power Frequency Current:

   a.   At 0.5 mA, shock is perceptible.

   b.   At 10 mA, a person may not be able to voluntarily let go of an energized electrical conductor or circuit part.

   c.   At about 40 mA, the shock, if lasting for 1 second or longer, can be fatal due to ventricular fibrillation.

   d.   Further increasing current leads to burns and cardiac arrest.

(2) Effects of Direct Current:

   a.   A dc current of 2 mA is perceptible.

   b.   A dc current of 40 mA is considered the threshold of the let-go current.

(3) Effects of Voltage. A voltage of 30 V rms, or 60 V dc, is considered safe, except when the skin is broken. The internal body resistance can be as low as 500 ohms, so fatalities can occur.

(4) Effects of Short Contact:

   a.   For contact less than 0.1 second and with currents just greater than 0.5 mA, ventricular fibrillation can occur only if the shock is during a vulnerable part of the cardiac cycle.

   b.   For contact of less than 0.1 second and with currents of several amperes, ventricular fibrillation can occur if the shock is during a vulnerable part of the cardiac cycle.

   c.   For contact of greater than 0.8 second and with currents just greater than 0.5 A, cardiac arrest (reversible) can occur.

   d.   For contact greater than 0.8 second and with currents of several amperes, burns and death are probable.

(5) Effects of Alternating Current at Frequencies Above 100 Hz. When the threshold of perception increases from 10 kHz to 100 kHz, the threshold of let-go current increases from 10 mA to 100 mA.

(6) Effects of Waveshape. Contact with voltages from phase controls usually causes effects between those of ac and dc sources.

(7) Effects of Capacitive Discharge:
   a.   A circuit of capacitance of 1 µF having a 10 kV capacitor charge can cause ventricular fibrillation.
   b.   A circuit of capacitance of 20 µF having a 10 kV capacitor charge can be dangerous and probably will cause ventricular fibrillation.

Section 340.5 is the only place in *NFPA 70E* to find quantitative data on the physiological effects of current on the human body. Research in this area was done by Charles F. Dalziel (inventor of the ground-fault circuit interrupter) and William B. Kouwenhoven (inventor of the cardiac defibrillator). Data presented by Dalziel is given in Commentary Table 340.1.

340.6 Hazards Associated with Power Electronic Equipment.

The employer and employees shall be aware of the hazards associated with the following:

(1) High voltages within the power supplies
(2) Radio frequency energy–induced high voltages
(3) Effects of RF fields in the vicinity of antennas and antenna transmission lines, which can introduce electrical shock and burns
(4) Ionizing (X-radiation) hazards from magnetrons, klystrons, thyratrons, cathode-ray tubes, and similar devices
(5) Nonionizing RF radiation hazards from the following:

   a.   Radar equipment
   b.   Radio communication equipment, including broadcast transmitters
   c.   Satellite–earth-transmitters
   d.   Industrial scientific and medical equipment
   e.   RF induction heaters and dielectric heaters
   f.   Industrial microwave heaters and diathermy radiators

Power electronic equipment commonly holds stored electrical energy. Filter boards and power boards in the equipment contain capacitors that could have stored energy in addition to the stored energy in the bus capacitors. Electromagnetic interference (EMI) filters must be discharged prior to servicing to prevent electric shock or arc flash hazard. Exhibit 340.1 shows a caution label requiring removal of power and a 10-minute waiting period before the enclosure is opened in order to allow time for stored energy to discharge.

Before accessing the interior of an uninterruptible power supply (UPS) unit, care must be exercised to ensure that each source of input and output power has been electrically or physically isolated. Batteries or capacitors located within the UPS unit might still be charged even after all apparent voltage sources have been isolated or confirmed to be de-energized.

COMMENTARY TABLE 340.1 *Quantitative Effects of Electric Current on Humans (Average Data)*

Effects	Direct Current (mA) DC 150 lbs	Direct Current (mA) DC 115 lbs	Alternating Current (mA) 60 Hz 150 lbs	Alternating Current (mA) 60 Hz 115 lbs	Alternating Current (mA) 10 kHz 150 lbs	Alternating Current (mA) 10 kHz 115 lbs
Slight sensation on hand	1	0.6	0.4	0.3	7	5
Perception threshold, median	5.2	3.5	1.1	0.7	12	8
Shock — not painful and no loss of muscular control	9	6	1.8	1.2	17	11
Painful shock — muscular control lost by 0.5%	62	41	9	6	55	37
Painful shock — let-go threshold, median	76	51	16	10.5	75	50
Painful and severe shock — breathing difficult, muscular control lost by 99.5%	90	60	23	15	94	63
Possible ventricular fibrillation:						
Three-second shocks	500	500	100	100		
Short shocks (T in seconds)			$165/(\sqrt{T})$	$165/(\sqrt{T})$		
High-voltage surges (Energy in joules, i.e., watt-seconds)	50 J	50 J	13.6 J	13.6 J		

Notes:

Derived from "Deleterious Effects of Electric Shock," Charles F. Dalziel, p. 24. Presented at a meeting of experts on electrical accidents and related matters, sponsored by the International Labour Office, World Health Office and International Electrotechnical Commission, Geneva, Switzerland, October 23–31, 1961. Refer to the study for definitions and details.

The data in the preceding table are based on limited experiments conducted on human subjects and animals in 1961. These figures may not be fully dependable due to lack of additional information and normal physiological differences between individuals. Electric current should probably be considered fatal at current values lower than indicated. From the *Health and Safety Manual*, LBNL/PPUB-3000, Environmental, Health and Safety (EH&S) Division of Lawrence Berkley National Laboratory, operated by the University of California for the U.S. Department of Energy, from experimental data derived from 115 subjects, information is as follows:

- The average threshold of perception of dc current in a 150 lb person (average healthy young male) is 5 mA.
- The threshold varies considerably from 2 mA to 10 mA dc.
- The average threshold for 60 Hz ac current in the 150 lb person is 1 mA.
- The threshold of perception increases with frequency so that at 100 kHz it is 150 mA (0.15 A).
- The threshold for a 115 lb person (average healthy young female) is around two-thirds of that for the 150 lb person (average healthy young male).

340.7 Specific Measures for Personnel Safety.

(A) Employer Responsibility. The employer shall be responsible for the following:

(1) Proper training and supervision by properly qualified personnel, including the following:

    a. Identification of associated hazards

    b. Strategies to reduce the risk associated with the hazards

    c. Methods of avoiding or protecting against the hazard

    d. Necessity of reporting any incident that resulted in, or could have resulted in, injury or damage to health

(2) Properly installed equipment

(3) Proper access to the equipment

(4) Availability of the correct tools for operation and maintenance

(5) Proper identification and guarding of dangerous equipment

(6) Provision of complete and accurate circuit diagrams and other published information to the employee prior to the employee starting work (The circuit diagrams should be marked to indicate the components that present an electrical hazard.)

(7) Maintenance of clear and clean work areas around the equipment to be worked on

(8) Provision of adequate and proper illumination of the work area

(B) Employee Responsibility. The employee shall be responsible for the following:

(1) Understanding the hazards associated with the work

(2) Being continuously alert and aware of the possible hazards

(3) Using the proper tools and procedures for the work

(4) Informing the employer of malfunctioning protective measures, such as faulty or inoperable enclosures and locking schemes

(5) Examining all documents provided by the employer relevant to the work, especially those documents indicating the location of components that present an electrical hazard

(6) Maintaining good housekeeping around the equipment and work space

(7) Reporting any incident that resulted in, or could have resulted in, injury or damage to health

(8) Using and appropriately maintaining the PPE and tools required to perform the work safely

Servicing damaged equipment may present additional hazards not associated with servicing properly functioning equipment. For example, stored energy in damaged equipment may not discharge as intended. High earth leakage currents to the equipment chassis or enclosure can occur if the equipment grounding method is damaged or improperly installed. Testing for the presence of voltage is recommended even after waiting the prescribed time.

Safety-Related Work Requirements: Research and Development Laboratories

Article 350

Summary of Article 350 Changes

350.2, Competent Person: Replaced *meeting* with *who meets; electrical hazard exposure* with *exposure to electrical hazards*; and *controls for mitigating those hazards* with *control methods to reduce the risk associated with*. These changes clarify and promote the consistent use of terminology associated with hazard and risk and risk assessment, and with risk management principles. (Hazards are identified, whereas risk is assessed and controlled by following a hierarchy of risk control methods.)

Article 350 addresses unique conditions that might exist in laboratory and research areas. Electrical installations in laboratory or research areas often contain custom or specially designed electrical equipment that may have unique electrical safety–related requirements.

350.1 Scope.

The requirements of this article shall apply to the electrical installations in those areas, with custom or special electrical equipment, designated by the facility management for research and development (R&D) or as laboratories.

This article covers all laboratory facilities, including those that exist in educational facilities. Research and development facilities, including those associated with institutions of higher learning, are also covered. An example of a laboratory covered by Article 350 is shown in Exhibit 350.1.

350.2 Definitions.

For the purposes of this article, the definitions that follow shall apply.

Competent Person. A person who meets all the requirements of *qualified person*, as defined in Article 100 in Chapter 1 of this standard and who, in addition, is responsible for all work activities or safety procedures related to custom or special equipment and has detailed knowledge regarding the exposure to electrical hazards, the appropriate control methods to reduce the risk associated with those hazards, and the implementation of those methods.

Field Evaluated. A thorough evaluation of nonlisted or modified equipment in the field that is performed by persons or parties acceptable to the authority having jurisdiction. The evaluation approval ensures that the equipment meets appropriate codes and standards, or is similarly found suitable for a specified purpose.

Equipment in laboratory or research areas is often custom designed. A product standard covering the special equipment might not be issued. Also, submitting the one-off piece of equipment to a formal listing process may not be necessary because the equipment is not

EXHIBIT 350.1

A laboratory covered by Article 350. (© Copyright Sandia Corporation. Reproduced by permission.)

available for sale or use outside of the company for which it is designed. Field evaluation of the custom equipment ensures that it is suitable for employee use within the intended application.

Laboratory. A building, space, room, or group of rooms intended to serve activities involving procedures for investigation, diagnostics, product testing, or use of custom or special electrical components, systems, or equipment.

Research and Development (R&D). An activity in an installation specifically designated for research or development conducted with custom or special electrical equipment.

350.3 Applications of Other Articles.

The electrical system for R&D and laboratory applications shall meet the requirements of the remainder of this document, except as amended by Article 350.

> Informational Note: Examples of these applications include low-voltage–high-current power systems; high-voltage–low-current power systems; dc power supplies; capacitors; cable trays for signal cables and other systems, such as steam, water, air, gas, or drainage; and custom-made electronic equipment.

350.4 Specific Measures and Controls for Personnel Safety.

Each laboratory or R&D system application shall be assigned a competent person as defined in this article to ensure the use of appropriate electrical safety-related work practices and controls.

As required in Chapter 1, the employer should provide written procedures or other instructions of the electrical safety program for the unique conditions encountered in laboratory and research areas. A competent person, not just qualified, is required to oversee the laboratory to ensure that employees follow the procedures.

350.5 Listing Requirements.

The equipment or systems used in the R&D area or in the laboratory shall be listed or field evaluated prior to use.

> Informational Note: Laboratory and R&D equipment or systems can pose unique electrical hazards that might require mitigation. Such hazards include ac and dc, low voltage and high amperage, high voltage and low current, large electromagnetic fields, induced voltages, pulsed power, multiple frequencies, and similar exposures.

This requirement is not intended to be applied to equipment under development. Unique equipment is often necessary to conduct research or to evaluate items under development, and listing of the equipment is often not possible. This equipment is permitted to be field evaluated (see the definition *Field Evaluated* in 350.2) by a party acceptable to the authority having jurisdiction. The use of custom-made equipment does not remove the employer's responsibility for providing electrical safety–related work practices.

Referenced Publications

Summary of Annex A Changes

A.1: Replaced the phrase *shall be* with *are to be* to remove a requirement from the annex. Updated all references to reflect the current revision dates.

A.2: Updated *NFPA 70®, National Electrical Code®*, reference to the current edition.

A.3.2: Updated references to the most current editions.

A.3.5: Deleted the reference to IEEE 1584, *Guide for Performing Arc Flash Calculations*, because it is no longer referenced in the requirements.

A.4: Deleted the reference to NFPA *101®, Life Safety Code®*, because it is not included within *NFPA 70E®, Standard for Electrical Safety in the Workplace®*.

Informative Annex A identifies publications that are directly referenced in the requirements of the 2015 edition of *NFPA 70E*. One example of a reference in a requirement is Table 130.7(C)(14), which contains mandatory references to ASTM standards on personal protective equipment (PPE).

The state of electrical safety is constantly moving forward. Existing documents are either deleted or revised periodically while new documents are issued. The user of *NFPA 70E* is encouraged to verify that the referenced document is the latest published edition, since the document's issue cycle may be different from that of *NFPA 70E*.

A.1 General.

This informative annex is not part of the requirements of this document and is included for information only. To the extent the documents or portions thereof listed in this informative annex are referenced within this standard, those documents are to be considered part of the requirements of this document in the section and manner in which they are referenced.

A.2 NFPA Publications.

National Fire Protection Association, 1 Batterymarch Park, Quincy, MA 02169-7471.

NFPA 70®, National Electrical Code®, 2014 edition.

A.3 Other Publications.

A.3.1 ANSI Publications. American National Standards Institute, Inc., 25 West 43rd Street, 4th Floor, New York, NY 10036.

ANSI/ASC A14.1, *American National Standard for Ladders — Wood — Safety Requirements,* 2007.

ANSI/ASC A14.3, *American National Standard for Ladders — Fixed — Safety Requirements,* 2008.

ANSI/ASC A14.4, *American National Standard Safety Requirements for Job-Made Ladders,* 2009.

ANSI/ASC A14.5, *American National Standard for Ladders — Portable Reinforced — Safety Requirements,* 2007.

ANSI Z87.1, *Practice for Occupational and Educational Eye and Face Protection,* 2010.

ANSI Z89.1, *Requirements for Protective Headwear for Industrial Workers,* 2009.

ANSI Z535, *Series of Standards for Safety Signs and Tags,* 2011.

A.3.2 ASTM Publications. ASTM International, 100 Barr Harbor Drive, P.O Box C700, West Conshohocken, PA 19428-2959.

ASTM D120, *Standard Specification for Rubber Insulating Gloves,* 2009.

ASTM D1048, *Standard Specification for Rubber Insulating Blankets,* 2012.

ASTM D1049, *Standard Specification for Rubber Covers,* 1998 (R 2010).

ASTM D1050, *Standard Specification for Rubber Insulating Line Hoses,* 2005 (R 2011).

ASTM D1051, *Standard Specification for Rubber Insulating Sleeves,* 2008.

ASTM F478, *Standard Specification for In-Service Care of Insulating Line Hose and Covers,* 2009.

ASTM F479, *Standard Specification for In-Service Care of Insulating Blankets,* 2006 (R 2011).

ASTM F496, *Standard Specification for In-Service Care of Insulating Gloves and Sleeves,* 2008.

ASTM F696, *Standard Specification for Leather Protectors for Rubber Insulating Gloves and Mittens,* 2006 (R 2011).

ASTM F711, *Standard Specification for Fiberglass-Reinforced Plastic (FRP) Rod and Tube Used in Live Line Tools,* 2002 (R 2007).

ASTM F712, *Standard Test Methods and Specifications for Electrically Insulating Plastic Guard Equipment for Protection of Workers,* 2006 (R 2011).

ASTM F855, *Standard Specification for Temporary Protective Grounds to Be Used on De-energized Electric Power Lines and Equipment,* 2009.

ASTM F887, *Standard Specification for Personal Climbing Equipment,* 2011.

ASTM F1116, *Standard Test Method for Determining Dielectric Strength of Dielectric Footwear,* 2003 (R 2008).

ASTM F1117, *Standard Specification for Dielectric Footwear,* 2003 (R 2008).

ASTM F1236, *Standard Guide for Visual Inspection of Electrical Protective Rubber Products,* 1996 (R 2012).

ASTM F1296, *Standard Guide for Evaluating Chemical Protective Clothing,* 2008.

ASTM F1449, *Standard Guide for Industrial Laundering of Flame, Thermal, and Arc Resistant Clothing,* 2008.

ASTM F1505, *Standard Specification for Insulated and Insulating Hand Tools,* 2010.

ASTM F1506, *Standard Performance Specification for Flame Resistant and Arc Rated Textile Materials for Wearing Apparel for Use by Electrical Workers Exposed to Momentary Electric Arc and Related Thermal Hazards,* 2010a.

ASTM F1742, *Standard Specification for PVC Insulating Sheeting,* 2003 (R 2011).

ASTM F1891, *Standard Specification for Arc and Flame Resistant Rainwear,* 2012.

ASTM F1959/F1959M, *Standard Test Method for Determining the Arc Rating of Materials for Clothing, 2012.*

ASTM F2178, *Standard Test Method for Determining the Arc Rating and Standard Specification for Eye or Face Protective Products, 2012.*

ASTM F2249, *Standard Specification for In-Service Test Methods for Temporary Grounding Jumper Assemblies Used on De-Energized Electric Power Lines and Equipment, 2003 (R 2009).*

ASTM F2412/F2320, *Standard Specification for Rubber Insulating Sheeting, 2011.*

ASTM F2412, *Standard Test Methods for Foot Protections, 2011.*

ASTM F2413, *Standard Specification for Performance Requirements for Protective (Safety) Toe Cap Footwear, 2011.*

ASTM F2522, *Standard Test Method for Determining the Protective Performance of a Shield Attached on Live Line Tools or on Racking Rods for Electric Arc Hazards, 2012.*

ASTM F2676, *Standard Test Method for Determining the Protective Performance of an Arc Protective Blanket for Electric Arc Hazards, 2009.*

ASTM F2677, *Standard Specification for Electrically Insulating Aprons, 2008a.*

ASTM F2757, *Standard Guide for Home Laundering Care and Maintenance of Flame, Thermal and Arc Resistant Clothing, 2009.*

A.3.3 ICRP Publications. International Commission on Radiological Protection, SE-171 16 Stockholm, Sweden.

ICRP Publication 33, *Protection Against Ionizing Radiation from External Sources Used in Medicine,* March 1981.

A.3.4 IEC Publications. International Electrotechnical Commission, 3, rue de Varembé, P.O. Box 131, CH-1211 Geneva 20, Switzerland.

IEC 60479, *Effects of Current Passing Through the Body.*

60479-1 Part 1: General aspects, 2005.
60479-1-1 Chapter 1: Electrical impedance of the human body.
60479-1-2 Chapter 2: Effects of ac in the range of 15 Hz to 100 Hz.
60479-2 Part 2: Special aspects, 2007.
60479-2-4 Chapter 4: Effects of ac with frequencies above 100 Hz.
60479-2-5 Chapter 5: Effects of special waveforms of current.
60479-2-6 Chapter 6: Effects of unidirectional single impulse currents of short duration.

A.3.5 IEEE Publications. Institute of Electrical and Electronics Engineers, IEEE Operations Center, 445 Hoes Lane, P.O. Box 1331, Piscataway, NJ 08855-1331.

IEEE C37.20.7, *Guide for Testing Metal-Enclosed Switchgear Rated up to 38 kV for Internal Arcing Faults, 2007/Corrigendum 1, 2010.*

•

A.4 References for Extracts in Mandatory Sections.

NFPA 70®, National Electrical Code®, 2014 edition.

•

Informational References

Summary of Annex B Changes

B.1: Updated all references to the most current editions.

B.1.7: Added new IEEE standards supporting the implementation of *NFPA 70E®, Standard for Electrical Safety in the Workplace®*, requirements. Moved IEEE papers that were previously located in the Other Publications subsection to a more appropriate location, the *IEEE* Publications subsection.

Informative Annex B identifies important publications that provide additional information to assist the user in understanding and applying the requirements of *NFPA 70E*. This informative annex provides further information on the publisher and edition of the documents referenced in the Informational Notes. These documents are for informational purposes only.

The state of electrical safety is constantly moving forward. Existing documents are either deleted or revised periodically while new documents are issued. The user of *NFPA 70E* is encouraged to verify that the referenced document is the latest published edition, since the document's issue cycle may be different from that of *NFPA 70E*.

B.1 Referenced Publications.

The following documents or portions thereof are referenced within this standard for informational purposes only and are thus not part of the requirements of this document unless also listed in Informative Annex A.

B.1.1 NFPA Publications. National Fire Protection Association, 1 Batterymarch Park, Quincy, MA 02169-7471.

NFPA 1, *Fire Code*, 2015 edition.
NFPA 70,® National Electrical Code®, 2014 edition.
NFPA 70B, *Recommended Practice for Electrical Equipment Maintenance,* 2013 edition.
NFPA 79, *Electrical Standard for Industrial Machinery*, 2015 edition.

B.1.2 ANSI Publications. American National Standards Institute, Inc., 25 West 43rd Street, 4th Floor, New York, NY 10036.

ANSI/AIHA Z10, *American National Standard for Occupational Health and Safety Management Systems, 2012*.

ANSI/ASSE Z244.1, *Control of Hazardous Energy — Lockout/Tagout and Alternative Methods*, 2003 (R 2008).

ANSI C84.1, *Electric Power Systems and Equipment – Voltage Ratings (60 Hz), 2011*.

ANSI/ISO 14001, *Environmental Management Systems — Requirements with Guidance for Use*, 2004/Corrigendum 1, 2009.

ANSI/NETA MTS, *Standard for Maintenance Testing Specifications for Electrical Power Distribution Equipment and Systems,* 2011.

ANSI Z535.4, *Product Safety Signs and Labels*, 2011.

B.1.3 ASTM Publications. ASTM International, 100 Barr Harbor Drive, P.O. Box C700, West Conshohocken, PA 19428-2959.

ASTM F496, *Standard Specification for In-Service Care of Insulating Gloves and Sleeves*, 2008.

ASTM F711, *Standard Specification for Fiberglass-Reinforced Plastic (FRP) Rod and Tube Used in Live Line Tools*, 2002 (R 2007).

ASTM F1449, *Standard Guide for Industrial Laundering of Flame, Thermal, and Arc Resistant Clothing, 2008*.

ASTM F1506, *Standard Performance Specification for Flame Resistant and Arc Rated Textile Materials for Wearing Apparel for Use by Electrical Workers Exposed to Momentary Electric Arc and Related Thermal Hazards*, 2010a.

ASTM F1959/F1959M, *Standard Test Method for Determining the Arc Rating of Materials for Clothing*, 2012.

ASTM F2249, *Standard Specification for In-Service Test Methods for Temporary Grounding Jumper Assemblies Used on De-Energized Electric Power Lines and Equipment*, 2003 (R 2009).

ASTM F2413, *Standard Specifications for Performance Requirements for Protective (Safety) Toe Cap Footwear,* 2011.

ASTM F2757, *Standard Guide for Home Laundering Care and Maintenance of Flame, Thermal and Arc Resistant Clothing, 2009*.

B.1.4 British Standards Institute, Occupational Health and Safety Assessment Series (OHSAS) Project Group Publications. British Standards Institute, American Headquarters, 12110 Sunset Hills Road, Suite 200, Reston VA 20190-5902.

BS OSHAS 18001, *Occupational Health and Safety Management Systems*, 2007.

B.1.5 CSA Publications. Canadian Standards Association, 5060 Spectrum Way, Mississauga, ON L4W 5N6, Canada.

CAN/CSA Z462, *Workplace Electrical Safety, 2012*.

CAN/CSA Z1000, *Occupational Health and Safety Management, 2006 (R 2011)*.

•

B.1.6 IEC Publications. International Electrotechnical Commission, 3, rue de Varembé, P.O. Box 131, CH-1211 Geneva 20, Switzerland.

IEC 60204-1 ed 5.1 Consol. with am 1, *Safety of Machinery — Electrical Equipment of Machines — Part 1: General Requirements*, 2009.

B.1.7 IEEE Publications. Institute of Electrical and Electronic Engineers, IEEE Operations Center, 445 Hoes Lane, P.O. Box 1331, Piscataway, NJ 08855-1331.

ANSI/IEEE C2, *National Electrical Safety Code*, 2012.

ANSI/IEEE C 37.20.6, *Standard for 4.76 kV to 38 kV-Rated Ground and Test Devices Used in Enclosures*, 2007.

•

IEEE 4, *Standard Techniques for High Voltage Testing*, 2013.

•

IEEE 450, *IEEE Recommended Practice for Maintenance, Testing, and Replacement of Vented Lead-Acid Batteries for Stationary Applications*, 2010.

•

IEEE 516, *Guide for Maintenance Methods on Energized Power Lines*, 2009.

IEEE 937, *Recommended Practice for Installation and Maintenance of Lead-Acid Batteries for Photovoltaic Systems*, 2007.

IEEE 946, *IEEE Recommended Practice for the Design of DC Auxiliary Power Systems for Generating Systems*, 2004.

IEEE 1106, *IEEE Recommended Practice for Installation, Maintenance, Testing, and Replacement of Vented Nickel-Cadmium Batteries for Stationary Applications*, 2005 (R 2011).

IEEE 1184, *IEEE Guide for Batteries for Uninterruptible Power Supply Systems*, 2006.

IEEE 1187, *Recommended Practice for Installation Design and Installation of Valve-Regulated Lead-Acid Storage Batteries for Stationary Applications*, 2002.

IEEE 1188, *IEEE Recommended Practice for Maintenance, Testing, and Replacement of Valve-Regulated Lead-Acid (VRLA) Batteries for Stationary Applications*, 2005 (R 2010).

•

IEEE 1491, *IEEE Guide for Selection and Use of Battery Monitoring Equipment in Stationary Applications*, 2012.

IEEE 1584™, *Guide for Performing Arc Flash Hazard Calculations*, 2002.

IEEE 1584a™, *Guide for Performing Arc Flash Hazard Calculations, Amendment 1*, 2004.

IEEE 1584b™, *Guide for Performing Arc Flash Hazard Calculations — Amendment 2: Changes to Clause 4*, 2011.

IEEE 1657, *Recommended Practice for Personnel Qualifications for Installation and Maintenance of Stationary Batteries*, 2009.

IEEE 3007.1, *IEEE Recommended Practice for the Operation and Management of Industrial and Commercial Power Systems*, 2010.

IEEE 3007.2, *IEEE Recommended Practice for the Maintenance of Industrial and Commercial Power Systems*, 2010.

IEEE 3007.3, *IEEE Recommended Practice for Electrical Safety in Industrial and Commercial Power Systems*, 2012.

Anderson, W. E., "Risk Analysis Methodology Applied to Industrial Machine Development," *IEEE Transactions on Industrial Applications*, Vol. 41, No. 1, January/February 2005, pp. 180–187.

Ammerman, R. F., Gammon, T., Sen, P. K., and Nelson, J. P., "DC-Arc Models and Incident-Energy Calculations," *IEEE Transactions on Industrial Applications*, Vol. 46, No. 5, 2010.

Doan, D. R, "Arc Flash Calculations for Exposures to DC Systems," *IEEE Transactions on Industrial Applications*, Vol 46, No. 6, 2010.

Doughty, R. L., T. E. Neal, and H. L. Floyd II, "Predicting Incident Energy to Better Manage the Electric Arc Hazard on 600 V Power Distribution Systems," *Record of Conference Papers IEEE IAS 45th Annual Petroleum and Chemical Industry Conference*, September 28–30, 1998.

Lee, R., "The Other Electrical Hazard: Electrical Arc Flash Burns," *IEEE Trans. Applications*, Vol. 1A-18, No. 3, May/June 1982.

B.1.8 ISA Publications. Instrumentation, Systems, and Automation Society, 67 Alexander Drive, Research Triangle Park, NC 27709.

ANSI/ISA 61010-1, *Safety Requirements for Electrical Equipment for Measurement, Control, and Laboratory Use, Part 1: General Requirements,* 2007.

B.1.9 ISEA Publications. International Safety Equipment Association, 1901 North Moore Street, Arlington, VA 22209-1762.

ANSI/ISEA Z358.1, *American National Standard for Emergency Eye Wash and Shower Equipment,* 2009.

B.1.10 ISO Publications. International Organization for Standardization, 1, Ch. de la Voie-Creuse, Case postale 56, CH-1211 Geneva 20, Switzerland.

ISO 14001, *Environmental Management Systems — Requirements with Guidance for Use,* 2004.

B.1.11 NIOSH Publications. National Institute for Occupational Safety and Health, Centers for Disease Control and Prevention, 1600 Clifton Road, Atlanta, GA 30333.

DHHS (NIOSH) Publication No. 94-110, *Applications Manual for the Revised NIOSH Lifting Equation,* 1994.

B.1.12 UL Publications. Underwriters Laboratories Inc., 333 Pfingsten Road, Northbrook, IL 60062-2096.

ANSI/UL 943, *Standard for Ground-Fault Circuit Interrupters,* 2006 (R 2012).

-

B.1.13 U.S. Government Publications. U.S. Government Printing Office, Washington, DC 20402.

-

Title 29, Code of Federal Regulations, Part 1910, Occupational Safety and Health Standards, Subpart S, Electrical, 1910.137, Personal Protective Equipment, and 1910.305(j)(7), Storage Batteries; and Part 1926, Safety and Health Regulations for Construction, Subpart K, Electrical, 1926.441, Batteries and Battery Charging.

B.1.14 Other Publications.

-

"DC Arc Hazard Assessment Phase II," Copyright Material, Kinectrics Inc., Report No. K-012623-RA-0002-R00.

Limits of Approach

This informative annex is not a part of the requirements of this NFPA document but is included for informational purposes only.

Informative Annex C provides information to illustrate approach boundaries. The information is intended to provide suggestions regarding workers' safe approach to each limit. The approach limits trigger the need for greater control of work performed inside that approach limit.

C.1 Preparation for Approach.

Observing a safe approach distance from exposed energized electrical conductors or circuit parts is an effective means of maintaining electrical safety. As the distance between a person and the exposed energized conductors or circuit parts decreases, the potential for electrical accident increases.

C.1.1 Unqualified Persons, Safe Approach Distance. Unqualified persons are safe when they maintain a distance from the exposed energized conductors or circuit parts, including the longest conductive object being handled, so that they cannot contact or enter a specified air insulation distance to the exposed energized electrical conductors or circuit parts. This safe approach distance is the limited approach boundary. Further, persons must not cross the arc flash boundary unless they are wearing appropriate personal protective clothing and are under the close supervision of a qualified person. Only when continuously escorted by a qualified person should an unqualified person cross the limited approach boundary. Under no circumstance should an unqualified person cross the restricted approach boundary, where special shock protection techniques and equipment are required.

According to 130.3(A), safety-related work practices must be used to safeguard employees from injury when the risk of exposure to electrical hazards or potential electrical hazards is unacceptable, and the work practices must be consistent with the nature and extent of the hazard. These work practices are used to protect employees from the risk associated with the four conditions of electrical hazards: arc flash, arc blast, thermal burn, and electrical shock. Section 110.2(D)(1)(a) indicates that a person may be judged to be qualified with regard to certain equipment and methods but still unqualified for others.

Even where under the close supervision of, or while continuously escorted by, a qualified person, a person considered unqualified to perform work must not be allowed to cross the arc flash boundary without first receiving the specific safety-related training to understand the hazard(s) involved and the appropriate use of the necessary personal protective equipment (PPE). Crossing the arc flash boundary could be seen as similar to an unqualified person crossing the limited approach boundary. See Section 130.4(C)(3).

C.1.2 Qualified Persons, Safe Approach Distance.

C.1.2.1 Determine the arc flash boundary and, if the boundary is to be crossed, appropriate arc-rated protective equipment must be utilized.

Qualified person(s) should do the following before crossing the arc flash boundary:

1. Perform an arc flash risk assessment.
2. Have a documented plan justifying the need to work within the arc flash boundary where the risk is considered unacceptable.
3. Have the plan and the risk assessment approved by authorized management.
4. Use PPE that is appropriate for the hazard and is rated for the incident energy level involved or the arc flash PPE category determined.

C.1.2.2 For a person to cross the limited approach boundary and enter the limited space, a person should meet the following criteria:

(1) Be qualified to perform the job/task
(2) Be able to identify the hazards and associated risks with the tasks to be performed

C.1.2.3 To cross the restricted approach boundary and enter the restricted space, qualified persons should meet the following criteria:

(1) Have an energized electrical work permit authorized by management
(2) Use personal protective equipment (PPE) that is rated for the voltage and energy level involved

(3) Minimize the likelihood of bodily contact with exposed energized conductors and circuit parts from inadvertent movement by keeping as much of the body out of the restricted space as possible and using only protected body parts in the space as necessary to accomplish the work
(4) Use insulated tools and equipment

(See Figure C.1.2.3.)

C.2 Basis for Distance Values in Tables 130.4(D)(a) and 130.4(D)(b).

Section C.2 illustrates how the information contained in Table 130.4(D)(a) is derived.

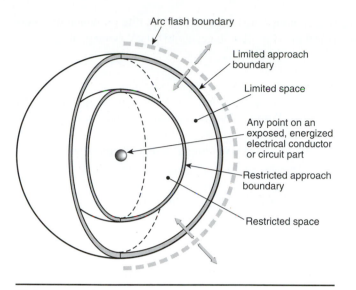

FIGURE C.1.2.3 Limits of Approach.

C.2.1 General Statement. Columns 2 through 4 of Table 130.4(D)(a) and Table 130.4(D)(b) show various distances from the exposed energized electrical conductors or circuit parts. They include dimensions that are added to a basic minimum air insulation distance. Those basic minimum air insulation distances for voltages 72.5 kV and under are based on IEEE 4, *Standard Techniques for High Voltage Testing,* Appendix 2B; and voltages over 72.5 kV are based on IEEE 516, *Guide for Maintenance Methods on Energized Power Lines.* The minimum air insulation distances that are required to avoid flashover are as follows:

 (1) ≤300 V: 1 mm (0 ft 0.03 in.)
 (2) >300 V to ≤750 V: 2 mm (0 ft 0.07 in.)
 (3) >750 V to ≤2 kV: 5 mm (0 ft 0.19 in.)
 (4) >2 kV to ≤15 kV: 39 mm (0 ft 1.5 in.)
 (5) >15 kV to ≤36 kV: 161 mm (0 ft 6.3 in.)
 (6) >36 kV to ≤48.3 kV: 254 mm (0 ft 10.0 in.)
 (7) >48.3 kV to ≤72.5 kV: 381 mm (1 ft 3.0 in.)
 (8) >72.5 kV to ≤121 kV: 640 mm (2 ft 1.2 in.)
 (9) >138 kV to ≤145 kV: 778 mm (2 ft 6.6 in.)
(10) >161 kV to ≤169 kV: 915 mm (3 ft 0.0 in.)
(11) >230 kV to ≤242 kV: 1.281 m (4 ft 2.4 in.)
(12) >345 kV to ≤362 kV: 2.282 m (7 ft 5.8 in.)
(13) >500 kV to ≤550 kV: 3.112 m (10 ft 2.5 in.)
(14) >765 kV to ≤800 kV: 4.225 m (13 ft 10.3 in.)

C.2.1.1 Column 1. The voltage ranges have been selected to group voltages that require similar approach distances based on the sum of the electrical withstand distance and an inadvertent movement factor. The value of the upper limit for a range is the maximum voltage for the highest nominal voltage in the range, based on ANSI C84.1, *Electric Power Systems and Equipment— Voltage Ratings (60 Hz).* For single-phase systems, select the range that is equal to the system's maximum phase-to-ground voltage multiplied by 1.732.

C.2.1.2 Column 2. The distances in column 2 are based on OSHA's rule for unqualified persons to maintain a 3.05 m (10 ft) clearance for all voltages up to 50 kV (voltage-to-ground), plus 100 mm (4.0 in.) for each 10 kV over 50 kV.

C.2.1.3 Column 3. The distances in column 3 are based on the following:

(1) ≤750 V: Use *NEC* Table 110.26(A)(1), Working Spaces, Condition 2, for the 151 V to 600 V range.
(2) >750 V to ≤145 kV: Use *NEC* Table 110.34(A), Working Space, Condition 2.
(3) >145 kV: Use OSHA's 3.05 m (10 ft) rules as used in Column 2.

C.2.1.4 Column 4. The distances in column 4 are based on adding to the flashover dimensions shown in C.2.1 the following inadvertent movement distance:

≤300 V: Avoid contact.

Based on experience and precautions for household 120/240-V systems:

>300 V to ≤750 V: Add 304.8 mm (1 ft 0 in.) for inadvertent movement.

These values have been found to be adequate over years of use in ANSI/IEEE C2, *National Electrical Safety Code,* in the approach distances for communication workers.

>72.5 kV: Add 304.8 mm (1 ft 0 in.) for inadvertent movement.

These values have been found to be adequate over years of use in ANSI/IEEE C2, *National Electrical Safety Code,* in the approach distances for supply workers.

•

Incident Energy and Arc Flash Boundary Calculation Methods

Summary of Annex D Changes

D.1: Rearranged Table D.1 according to the calculation method. Replaced *Ralph Lee paper* with *Lee, "The Other Electrical Hazard: Electrical Arc Flash Burns"*; replaced *Doughty/Neal paper* with *Doughty, et al., "Predicting Incident Energy to Better Manage the Electrical Arc Hazard on 600 V Power Distribution Systems"*; replaced *IEEE Std. 1584* with *IEEE 1584, Guide for Performing Arc Flash Calculations;* and replaced *ANSI/IEEE C2 NESC, Section 410, Table 410-1 and Table 410-2* with *Doan, "Arc Flash Calculations for Exposure to DC Systems."*

D.2, D.3, D.4, D.5: Sections D.2 through D.8 were renumbered and renamed to span from D.2 through D.5. Redundant text deleted to provide clarity and consistency with other safety standards that address hazard, risk, and risk assessment.

D.5.3: Subsection added to assist in calculating short circuit currents for stationary battery systems.

This informative annex is not a part of the requirements of this NFPA document but is included for informational purposes only.

Informative Annex D illustrates how the arc flash boundary and incident energy might be calculated. These examples are not intended to limit the choice of calculation methods; the method chosen should be applicable to the situation. All the publicly known methods of calculating the arc flash incident energy and arc flash boundary produce results that are estimates of the actual values. The thermal hazard associated with an arcing fault is very complex, with many variable attributes having an impact on the calculation. NFPA and IEEE (Institute of Electrical and Electronic Engineers) are sponsoring a joint endeavor, the IEEE/NFPA Arc Flash Phenomena Collaborative Research Project. The intent of the project is to produce the data necessary to further understand arc flash phenomena, which will help improve worker safety. The results of this project have not yet been issued but could have a significant impact on hazard assessment methods used today.

Workers who might be exposed to an arcing fault must wear arc-rated clothing or use other equipment to avoid a thermal injury. Protecting a worker from the thermal effects of an arcing fault does not necessarily protect the worker from injury. An arcing fault exhibits characteristics of other hazards. For instance, the arc generates a significant pressure wave. A worker could be injured by the pressure differential developed between the outside and inside of the worker's body. The calculations illustrated in this informative annex do not offer protection from the effects of any pressure wave.

If the calculations illustrated in this informative annex (or from any other method) indicate that incident energy is 40 cal/cm² or greater, the pressure wave might be hazardous.

D.1 Introduction.

Informative Annex D summarizes calculation methods available for calculating arc flash boundary and incident energy. It is important to investigate the limitations of any methods to be used. The limitations of methods summarized in Informative Annex D are described in Table D.1.

Table D.1 identifies the source for and limitations of various methods of performing calculations to determine arc flash boundary and incident energy. The table identifies which section of Informative Annex D covers each calculation method.

Table D.1   *Limitation of Calculation Methods*

Section	Source	Limitations/Parameters
D.2	Lee, "The Other Electrical Hazard: Electrical Arc Flash Burns"	Calculates incident energy and arc flash boundary for arc in open air; conservative over 600 V and becomes more conservative as voltage increases
D.3	Doughty, et al., "Predicting Incident Energy to Better Manage the Electrical Arc Hazard on 600 V Power Distribution Systems"	Calculates incident energy for three-phase arc on systems rated 600 V and below; applies to short-circuit currents between 16 kA and 50 kA
D.4	IEEE *1584, Guide for Performing Arc Flash Calculations*	Calculates incident energy and arc flash boundary for: 208 V to 15 kV; three-phase; 50 Hz to 60 Hz; 700 A to 106,000 A short-circuit current; and 13 mm to 152 mm conductor gaps
D.5	Doan, "Arc Flash Calculations for Exposure to DC Systems"	Calculates incident energy for dc systems rated up to 1000 V dc

D.2 Ralph Lee Calculation Method.

D.2.1 Basic Equations for Calculating Arc Flash Boundary Distances. The short-circuit symmetrical ampacity, I_{sc}, from a bolted three-phase fault at the transformer terminals is calculated with the following formula:

$$I_{sc} = \left\{ \left[MVA\,\text{Base} \times 10^6 \right] \div \left[1.732 \times V \right] \right\} \times \left\{ 100 \div \%Z \right\} \qquad \textbf{[D.2.1(a)]}$$

where I_{sc} is in amperes, V is in volts, and $\%Z$ is based on the transformer *MVA*.

A typical value for the maximum power, P (in MW) in a three-phase arc can be calculated using the following formula:

$$P = \left[\text{maximum bolted fault, in } MVA_{bf}\right] \times 0.707^2 \qquad \textbf{[D.2.1(b)]}$$

$$P = 1.732 \times V \times I_{sc} \times 10^{-6} \times 0.707^2 \qquad \textbf{[D.2.1(c)]}$$

The arc flash boundary distance is calculated in accordance with the following formulae:

$$D_c = \left[2.65 \times MVA_{bf} \times t\right]^{\frac{1}{2}} \qquad \textbf{[D.2.1(d)]}$$

$$D_c = \left[53 \times MVA \times t\right]^{\frac{1}{2}} \qquad \textbf{[D.2.1(e)]}$$

where:

D_c	=	distance in feet of person from arc source for a just curable burn (that is, skin temperature remains less than 80°C).
MVA_{bf}	=	bolted fault MVA at point involved.
MVA	=	MVA rating of transformer. For transformers with MVA ratings below 0.75 MVA, multiply the transformer MVA rating by 1.25.
t	=	time of arc exposure in seconds.

The clearing time for a current-limiting fuse is approximately ¼ cycle or 0.004 second if the arcing fault current is in the fuse's current-limiting range. The clearing time of a 5-kV and 15-kV circuit breaker is approximately 0.1 second or 6 cycles if the instantaneous function is installed and operating. This can be broken down as follows: actual breaker time (approximately 2 cycles), plus relay operating time of approximately 1.74 cycles, plus an additional safety margin of 2 cycles, giving a total time of approximately 6 cycles. Additional time must be added if a time delay function is installed and operating.

The formulas used in this explanation are from Ralph Lee, "The Other Electrical Hazard: Electrical Arc Flash Burns," in *IEEE Trans. Industrial Applications*. The calculations are based on the worst-case arc impedance. *(See Table D.2.1.)*

Table D.2.1 *Flash Burn Hazard at Various Levels in a Large Petrochemical Plant*

(1)	*(2)*	*(3)*	*(4)*	*(5)*	*(6)*	*(7)*	
						*Arc Flash Boundary Typical Distance**	
Bus Nominal Voltage Levels	*System (MVA)*	*Transformer (MVA)*	*System or Transformer (% Z)*	*Short-Circuit Symmetrical (A)*	*Clearing Time of Fault (cycles)*	*SI*	*U.S.*
230 kV	9000		1.11	23,000	6.0	15 m	49.2 ft
13.8 kV	750		9.4	31,300	6.0	1.16 m	3.8 ft
Load side of all 13.8-V fuses	750		9.4	31,300	1.0	184 mm	0.61 ft
4.16 kV		10.0	5.5	25,000	6.0	2.96 m	9.7 ft
4.16 kV		5.0	5.5	12,600	6.0	1.4 m	4.6 ft
Line side of incoming 600-V fuse		2.5	5.5	44,000	60.0–120.0	7 m–11 m	23 ft–36 ft
600-V bus		2.5	5.5	44,000	0.25	268 mm	0.9 ft
600-V bus		1.5	5.5	26,000	6.0	1.6 m	5.4 ft
600-V bus		1.0	5.57	17,000	6.0	1.2 m	4 ft

*Distance from an open arc to limit skin damage to a curable second degree skin burn [less than 80°C (176°F) on skin] in free air.

D.2.2 Single-Line Diagram of a Typical Petrochemical Complex. The single-line diagram *(see Figure D.2.2)* illustrates the complexity of a distribution system in a typical petrochemical plant.

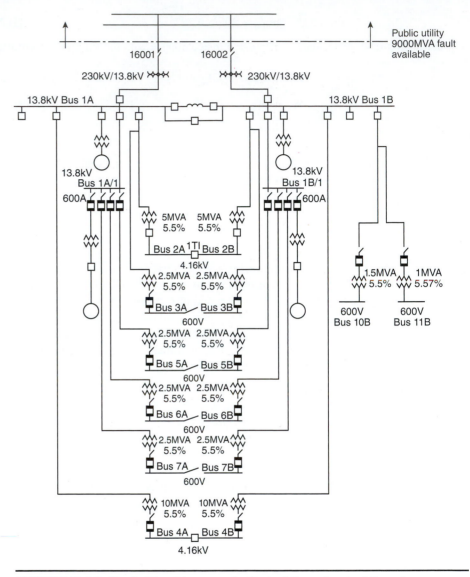

FIGURE D.2.2 *Single-Line Diagram of a Typical Petrochemical Complex.*

D.2.3 Sample Calculation. Many of the electrical characteristics of the systems and equipment are provided in Table D.2.1. The sample calculation is made on the 4160-volt bus 4A or 4B. Table D.2.1 tabulates the results of calculating the arc flash boundary for each part of the system. For this calculation, based on Table D.2.1, the following results are obtained:

(1) Calculation is made on a 4160-volt bus.
(2) Transformer *MVA* (and base *MVA*) = 10 *MVA*.
(3) Transformer impedance on 10 *MVA* base = 5.5 percent.
(4) Circuit breaker clearing time = 6 cycles.

Using Equation D.2.1(a), calculate the short-circuit current:

$$I_{sc} = \left\{ \left[MVA\ \text{Base} \times 10^6 \right] \div \left[1.732 \times V \right] \right\} \times \left\{ 100 \div \%Z \right\}$$
$$= \left\{ \left[10 \times 10^6 \right] \div \left[1.732 \times 4160 \right] \right\} \times \left\{ 100 \div 5.5 \right\}$$
$$= 25,000\ \text{amperes}$$

Using Equation D.2.1(b), calculate the power in the arc:

$$P = 1.732 \times 4160 \times 25,000 \times 10^{-6} \times 0.707^2$$
$$= 91\ \text{MW}$$

Using Equation D.2.1(d), calculate the second degree burn distance:

$$D_c = \left\{ 2.65 \times \left[1.732 \times 25,000 \times 4160 \times 10^{-6} \right] \times 0.1 \right\}^{\frac{1}{2}}$$
$$= 6.9\ \text{or}\ 7.00\ \text{ft}$$

Or, using Equation D.2.1(e), calculate the second degree burn distance using an alternative method:

$$D_c = \left[53 \times 10 \times 0.1 \right]^{\frac{1}{2}}$$
$$= 7.28\ \text{ft}$$

D.2.4 Calculation of Incident Energy Exposure Greater Than 600 V for an Arc Flash Hazard Analysis. The equation that follows can be used to predict the incident energy produced by a three-phase arc in open air on systems rated above 600 V. The parameters required to make the calculations follow.

(1) The maximum bolted fault, three-phase short-circuit current available at the equipment.
(2) The total protective device clearing time (upstream of the prospective arc location) at the maximum short-circuit current. If the total protective device clearing time is longer than 2 seconds, consider how long a person is likely to remain in the location of the arc flash. It is likely that a person exposed to an arc flash will move away quickly if it is physically possible, and 2 seconds is a reasonable maximum time for calculations. A person in a bucket truck or a person who has crawled into equipment will need more time to move away. Sound engineering judgment must be used in applying the 2-second maximum clearing time, since there could be circumstances where an employee's egress is inhibited.

REACTION TIME

Reaction time is a complicated subject. Often standard numbers are used without a good comprehension of where the numbers derive from, how they were acquired, or the variables that shaped them. A reaction time of 1.5 seconds is commonly quoted by experts in regard to automobile accidents. In regard to arc flash incidents, 2 seconds is a reasonable maximum time to use in calculating the arc flash incident energy, provided the worker's egress is not inhibited. If the total clearing time of the upstream overcurrent protective device is greater than 2 seconds or egress is restricted, then additional time may be needed to exit the arc flash boundary.

Sound engineering judgment must be used in determining if the 2-second exposure time is applicable. Reaction time can be broken down into different components or categories, such as perception, decision, and motor response times, each having fairly dissimilar properties.

MENTAL PROCESSING TIME

Mental processing time is the period of time a worker takes to recognize that an event has happened (perception time) and to decide upon a response (decision time). For example, it might be the time it takes a worker to realize that an arc flash event is underway and decide what action to take. Perception and decision time can be further broken down into subcategories in order to better understand the hazard response.

Perception time can be broken down into the following two subcategories:

- *Sensation time.* The time it takes to detect the sensory input from an event is the sensation time. The greater the signal intensity, the better the visibility and the faster the reaction time. Since reaction times can be faster for acoustic signals than for visual signals, the sound wave associated with an arc blast might lead to a faster response time.
- *Recognition time.* The time needed to understand the meaning of the event is the recognition time. In some cases, an extremely fast automatic response may kick in; while in others, a controlled response, which can take substantial time, may happen. Therefore, training can help to decrease the response time.

Decision time can be broken down into the following two subcategories:

- *Situational awareness time.* The time it takes to recognize and interpret the event, extract its meaning, and possibly extrapolate it into the future is the situational awareness time. Again, practice or rehearsal might decrease response time.
- *Response selection time.* The time needed to determine which reaction to make and to mentally encode the movement is the response selection time. Generally, response selection time slows when more than one event is perceived or when more than one response is possible (that is, when choice is involved). As with other categories here, practice or rehearsal might decrease response time.

The following additional factors might also have an impact on response time to an event:

- *Movement time.* A number of factors affect movement time, such as the number of exit passageways, length of the passageway, and any obstacles that might be in the way. As a general rule, the greater complexity of the movement, the longer it takes; yet practice or rehearsal in general can lower movement times. An emotional stimulus can accelerate gross motor movements but inhibit fine detail movement.
- *Expectation.* If an event is expected, practiced for, and rehearsed, it may be possible to get the reaction time down to around 1.4 seconds when the movement time is around 0.7 seconds. Reaction time for an unexpected event could be around 1.75 seconds, including 0.7 seconds of travel time. Extra time is needed to interpret and decide on an appropriate response where the event is a complete surprise. In this case a best estimate might be around 2.0 seconds, including 1.2 seconds for perception and decision, and 0.8 seconds for movement. It may be necessary to study the situation to determine the appropriate reaction time.
- *Other factors.* Factors such as perceived urgency, the complexity of the task, whether the employee is holding tools, age, and gender can affect the overall response time to an event. The more complex a worker's task when an event happens, the longer the anticipated response time.

MINIMIZING EXPOSURE

Overcurrent protective device operating time depends on whether the device is current limiting or not. Depending on the available fault current, an overcurrent protective device might be expected to extinguish an arcing current in anywhere from ¼ cycle to 30 cycles. Workers cannot outrun an arc flash event or minimize their exposure time for such short intervals, but they might be able to minimize their time inside the arc flash boundary when extended tripping times are involved.

Where the tripping time of the overcurrent protective device is longer than the reaction time of the employee, the employee may be able to increase his or her distance from the arcing event or exit the arc flash boundary. Doubling the distance between the source of an arc flash event and the employee decreases the incident energy level by around four (it is an inverse square type of relationship).

If the overcurrent protective device takes longer than 2 seconds to trip in response to an arcing current, consideration should be made into how long a person is likely to remain within the arc flash boundary. The physical layout needs to be considered where exposure time is used to calculate incident energy level. Anticipated exposure time might have to be increased where a narrow aisle is the only means of egress. Some other factors that should be considered are whether it is possible for an employee to get stuck in a position and not escape within the anticipated exposure time; whether the employee has studied the exit routes; whether the event is in an enclosed area or an outside switchyard; and whether the employee is in a bucket, aerial lift, or on a ladder. Field staff should evaluate and verify that, in the event of arc flash, they can exit the arc flash boundary within the exposure time assumed. This discussion demonstrates the basic principles and thought process involved when using exposure time as a basis for calculating incident energy. Each type of reaction time needs to be studied, because each type of reaction has its own eccentricities.

(3) The distance from the arc source.

(4) Rated phase-to-phase voltage of the system.

$$E = \frac{793 \times F \times V \times t_A}{D^2} \qquad \text{[D.2.4(4)]}$$

where:

E = incident energy, cal/cm^2
F = bolted fault short-circuit current, kA
V = system phase-to-phase voltage, kV
t_A = arc duration, sec
D = distance from the arc source, in.

D.3 Doughty Neal Paper.

D.3.1 Calculation of Incident Energy Exposure. The following equations can be used to predict the incident energy produced by a three-phase arc on systems rated 600 V and below. The results of these equations might not represent the worst case in all situations. It is essential that the equations be used only within the limitations indicated in the definitions of the variables shown under the equations. The equations must be used only under qualified engineering supervision.

Informational Note: Experimental testing continues to be performed to validate existing incident energy calculations and to determine new formulas.

The parameters required to make the calculations follow.

(1) The maximum bolted fault, three-phase short-circuit current available at the equipment and the minimum fault level at which the arc will self-sustain. (Calculations should be made using the maximum value, and then at lowest fault level at which the arc is self-sustaining. For 480-volt systems, the industry accepted minimum level for a sustaining arcing fault is 38 percent of the available bolted fault, three-phase short-circuit current. The highest incident energy exposure could occur at these lower levels where the overcurrent device could take seconds or minutes to open.)

(2) The total protective device clearing time (upstream of the prospective arc location) at the maximum short-circuit current, and at the minimum fault level at which the arc will sustain itself.

(3) The distance of the worker from the prospective arc for the task to be performed.

Typical working distances used for incident energy calculations are as follows:

(1) Low voltage (600 V and below) MCC and panelboards — 455 mm (18 in.)
(2) Low voltage (600 V and below) switchgear — 610 mm (24 in.)
(3) Medium voltage (above 600 V) switchgear — 910 mm (36 in.)

D.3.2 Arc in Open Air. The estimated incident energy for an arc in open air is as follows:

$$E_{MA} = 5271 D_A^{-1.9593} t_A \begin{bmatrix} 0.0016F^2 \\ -0.0076F \\ +0.8938 \end{bmatrix} \qquad \text{[D.3.2(a)]}$$

where:

E_{MA} = maximum open arc incident energy, cal/cm^2
D_A = distance from arc electrodes, in. (for distances 18 in. and greater)
t_A = arc duration, sec
F = short-circuit current, kA (for the range of 16 kA to 50 kA)

Sample Calculation: Using Equation D.3.2(a), calculate the maximum open arc incident energy, cal/cm^2, where D_A = 18 in., t_A = 0.2 second, and F = 20 kA.

$$
\begin{aligned}
E_{MA} &= 5271 D_A^{-1.9593} t_A [0.0016F^2 - 0.0076F + 0.8938] \\
&= 5271 \times .0035 \times 0.2 [0.0016 \times 400 - 0.0076 \times 20 + 0.8938] \\
&= 3.69 \times [1.381] \\
&= 21.33 \text{ J/ cm}^2 \left(5.098 \text{ cal/cm}^2\right) \qquad \text{[D.3.2(b)]}
\end{aligned}
$$

D.3.3 Arc in a Cubic Box. The estimated incident energy for an arc in a cubic box (20 in. on each side, open on one end) is given in the equation that follows. This equation is applicable to arc flashes emanating from within switchgear, motor control centers, or other electrical equipment enclosures.

$$E_{MB} = 1038.7 D_B^{-1.4738} t_A \begin{bmatrix} 0.0093F^2 \\ -0.3453F \\ +5.9675 \end{bmatrix} \qquad \text{[D.3.3(a)]}$$

where:

E_{MB} = maximum 20 in. cubic box incident energy, cal/cm^2
D_B = distance from arc electrodes, in. (for distances 18 in. and greater)
t_A = arc duration, sec
F = short-circuit current, kA (for the range of 16 kA to 50 kA)

Sample Calculation: Using Equation D.3.3(a), calculate the maximum 20 in. cubic box incident energy, cal/cm^2, using the following:

(1) $D_B = 18$ in.
(2) $t_A = 0.2$ sec
(3) $F = 20$ kA

$$E_{MB} = 1038.7 D_B^{-1.4738} t_A \left[0.0093 F^2 - 0.3453 F + 5.9675 \right]$$
$$= 1038 \times 0.0141 \times 0.2 \left[0.0093 \times 400 - 0.3453 \times 20 + 5.9675 \right]$$
$$= 2.928 \times \left[2.7815 \right]$$
$$= 34.1 \text{ J/cm}^2 \left(8.144 \text{ cal/cm}^2 \right) \qquad \text{[D.3.3(b)]}$$

D.3.4 Reference. The equations for this section were derived in the IEEE paper by R. L. Doughty, T. E. Neal, and H. L. Floyd, II, "Predicting Incident Energy to Better Manage the Electric Arc Hazard on 600 V Power Distribution Systems."

.

D.4 IEEE 1584 Calculation Method.

D.4.1 Basic Equations for Calculating Incident Energy and Arc Flash Boundary. This section provides excerpts from IEEE 1584, *IEEE Guide for Performing Arc Flash Hazard Calculations,* for estimating incident energy and arc flash boundaries based on statistical analysis and curve fitting of available test data. An IEEE working group produced the data from tests it performed to produce models of incident energy.

The complete data, including a spreadsheet calculator to solve the equations, can be found in the IEEE 1584, *Guide for Performing Arc Flash Hazard Calculations.* Users are encouraged to consult the latest version of the complete document to understand the basis, limitation, rationale, and other pertinent information for proper application of the standard. It can be ordered from the Institute of Electrical and Electronics Engineers, Inc., 445 Hoes Lane, P.O. Box 1331, Piscataway, NJ 08855-1331.

D.4.1.1 System Limits. An equation for calculating incident energy can be empirically derived using statistical analysis of raw data along with a curve-fitting algorithm. It can be used for systems with the following limits:

(1) 0.208 kV to 15 kV, three-phase
(2) 50 Hz to 60 Hz
(3) 700 A to 106,000 A available short-circuit current
(4) 13 mm to 152 mm conductor gaps

For three-phase systems in open-air substations, open-air transmission systems, and distribution systems, a theoretically derived model is available. This theoretically derived model is intended for use with applications where faults escalate to three-phase faults. Where such an escalation is not possible or likely, or where single-phase systems are encountered, this equation will likely provide conservative results.

The results of the IEEE/NFPA Arc Flash Phenomena Collaborative Research Project have not yet been issued, but they could have a significant impact on hazard assessment methods used today. Since IEEE periodically revises its documents, a prudent action for the *NFPA 70E* user would be to consult the latest adopted version or amendment of IEEE 1584, *IEEE Guide for Performing Arc Flash Hazard Calculations.*

D.4.2 Arcing Current. To determine the operating time for protective devices, find the predicted three-phase arcing current.

For applications with a system voltage under 1 kV, solve Equation D.4.2(a) as follows:

$$\lg I_a = K + 0.662 \lg I_{bf} + 0.0966V$$
$$+ 0.000526G + 0.5588V\left(\lg I_{bf}\right)$$
$$- 0.00304G\left(\lg I_{bf}\right) \qquad \textbf{[D.4.2(a)]}$$

where:

$\lg$ = the $\log_{10}$
I_a = arcing current, kA
K = −0.153 for open air arcs; −0.097 for arcs-in-a-box
I_{bf} = bolted three-phase available short-circuit current (symmetrical rms), kA
V = system voltage, kV
G = conductor gap, mm (*see Table D.4.2*)

For systems greater than or equal to 1 kV, use Equation D.4.2(b):

$$\lg I_a = 0.00402 + 0.983 \lg I_{bf} \qquad \textbf{[D.4.2(b)]}$$

This higher voltage formula is used for both open-air arcs and for arcs-in-a-box. Convert from lg:

$$I_a = 10^{\lg I_a} \qquad \textbf{[D.4.2(c)]}$$

Use 0.85 I_a to find a second arc duration. This second arc duration accounts for variations in the arcing current and the time for the overcurrent device to open. Calculate the incident energy using both arc durations (I_a and 0.85 I_a), and use the higher incident energy.

Table D.4.2 *Factors for Equipment and Voltage Classes*

System Voltage (kV)	Type of Equipment	Typical Conductor Gap (mm)	Distance Exponent Factor x
0.208–1	Open air	10–40	2.000
	Switchgear	32	1.473
	MCCs and panels	25	1.641
	Cables	13	2.000
>1–5	Open air	102	2.000
	Switchgear	13–102	0.973
	Cables	13	2.000
>5–15	Open air	13–153	2.000
	Switchgear	153	0.973
	Cables	13	2.000

D.4.3 Incident Energy at Working Distance — Empirically Derived Equation. To determine the incident energy using the empirically derived equation, determine the $\log 10$ of the normalized incident energy. The following equation is based on data normalized for an arc time of 0.2 second and a distance from the possible arc point to the person of 610 mm:

$$\lg E_n = k_1 + k_2 + 1.081 \lg l_a + 0.0011 G \qquad \text{[D.4.3(a)]}$$

where:

E_n = incident energy, normalized for time and distance, J/cm²

k_1 = −0.792 for open air arcs

   = −0.555 for arcs-in-a-box

k_2 = 0 for ungrounded and high-resistance grounded systems

   = −0.113 for grounded systems

G = conductor gap, mm (see *Table D.4.2*)

Then,

$$E_n = 10^{\lg E_n} \qquad \text{[D.4.3(b)]}$$

Converting from normalized:

$$E = 4.184 C_f E_n \left(\frac{t}{0.2}\right)\left(\frac{610^x}{D^x}\right) \qquad \text{[D.4.3(c)]}$$

where:

E = incident energy, J/cm²

C_f = calculation factor

   = 1.0 for voltages above 1 kV.

   = 1.5 for voltages at or below 1 kV.

E_n = incident energy normalized.

t = arcing time, sec.

x = distance exponent from Table D.4.2.

D = distance, mm, from the arc to the person (working distance). See Table D.4.3 for typical working distances.

Table D.4.3 Typical Working Distances

Classes of Equipment	Typical Working Distance* (mm)
15-kV switchgear	910
5-kV switchgear	910
Low-voltage switchgear	610
Low-voltage MCCs and panelboards	455
Cable	455
Other	To be determined in field

* Typical working distance is the sum of the distance between the worker and the front of the equipment and the distance from the front of the equipment to the potential arc source inside the equipment.

If the arcing time, t, in Equation D.4.3(c) is longer than 2 seconds, consider how long a person is likely to remain in the location of the arc flash. It is likely that a person exposed to an arc flash will move away quickly if it is physically possible, and 2 seconds is a reasonable maximum time for calculations. Sound engineering judgment should be used in applying the 2-second maximum clearing time, because there could be circumstances where an employee's egress is inhibited. For example, a person in a bucket truck or a person who has crawled into equipment will need more time to move away.

Refer to the commentary that follows D.2.4(2) regarding maximum clearing time for use in calculations when an employee's egress path is not inhibited.

When using the empirical or theoretical method to determine arc flash incident energy, a calculation factor (C_f) is used in equations D.4.3(c) and D.4.5(a). The calculation factors (1.0 or 1.5) were chosen to give a 95 percent numeral confidence level, based on the use of the following personal protective equipment (PPE) levels — 1.2 cal/cm², 8 cal/cm², 25 cal/cm², 40 cal/cm², and 100 cal/cm². See IEEE 1584.

D.4.4 Incident Energy at Working Distance — Theoretical Equation. The following theoretically derived equation can be applied in cases where the voltage is over 15 kV or the gap is outside the range:

$$E = 2.142 \times 10^6 VI_{bf}\left(\frac{t}{D^2}\right)$$ **[D.4.4]**

where:

E = incident energy, J/cm²
V = system voltage, kV
I_{bf} = available three-phase bolted fault current
t = arcing time, sec
D = distance (mm) from the arc to the person (working distance)

For voltages over 15 kV, arcing fault current and bolted fault current are considered equal.

D.4.5 Arc Flash Boundary. The arc flash boundary is the distance at which a person is likely to receive a second degree burn. The onset of a second degree burn is assumed to be when the skin receives 5.0 J/cm² of incident energy.

For the empirically derived equation,

$$D_B = \left[4.184 C_f E_n \left(\frac{t}{0.2}\right)\left(\frac{610^x}{E_B}\right)\right]^{\frac{1}{x}}$$ **[D.4.5(a)]**

For the theoretically derived equation,

$$D_B = \sqrt{2.142 \times 10^6 VI_{bf}\left(\frac{t}{E_B}\right)}$$ **[D.4.5(b)]**

where:

D_B = distance (mm) of the arc flash boundary from the arcing point
C_f = calculation factor
   = 1.0 for voltages above 1 kV
   = 1.5 for voltages at or below 1 kV
E_n = incident energy normalized
t = time, sec

x =   distance exponent from Table D.4.2

E_B =   incident energy in J/cm$_2$ at the distance of the arc flash boundary

V =   system voltage, kV

I_{bf} =   bolted three-phase available short-circuit current

> Informational Note: These equations could be used to determine whether selected personal protective equipment (PPE) is adequate to prevent thermal injury at a specified distance in the event of an arc flash.

D.4.6 Current-Limiting Fuses. The formulas in this section were developed for calculating arc flash energies for use with current-limiting Class L and Class RK1 fuses. The testing was done at 600 V and at a distance of 455 mm, using commercially available fuses from one manufacturer. The following variables are noted:

I_{bf} = available three-phase bolted fault current (symmetrical rms), kA

E = incident energy, J/cm^2

(A) Class L Fuses 1601 A through 2000 A. Where $I_{bf}<$ 22.6 kA, calculate the arcing current using Equation D.4.2(a), and use time-current curves to determine the incident energy using Equations D.4.3(a), D.4.3(b), and D.4.3(c).
     Where 22.6 kA $\leq I_{bf} \leq$ 65.9 kA,

$$E = 4.184\left(-0.1284I_{bf} + 32.262\right) \qquad \textbf{[D.4.6(a)]}$$

Where 65.9 kA $< I_{bf} \leq$ 106 kA,

$$E = 4.184\left(-0.5177I_{bf} + 57.917\right) \qquad \textbf{[D.4.6(b)]}$$

Where $I_{bf}>$ 106 kA, contact the manufacturer.

(B) Class L Fuses 1201 A through 1600 A. Where $I_{bf}<$ 15.7 kA, calculate the arcing current using Equation D.4.2(a), and use time-current curves to determine the incident energy using Equations D.4.3(a), D.4.3(b), and D.4.3(c).
     Where 15.7 kA $\leq I_{bf} \leq$ 31.8 kA,

$$E = 4.184\left(-0.1863I_{bf} + 27.926\right) \qquad \textbf{[D.4.6(c)]}$$

Where 44.1 kA $\leq I_{bf} \leq$ 65.9 kA,

$$E = 12.3 \text{ J/ cm}^2 \left(2.94 \text{ cal/cm}^2\right) \qquad \textbf{[D.4.6(e)]}$$

Where 65.9 kA $< I_{bf} \leq$ 106 kA,

$$E = 4.184\left(-0.0631I_{bf} + 7.0878\right) \qquad \textbf{[D.4.6(f)]}$$

Where $I_{bf}>$ 106 kA, contact the manufacturer.

(C) Class L Fuses 801 A through 1200 A. Where $I_{bf}<$ 15.7 kA, calculate the arcing current using Equation D.4.2(a), and use time-current curves to determine the incident energy per Equations D.4.3(a), D.4.3(b), and D.4.3(c).
     Where 15.7 kA $\leq I_{bf} \leq$ 22.6 kA,

$$E = 4.184\left(-0.1928I_{bf} + 14.226\right) \qquad \textbf{[D.4.6(g)]}$$

Where $22.6 \text{ kA} < I_{bf} \le 44.1 \text{ kA}$,

$$E = 4.184 \left(0.0143 I_{bf}^{2} - 1.3919 I_{bf} + 34.045 \right) \qquad \textbf{[D.4.6(h)]}$$

Where $44.1 \text{ kA} < I_{bf} \le 106 \text{ kA}$,

$$E = 1.63 \qquad \textbf{[D.4.6(i)]}$$

Where $I_{bf} > 106 \text{ kA}$, contact the manufacturer.

(D) Class L Fuses 601 A through 800 A. Where $I_{bf} < 15.7 \text{ kA}$, calculate the arcing current using Equation D.4.2(a), and use time-current curves to determine the incident energy using Equations D.4.3(a), D.4.3(b), and D.4.3(c).

Where $15.7 \text{ kA} \le I_{bf} \le 44.1 \text{ kA}$,

$$E = 4.184 \left(-0.0601 I_{bf} + 2.8992 \right) \qquad \textbf{[D.4.6(j)]}$$

Where $44.1 \text{ kA} < I_{bf} \le 106 \text{ kA}$,

$$E = 1.046 \qquad \textbf{[D.4.6(k)]}$$

Where $I_{bf} > 106 \text{ kA}$, contact the manufacturer.

(E) Class RK1 Fuses 401 A through 600 A. Where $I_{bf} < 8.5 \text{ kA}$, calculate the arcing current using Equation D.4.2(a), and use time-current curves to determine the incident energy using Equations D.4.3(a), D.4.3(b), and D.4.3(c).

Where $8.5 \text{ kA} \le I_{bf} \le 14 \text{ kA}$,

$$E = 4.184 \left(-3.0545 I_{bf} + 43.364 \right) \qquad \textbf{[D.4.6(l)]}$$

Where $14 \text{ kA} < I_{bf} \le 15.7 \text{ kA}$,

$$E = 2.510 \qquad \textbf{[D.4.6(m)]}$$

Where $15.7 \text{ kA} < I_{bf} \le 22.6 \text{ kA}$,

$$E = 4.184 \left(-0.0507 I_{bf} + 1.3964 \right) \qquad \textbf{[D.4.6(n)]}$$

Where $22.6 \text{ kA} < I_{bf} \le 106 \text{ kA}$,

$$E = 1.046 \qquad \textbf{[D.4.6(o)]}$$

Where $I_{bf} > 106 \text{ kA}$, contact the manufacturer.

(F) Class RK1 Fuses 201 A through 400 A. Where $I_{bf} < 3.16 \text{ kA}$, calculate the arcing current using Equation D.4.2(a), and use time-current curves to determine the incident energy using Equations D.4.3(a), D.4.3(b), and D.4.3(c).

Where $3.16 \text{ kA} \le I_{bf} \le 5.04 \text{ kA}$,

$$E = 4.184 \left(-19.053 I_{bf} + 96.808 \right) \qquad \textbf{[D.4.6(p)]}$$

Where $5.04 \text{ kA} < I_{bf} \le 22.6 \text{ kA}$,

$$E = 4.184 \left(-0.0302 I_{bf} + 0.9321 \right) \qquad \textbf{[D.4.6(q)]}$$

Where $22.6 \text{ kA} < I_{bf} \le 106 \text{ kA}$,

$$E = 1.046 \qquad \textbf{[D.4.6(r)]}$$

Where $I_{bf} > 106$ kA, contact the manufacturer.

(G) Class RK1 Fuses 101 A through 200 A. Where $I_{bf} < 1.16$ kA, calculate the arcing current using Equation D.4.2(a), and use time-current curves to determine the incident energy using Equations D.4.3(a), D.4.3(b), and D.4.3(c).

Where 1.16 kA $\leq I_{bf} \leq 1.6$ kA,

$$E = 4.184\left(-18.409I_{bf} + 36.355\right)$$ **[D.4.6(s)]**

Where 1.6 kA $< I_{bf} \leq 3.16$ kA,

$$E = 4.184\left(-4.2628I_{bf} + 13.721\right)$$ **[D.4.6(t)]**

Where 3.16 kA $< I_{bf} \leq 106$ kA,

$$E = 1.046$$ **[D.4.6(u)]**

Where $I_{bf} > 106$ kA, contact the manufacturer.

(H) Class RK1 Fuses 1 A through 100 A. Where $I_{bf} < 0.65$ kA, calculate the arcing current using Equation D.4.2(a), and use time-current curves to determine the incident energy using Equations D.4.3(a), D.4.3(b), and D.4.3(c).

Where 0.65 kA $\leq I_{bf} \leq 1.16$ kA,

$$E = 4.184\left(-11.176I_{bf} + 13.565\right)$$ **[D.4.6(v)]**

Where 1.16 kA $< I_{bf} \leq 1.4$ kA,

$$E = 4.184\left(-1.4583I_{bf} + 2.2917\right)$$ **[D.4.6(w)]**

Where 1.4 kA $< I_{bf} \leq 106$ kA,

$$E = 1.046$$ **[D.4.6(x)]**

Where $I_{bf} > 106$ kA, contact the manufacturer.

D.4.7 Low-Voltage Circuit Breakers. The equations in Table D.4.7 can be used for systems with low-voltage circuit breakers. The results of the equations will determine the incident energy and arc flash boundary when I_{bf} is within the range as described. Time-current curves for the circuit breaker are not necessary within the appropriate range.

When the bolted fault current is below the range indicated, calculate the arcing current using Equation D.4.2(a), and use time-current curves to determine the incident energy using Equations D.4.3(a), D.4.3(b), and D.4.3(c).

The range of available three-phase bolted fault currents is from 700 A to 106,000 A. Each equation is applicable for the following range:

$$I_1 < I_{bf} < I_2$$

Table D.4.7 *Incident Energy and Arc Flash Protection Boundary by Circuit Breaker Type and Rating*

Rating (A)	Breaker Type	Trip Unit Type	480 V and Lower		575 V–600 V	
			Incident Energy (J/cm²)[a]	Arc Flash Boundary (mm)[a]	Incident Energy (J/cm²)[a]	Arc Flash Boundary (mm)[a]
100–400	MCCB	TM or M	$0.189\,I_{bf} + 0.548$	$9.16\,I_{bf} + 194$	$0.271\,I_{bf} + 0.180$	$11.8\,I_{bf} + 196$
600–1200	MCCB	TM or M	$0.223\,I_{bf} + 1.590$	$8.45\,I_{bf} + 364$	$0.335\,I_{bf} + 0.380$	$11.4\,I_{bf} + 369$
600–1200	MCCB	E, LI	$0.377\,I_{bf} + 1.360$	$12.50\,I_{bf} + 428$	$0.468\,I_{bf} + 4.600$	$14.3\,I_{bf} + 568$
1600–6000	MCCB or ICCB	TM or E, LI	$0.448\,I_{bf} + 3.000$	$11.10\,I_{bf} + 696$	$0.686\,I_{bf} + 0.165$	$16.7\,I_{bf} + 606$
800–6300	LVPCB	E, LI	$0.636\,I_{bf} + 3.670$	$14.50\,I_{bf} + 786$	$0.958\,I_{bf} + 0.292$	$19.1\,I_{bf} + 864$
800–6300	LVPCB	E, LS[b]	$4.560\,I_{bf} + 27.230$	$47.20\,I_{bf} + 2660$	$6.860\,I_{bf} + 2.170$	$62.4\,I_{bf} + 2930$

MCCB: Molded-case circuit breaker.

TM: Thermal-magnetic trip units.

M: Magnetic (instantaneous only) trip units.

E: Electronic trip units have three characteristics that may be used separately or in combination: L: Long time, S: Short time, I: Instantaneous.

ICCB: Insulated-case circuit breaker.

LVPCB: Low-voltage power circuit breaker.

[a] I_{bf} is in kA; working distance is 455 mm (18 in.).

[b] Short-time delay is assumed to be set at maximum.

where:

I_1 = minimum available three-phase, bolted, short-circuit current at which this method can be applied. I_1 is the lowest available three-phase, bolted, short-circuit current level that causes enough arcing current for instantaneous tripping to occur, or, for circuit breakers with no instantaneous trip, that causes short-time tripping to occur.

I_2 = interrupting rating of the circuit breaker at the voltage of interest.

To find I_1, the instantaneous trip (I_t) of the circuit breaker must be found. I_t can be determined from the time-current curve, or it can be assumed to be 10 times the rating of the circuit breaker for circuit breakers rated above 100 amperes. For circuit breakers rated 100 amperes and below, a value of $I_t = 1300$ A can be used. When short-time delay is utilized, I_t is the short-time pickup current.

The corresponding bolted fault current, I_{bf}, is found by solving the equation for arc current for box configurations by substituting I_t for arcing current. The 1.3 factor in Equation D.4.7(b) adjusts current to the top of the tripping band.

$$\lg(1.3I_t) = 0.084 + 0.096V + 0.586\left(\lg I_{bf}\right) + 0.559V\left(\lg I_{bf}\right) \qquad \textbf{[D.4.7(a)]}$$

At 600 V,

$$\lg I_1 = 0.0281 + 1.09\,\lg(1.3I_t) \qquad \textbf{[D.4.7(b)]}$$

At 480 V and lower,

$$\lg I_1 = 0.0407 + 1.17\,\lg(1.3I_t) \qquad \textbf{[D.4.7(c)]}$$

$$I_{bf} = I_1 = 10^{\lg I_1} \qquad \textbf{[D.4.7(d)]}$$

D.4.8 References. The complete data, including a spreadsheet calculator to solve the equations, can be found in IEEE 1584, *Guide for Performing Arc Flash Hazard Calculations*. IEEE publications are available from the Institute of Electrical and Electronics Engineers, 445 Hoes Lane, P.O. Box 1331, Piscataway, NJ 08855-1331, USA (http://standards.ieee.org/).

D.5 Direct-Current Incident Energy Calculations.

•

D.5.1 Maximum Power Method. The following method of estimating dc arc flash incident energy that follows was published in the *IEEE Transactions on Industry Applications* (*see reference 2, which follows*). This method is based on the concept that the maximum power possible in a dc arc will occur when the arcing voltage is one-half the system voltage. Testing completed for Bruce Power (*see reference 3, which follows*) has shown that this calculation is conservatively high in estimating the arc flash value. This method applies to dc systems rated up to 1000 V.

$$I_{arc} = 0.5 \times I_{bf}$$

$$IE_m = 0.01 \times V_{sys} \times I_{arc} \times T_{arc} / D^2$$

where:

I_{arc} = arcing current amperes

I_{bf} = system bolted fault current amperes

IE_m = estimated dc arc flash incident energy at the maximum power point cal/cm$_2$

V_{sys} = system voltage volts

T_{arc} = arcing time sec

D = working distance cm

For exposures where the arc is in a box or enclosure, it would be prudent to use a multiplying factor of 3 for the resulting incident energy value.

This maximum power method to calculate the arc flash incident energy for a direct current power source uses principles similar to those used by Ralph Lee in his paper, "The Other Electrical Hazard: Electrical Arc Blast Burns," in *IEEE Transactions on Industrial Applications*, Volume 1A-18, Issue 3, page 246, 1987.

Without having a full appreciation of the intricacies involved, it is possible to misapply the maximum power method equations. Key to the use of the method is the appropriate determination of arcing time. Under transient conditions, such as at the initiation of an arcing fault, the current in a dc circuit does not change instantaneously. The rate of change is dependent on the L/R time constant of the circuit. The longer the time constant, the longer the circuit will take to reach its new steady state value. The rms current value during this period of time is dependent on the time constant. For devices that trip in response to rms current values, the L/R time constant can affect the total clearing time of the overcurrent protective device used to protect the circuit. The manufacturer of the overcurrent protective device may need to be contacted to obtain an appropriate time–current characteristic curve.

D.5.2 Detailed Arcing Current and Energy Calculations Method. A thorough theoretical review of dc arcing current and energy was published in the *IEEE Transactions on Industry Applications.* Readers are advised to refer to that paper (*see reference 1*) for those detailed calculations.

References:

1. "DC-Arc Models and Incident-Energy Calculations," Ammerman, R.F.; et al.; *IEEE Transactions on Industry Applications*, Vol. 46, No. 5.

2. "Arc Flash Calculations for Exposures to DC Systems," Doan, D.R., *IEEE Transactions on Industry Applications,* Vol. 46, No. 6.

3. "DC Arc Hazard Assessment Phase II", Copyright Material, Kinectrics Inc., Report No. K-012623-RA-0002-R00.

D.5.3 Short Circuit Current. The determination of short circuit current is necessary in order to use Table 130.7(C)(15)(B). The arcing current is calculated at 50 percent of the dc short-circuit value. The current that a battery will deliver depends on the total impedance of the short-circuit path. A conservative approach in determining the short-circuit current that the battery will deliver at 25°C is to assume that the maximum available short-circuit current is 10 times the 1 minute ampere rating (to 1.75 volts per cell at 25°C and the specific gravity of 1.215) of the battery. A more accurate value for the short-circuit current for the specific application can be obtained from the battery manufacturer.

Reference

1. IEEE 946, *Recommended Practice for the Design of DC Auxiliary Powers Systems for Generating Stations.*

Electrical Safety Program

Summary of Annex E Changes

Informative Annex E: Updated to correlate with the redefined hazard and risk terminology for consistency with the definitions of *hazard, risk,* and *risk assessment,* and with risk management principles (hazards are identified; risk is assessed and controlled by following a hierarchy of risk control methods). Modifications were made in E.2 to provide a more logical sequence of safety program controls.

This informative annex is not a part of the requirements of this NFPA document but is included for informational purposes only.

Informative Annex E provides information that could be used as the groundwork for an electrical safety program.

(See 110.1, Electrical Safety Program.)

E.1 Typical Electrical Safety Program Principles.

Electrical safety program principles include, but are not limited to, the following:

(1) Inspecting and evaluating the electrical equipment
(2) Maintaining the electrical equipment's insulation and enclosure integrity
(3) Planning every job and document first-time procedures
(4) De-energizing, if possible *(see 120.1)*
(5) Anticipating unexpected events
(6) Identifying the electrical hazards and reduce the associated risk
(7) Protecting employees from shock, burn, blast, and other hazards due to the working environment
(8) Using the right tools for the job
(9) Assessing people's abilities
(10) Auditing the principles

E.2 Typical Electrical Safety Program Controls.

Electrical safety program controls can include, but are not limited to, the following:

(1) The employer develops programs, including training, and the employees apply them.

.

(2) Employees are to be trained to be qualified for working in an environment influenced by the presence of electrical energy.

(3) Procedures are to be used to identify the electrical hazards and to develop plans to eliminate those hazards or to control the associated risk for those hazards that cannot be eliminated.

(4) Every electrical conductor or circuit part is considered energized until proved otherwise.

(5) De-energizing an electrical conductor or circuit part and making it safe to work on is, in itself, a potentially hazardous task.

(6) Tasks to be performed on or near exposed energized electrical conductors and circuit parts are to be identified and categorized.

(7) Precautions appropriate to the working environment are to be determined and taken.

(8) A logical approach is to be used to determine the associated risk of each task.

E.3 Typical Electrical Safety Program Procedures.

Electrical safety program procedures can include, but are not limited to determination and assessment of the following:

(1) Purpose of task

(2) Qualifications and number of employees to be involved

(3) Identification of hazards and assessment of risks of the task

(4) Limits of approach

(5) Safe work practices to be used

(6) Personal protective equipment (PPE) involved

(7) Insulating materials and tools involved

(8) Special precautionary techniques

(9) Electrical single-line diagrams

(10) Equipment details

(11) Sketches or photographs of unique features

(12) Reference data

Risk Assessment Procedure

Summary of Annex F Changes

Replaced the term *probability* with *likelihood* to provide clarity and promote consistent use of the term. (*Likelihood* can refer to the chance of something happening and includes a probability or a frequency over a given time period.)

Changed the title of Informative Annex F from *Hazard Analysis, Risk Estimation, and Risk Evaluation* to *Risk Assessment Procedure*. (Risk assessment includes hazard analysis, risk estimation, and risk evaluation.)

Replaced the term *harm* with the phrase *injury or damage to health*.

This informative annex is not a part of the requirements of this NFPA document but is included for informational purposes only.

This informative annex deals with risk assessment, which is composed of risk estimation and risk evaluation. The procedures identified involve an iterative process of risk reduction until an acceptable risk level is attained. The hierarchy of health and safety controls, as detailed in the commentary to Informative Annex P and in this informative annex, is applied from the highest level to the lowest level of controls in order to reduce the risk to an acceptable level for the task at hand. This hierarchy provides an orderly method for establishing the most effective or feasible way to reduce the risk associated with a hazard. The highest level of control is elimination, and the lowest level of control before an event happens is personal protective equipment (PPE). PPE is the last line of defense in protecting an employee from an electrical hazard before an incident happens. After an event happens, mitigation can be used to minimize the extent of the injury, but by this time most of the damage is already done.

Electrical safety issues (not health issues) are the focus of this standard. However, from an overall safety point of view, both health and safety issues are important. Electrical safety issues need to be assessed and prioritized. Electrical safety issues include the potential electrical hazards of electrical shock, arc flash, arc blast, and burns from hot electrical equipment associated with a particular task, risks associated with the hazards, electrical safety management system deficiencies, the opportunities for improvement, and the appropriate PPE necessary for the assigned task.

F.1 Risk Assessment (General).

This informative annex provides guidance regarding a qualitative approach for risk assessment, including risk estimation and risk evaluation, which can be helpful in determining the protective measures that are required to reduce the likelihood of injury or damage to health occurring in the circumstances under consideration. To receive the full benefit of completing the risk assessment process the relationships between the source or cause of risk and the effects of the hierarchy of controls on those causes must be understood. This informative annex is intended to provide guidance.

Risk assessment is an analytical process consisting of a number of discrete steps intended to ensure that hazards are properly identified and analyzed with regard to their severity and the likelihood of their occurrence. Once hazards have been identified and analyzed, the risk associated with those hazards can be estimated using the parameters outlined in F.2.1. Appropriate protective measures can then be implemented and evaluated in order to determine if adequate risk reduction has been achieved.

Risk assessment includes a comprehensive review of the hazards, the associated foreseeable tasks, and the protective measures that are required in order to maintain a tolerable level of risk, including the following:

(1) Identifying and analyzing electrical hazards
(2) Identifying tasks to be performed
(3) Documenting hazards associated with each task
(4) Estimating the risk for each hazard/task pair
(5) Determining the appropriate protective measures needed to adequately reduce the level of risk

Figure F.1(a) illustrates the steps to be taken and the decisions to be considered when performing an electrical work risk assessment. See 110.1 for a hazard and risk evaluation procedure. Figure F.1(b) illustrates in more detail the steps of the risk analysis, assessment, and evaluation process.

In a general sense, risk can be described as the potential that a chosen action or activity (including inaction) will lead to some type of loss or injury. For the purpose of electrical safety, risk is defined as a combination of the likelihood of occurrence of injury or damage to health and the severity of injury or damage to health that results from a hazard. The type of injury of concern is damage to health (either direct or indirect) that can be caused by contact with exposed energized conductors and circuit parts, and the injury (either direct or indirect) that can be caused when an arc flash occurs.

Risk assessment is a step in a risk management procedure that involves risk estimation and risk evaluation. Risk assessment is the determination of a quantitative or a qualitative value of risk related to an actual situation or task involving a recognized hazard. Of the two basic methods of risk assessment — the quantitative approach and the qualitative approach — this informative annex provides guidance only on the qualitative approach.

F.1.1 Responsibility. Electrical system designers, constructors, and users have responsibilities for defining and achieving tolerable risk. The supplier and the user either separately or jointly identify hazards, estimate risks, and reduce risks to a tolerable level within the scope of their respective work activities. Although the responsibilities of the supplier and the user differ over the life cycle of the electrical equipment, each entity should use the risk assessment process.

In general, the electrical system supplier is responsible for the design, construction, and information for operation and maintenance of the electrical system, while the user is responsible for the operation and maintenance of the electrical system.

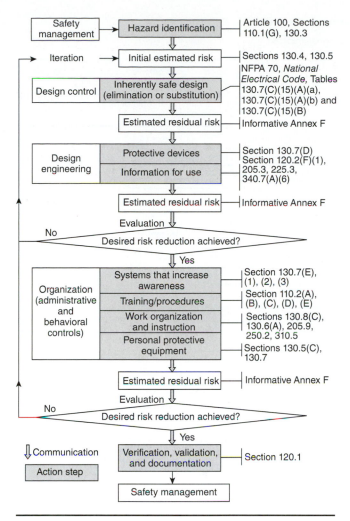

FIGURE F.1(a) *Risk Assessment Process.*

Suppliers and users should involve qualified personnel in meeting their respective responsibilities. The supplier and the user should ensure compliance with the regulations and standards applicable to their work activity. This could include regulations and standards for a specific location, a specific application, or both.

Suppliers and users of electrical systems are responsible for attaining tolerable risk levels for the anticipated tasks to be performed. Engineers (designers) and contractors (constructors) are specifically identified as being under the supplier category, while electrical equipment suppliers and manufacturers are implied to be in the supplier category.

F.2 Risk Assessment.

Several methods are available for qualitatively determining risk, such as risk assessment matrices, risk ranking, or risk scoring systems. The application of a risk matrix is a useful method in many situations. The more levels in a matrix, the more complicated the matrix is to use. However, a matrix with too few levels may not be suitable for the task. The method identified in Section F.2 can be considered a risk scoring method and may be used as the basis to form a risk register. Even though the method used in Section F.2 uses parameter estimates (numbers), it is not a quantitative method of analysis because it does not use measureable, objective data to determine likelihood of loss and associated risk.

Risk Assessment Process

Note: Italicized text represents information used during the risk assessment process.

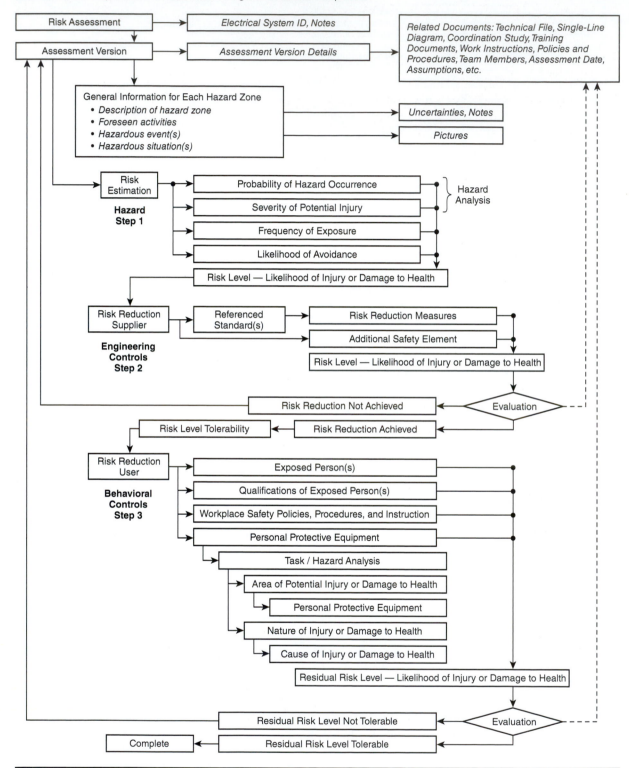

FIGURE F.1(b) *Detailed Risk Assessment Process.*

F.2.1 Initial Risk Estimation. An initial estimation of risk should be carried out for each hazard. Risk related to the identified hazard should be derived by using the risk parameters that are shown in Figure F.2.1 including the following:

(1) Severity of injury or damage to health (Se)
(2) Likelihood of occurrence of that injury or damage to health, which is a function of all of the following:

    a. Frequency and duration of the exposure of persons to the hazard (Fr)
    b. Likelihood of occurrence of a hazardous event (Pr)
    c. Likelihood of avoiding or limiting the injury or damage to health (Av)

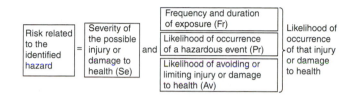

FIGURE F.2.1   *Elements of Risk.*

F.2.2 Parameters Used in Risk Estimation. In preparation for the risk assessment, parameters are estimated and can be entered into Table F.2.5. These parameters should be based on worst-case considerations for the electrical system. Different risk reduction strategies can be implemented for each hazard. The risk estimation stage is the only one at which hazards can be eliminated, thus avoiding the need for additional protective measures, such as safeguarding or complementary protective measures.

The risk related to an identified hazard may be thought of as being composed of the severity of the possible damage to health (severity of the injury) and the likelihood of occurrence of that injury. These components can be looked at and evaluated separately. The user can then determine which risk method to use, addition or multiplication. The commentary on this section shows one example in which likelihood and severity are multiplied. The user could just as easily have added the components.

In the model shown in this informative annex, the severity of the injury (Se) is determined by the possible injury to the affected workers from Table F.2.3. Three independent factors or parameters are to be estimated independently and then added together to determine the likelihood of the occurrence of that harm as part of a risk register. These independent factors are as follows:

1. The frequency and duration of the exposure (Fr)
2. The likelihood of occurrence of the hazardous event (Pr)
3. The likelihood of avoiding or limiting the injury (Av)

In determining Fr, the average interval between exposures should be able to be estimated, and therefore the average frequency of access to the hazard can be determined. The duration of the hazard may be less than or more than 10 minutes, and the user of this standard may want to create columns for such categories instead of using the values given for all durations.

Table F.2.4.1 is used to determine the value of Fr. The value of Pr is selected from Table F.2.4.2, and the value of Av is determined from Table F.2.4.3. The values of the four parameters (Se, Fr, Pr, and Av) are then entered into Table F.2.5 in their respective columns and can be used as the basis for developing a risk register.

From this, there are three possible ways to evaluate the values of the parameters:

• One way is to look at the values separately.
• A second way is to perform an algebraic summation:

$$R = Se + Fr + Pr + Av$$

• The third way is to sum up the individual values for likelihood of occurrence of the injury:

$$Po = Fr + Pr + Av$$

Then multiply that value by the severity of the injury (Se), so that the risk is determined as a product:

$$R = Se \times Po$$

The risk number(s) determined would then have to be compared with a predetermined risk number value(s), above which additional safety measures would be required — such as training, procedures, and PPE. This third method is suitable for use in a risk register.

F.2.3 Severity of the Possible Injury or Damage to Health (Se). Severity of injuries or damage to health can be estimated by taking into account reversible injuries, irreversible injuries, and death. Typically, the types of hazards to be considered include, but are not limited to, shock and electrocution, burns, and impact. Choose the appropriate value of severity from Table F.2.3, based on the consequences of an injury, as follows:

(1) 8: a fatal or a significant irreversible injury, such that it will be very difficult to continue the same work after healing, if at all

(2) 6: a major or irreversible injury, in such a way that it can be possible to continue the same work after healing and can also include a severe major but reversible injury such as broken limbs

(3) 3: a reversible injury, including severe lacerations, stabbing, and severe bruises, that requires attention from a medical practitioner

(4) 1: a minor injury, including scratches and minor bruises that require attention by first aid.

Select the appropriate value for severity of the possible injury or damage to health (Se) from Table F.2.3 and insert in the Se column in Table F.2.5.

TABLE F.2.3   Severity of the Possible Injury or Damage to Health (Se) Classification

Severity of Injury or Damage to Health	Se Value
Irreversible — trauma, death	8
Permanent — skeletal damage, blindness, hearing loss, third degree burns	6
Reversible — minor impact, hearing damage, second degree burns	3
Reversible — minor laceration, bruises, first degree burns	1

F.2.4 Likelihood of Occurrence of Injury or Damage to Health. Each of the three parameters of likelihood of occurrence of injury or damage to health (that is, Fr, Pr, and Av) should be estimated independently of each other. A worst-case assumption needs to be used for each parameter to ensure that the protective measures, determined during risk evaluation, will provide adequate risk reduction. Generally, the use of a form of hazard/task–based evaluation is strongly recommended to ensure that proper consideration is given to the estimation of the likelihood of occurrence of injury or damage to health.

F.2.4.1 Frequency and Duration of Exposure (Fr). The following aspects should be considered to determine the level of exposure:

(1) Need for access to the hazard zone based on all modes of use; for example, normal operation and maintenance

(2) Nature of access, for example, examination, repair, and troubleshooting

It should then be possible to estimate the average interval between exposures and, thus, the average frequency of access.

This factor does not include consideration of the failure of the short-circuit interruption device(s) or the failure to use the appropriate PPE.

Select the appropriate row for frequency and duration of exposure (Fr) from Table F.2.4.1. Insert the appropriate number under the Fr column in Table F.2.5.

TABLE F.2.4.1 Frequency and Duration of Exposure (Fr) Classification

Frequency of Exposure	Fr Value (for Duration >10 min)
≤ 1 per hour	5
> 1 per hour to ≤ 1 per day	5
> 1 per day to ≤ 1 every 2 weeks	4
> 1 every 2 weeks to ≤ 1 per year	3
> 1 per year	2

F.2.4.2 Likelihood of Occurrence of a Hazardous Event (Pr). The occurrence of a hazardous event influences the likelihood of the occurrence of injury or damage to health. The possibility of the hazardous event occurring should describe the likelihood of the event materializing during the use or foreseeable misuse, or both, of the electrical system or process. Subjectivity may have a substantial impact on the result of the risk assessment. The use of subjective information should be minimized as far as reasonably practicable.

The likelihood of occurrence of the hazardous event should be estimated independently of other related parameters (Fr and Av) and will typically be based on the results of the completed study of the arc flash potential. The worst-case scenario should be used for this parameter to ensure that short-circuit interruption device(s) have, where practicable, been properly selected and installed and will provide adequate protection.

Elements of the electrical system that are intended to ensure an intrinsically safe design shall be taken into consideration in the determination of the likelihood of the hazardous event(s). These can include, but are not limited to, the mechanical structure, electrical devices, and electronic controls integral to the system, the process, or both at the time of the analysis. Types of components that could contribute to an inherently safe design include, but are not limited to, current-limiting devices and ground-fault circuit interrupters.

This parameter can be estimated by taking into account the following factors:

(1) The predictability of the performance of component parts of the electrical system relevant to the hazard in different modes of use (e.g., normal operation, maintenance, fault finding).

At this point in the risk assessment process, the protective effect of any personal protective equipment (PPE) and other protective measures should not be taken into account. This is necessary in order to estimate the amount of risk that will be present if the PPE and other protective measures are not in place at the time of the exposure. In general terms, it must be considered whether the electrical system being assessed has the propensity to act in an unexpected manner. The electrical system performance will vary from very predictable to not predictable. Unexpected events cannot be discounted until it can be clearly demonstrated that the electrical system will perform as expected.

> Informational Note: Predictability is often linked to the complexity of the electrical system and the characteristics of the energy supply.

(2) The specified or foreseeable characteristics of human behavior with regard to interaction with the component parts of the machine relevant to the hazard, which can be characterized by one or both of the following:

    a.   Stress (e.g., due to time constraints, work task, perceived damage limitation)

    b.   Lack of awareness of information relevant to the hazard

Human behavior will be influenced by factors such as skills, training, experience, and complexity of the machine or the process.

These attributes are not usually directly under the influence of the electrical system designer, but a task analysis will reveal activities in which total awareness of all issues, including unexpected outcomes, cannot be reasonably assumed. "Very high" likelihood of occurrence of a hazardous event should be selected to reflect normal workplace constraints and worst-case considerations. Positive reasons (e.g., well-defined application and a high level of user competence) are required for any lower values to be used.

Any required or assumed skills, knowledge, and so forth, should be stated in the information for use.

Select the appropriate row for likelihood of occurrence of a hazardous event (Pr) from Table F.2.4.2.

Indicate the appropriate risk level under the Pr column in Table F.2.5.

TABLE F.2.4.2 *Likelihood of a Hazardous Event (Pr) Classification*

Likelihood of a Hazardous Event	*Pr Value*
Very high	5
Likely	4
Possible	3
Rare	2
Negligible	1

When determining the likelihood of the occurrence of a hazardous event (Pr), it might be helpful to ask the following types of questions. These questions are an indication of the types of questions that might be considered. They are not intended to be a complete or accurate list of the actual questions that should be asked or investigations undertaken. The following are general questions to determine the likelihood of an event (risk):

1.   Has the equipment been installed in accordance with *NFPA 70®, National Electrical Code® (NEC®)*, or other appropriate installation standard?

2.   Has the equipment been maintained and tested in accordance with the manufacturer's instructions, NFPA 70B, *Recommended Practice for Electrical Equipment Maintenance* (2013), or ANSI/NETA MTS, *Standard for Maintenance Testing Specifications* (2011)?

3. At what point in its rated life is the equipment?

4. Have all connections been verified to be appropriately tightened (torqued) in accordance with the manufacturer's requirements or appropriate industry standard?

5. Is there any visual indication of overheating?

6. Is any component, device, or equipment loose?

7. Has any component, device, or equipment been damaged?

8. Has a circuit breaker (CB) calibration sampling program been instituted for the facility?

The following are enclosure questions:

9. Have all openings in the enclosure been closed in accordance with the manufacturer's instructions, the *NEC®*, or other appropriate installation standard?

10. Do all enclosure doors operate and latch properly?

11. Does the enclosure have all of its bolts and screws installed?

12. Does the equipment or enclosure have ventilation openings?

13. Is the enclosure arc rated?

14. Are there openings in the enclosure that rodents or other vermin could enter?

15. Was the right type of enclosure used for the installation?

16. Is there an indication of moisture in the equipment?

17. Has the enclosure been examined for dust, dirt, soot, or grease?

18. Is there any indication of overheating of the bus work, etc., in the enclosure, such as discoloration?

The following are circuit breaker (CB) condition questions:

19. Has the CB periodically been operated in accordance with the manufacturer's instructions or in accordance with standard(s) requirements, so as to wipe the mechanical mechanisms to ensure that the mechanical mechanism has not seized?

20. Has the CB been applied within its marked rating?

21. Has the right type of CB been used?

22. Have the proper conductor types and sizes been used to connect to the CB?

23. Has the CB interrupted repeated high-fault currents?

24. Has the CB been checked for burn marks?

25. Has the CB operating temperature been checked under normal use conditions?

26. Have the CB surfaces been examined for dust, dirt, soot, grease, or moisture? If any was found, have the CB surfaces been appropriately cleaned?

27. Has the CB been examined for cracks?

28. Have all electrical connections to the CB been checked to be certain that they are clean and secure?

29. Is there any indication of discoloration of the CB's molded case, discoloration or flaking of external metal parts, or melting or blistering of adjacent wire insulation?

30. Is there any evidence of overheating or melting of the arc chute vent or area surrounding the vents?

31. Is there evidence of overheating or case blistering?

32. If the CB has interchangeable trip units, have the trip units been visually checked for overheating or looseness?

33. Have mechanical operation tests been performed on the CB and proper contact operation verified?

34. Have insulation resistance and/or individual pole resistance (millivolt drop) tests been performed on the CB?

35. Have inverse-time and/or instantaneous overcurrent trip tests been conducted on the CB?
36. Has a rated hold-in test of the CB been conducted?
37. Have any accessory devices involved with the CB been tested?
38. What is the ampere rating of the CB involved?

F.2.4.3 Likelihood of Avoiding or Limiting Injury or Damage to Health (Av). This parameter can be estimated by taking into account aspects of the electrical system design and its intended application that can help to avoid or limit the injury or damage to health from a hazard, including the following examples:

(1) Sudden or gradual appearance of the hazardous event; for example, an explosion caused by high fault values under short-circuit conditions.
(2) Spatial possibility to withdraw from the hazard.
(3) Nature of the component or system; for example, the use of touch-safe components, which reduce the likelihood of contact with energized parts. Working in close proximity to high voltage can increase the likelihood of personnel being exposed to hazards due to approach to live parts.
(4) Likelihood of recognition of a hazard; for example, as an electrical hazard, a copper bar does not change its appearance, whether it is under voltage or not. To recognize the presence of the hazard, an instrument is needed to establish whether or not electrical equipment is energized; thus, both inadvertent and intentional contact need to be considered.

Select the appropriate row for likelihood of avoiding or limiting injury or damage to health (Av) from Table F.2.4.3.
Insert the appropriate value for risk level in the Av column in Table F.2.5.

TABLE F.2.4.3 Likelihood of Avoiding or Limiting Injury or Damage to Health (Av) Classification

Likelihood of Avoiding or Limiting Injury or Damage to Health	Av Value
Impossible	5
Rare	3
Probable	1

F.2.5 Risk Level and Likelihood of Injury or Damage to Health. Once the parameters for each hazard under consideration have been entered in Table F.2.5, the information can be used in the first step of the risk assessment process as outlined in Figure F.1(a).

TABLE F.2.5 Parameters for Determining Risk Levels and Likelihood of Injury or Damage to Health (See Figure F.2.1)

Zone No.	Hazard	Se	Fr	Pr	Av

COMMENTARY TABLE F.1 *Risk Register (Based on Table F.2.5)*

Scenario No.	Hazard	Severity Se (Table F.2.3)	Probability of Occurrence of Harm, Po = (Fr + Pr + Av)				Risk Score (R)
			Fr (Table F.2.4.1)	Pr (Table F.2.4.2)	Av (Table F.2.4.3)	Total	Se × Po

RISK REGISTER

In order to make the risk estimation table easier to use, a risk register based on Table F.2.5 is suggested. (See commentary Table F.1.)

A risk register is a tool that can be tailored for each use but should contain some of the following:

- In its heading, the name of the equipment involved and its mark number, or the applicable facility, the applicable project, or any other appropriate designation
- Information on who created the register and who reviewed it
- The date created, the date modified, and the date reviewed
- As individual column headings, a scenario number or a reference number for tracking
- The hazard or risk involved; the severity; the likelihood of occurrence of the injury, its constituent parts and the sum of the constituent parts; and the risk score, which can be obtained by multiplying the severity rating by the likelihood of occurrence rating
- Information about consequences, control measures, control scores, and a revised evaluation based on the corrective measures taken

When using a risk register, the risk score at which corrective action should be taken (i.e., how much risk is acceptable) must be determined.

RISK ASSESSMENT MATRICES

A matrix is an ordered presentation of data in a display of rows and columns that defines the cells of the array. A risk assessment matrix, suitable for use by people with little or no formal training in statistics, is a visual tool used to characterize risk. A risk assessment matrix is a simple table that groups risk based on severity and likelihood. It can be used to assess the need for remedial action, such as the use of PPE for a given task, and to prioritize safety issues.

The risk assessment matrix that appears in this informational annex displays the two factors that have to be considered when evaluating risk: likelihood of the occurrence of a hazardous event or exposure over a specific period of time and the severity of the injury that can result. Therefore, the likelihood will be the general row heading, and the severity will be the general column heading. The columns and rows must then be assigned values. The matrix will become cumbersome if too many columns and rows are used, and it will fail to impart the necessary information if too few are used.

In building the column headings for a matrix, the electrical hazards — shock, arc flash, and arc blast — are causes used to determine values of the severity of injury; depending on circumstances and conditions, the effects can range from minor injury to death. Practically speaking, the maximum number of categories for a useful matrix is five. Five categories for the severity of the injury will be used: catastrophic, critical, medium, minor, and slight.

For the rows, values will be assigned on a continuum of the likelihood of occurrence, from *certain to happen* to *extremely unlikely to happen*. Again, keeping to five categories,

COMMENTARY TABLE F.2 *Title and Risk Categories for Use in Risk Assessment Matrix*

Likelihood (Probability) of Occurrence	
Definite	Almost certain of happening
Likely	Can happen at any time
Occasional	Occurs sporadically, from time to time
Seldom	Remote possibility; could happen sometime; most likely will not happen
Unlikely	Rare and exceptional for all practical purposes; can assume it will not happen

Severity of Injury	
Catastrophic	Death or permanent total disability (PTD)
Critical	Permanent partial disability (PPD) or temporary total disability (TTD) 3 months or longer
Medium	Medical treatment and lost work injury (LWI)
Minor	Minor medical treatment possible
Slight	First aid or minor treatment

Risk	

Extreme (E)	Intolerable risk
	    Do not proceed
	    Immediately introduce further controls
	    Detailed action and plan required
	    Color code red

High (H)	Unsupportable risk
	    Review and introduce additional controls
	    Requires senior management attention
	    Color code orange

Moderate (M)	Tolerable risk
	    Incorporates some level of risk that is unlikely to occur
	    Specific management responsibility
	    Consider additional controls
	    Take remedial action at appropriate time
	    Color code yellow

Low (L)	Supportable risk
	    Monitor and maintain controls in place
	    Manage by routine
	    Procedures
	    Little or no impact
	    Color code green

the following headings for the likelihood of occurrence will be used: definite, likely, occasional, seldom, and unlikely.

The categories and their suggested meanings are contained in Commentary Table F.2.

Exhibit F.1 and Exhibit F.2 show two different forms for a risk assessment matrix that display the same general information in different manners. From left to right and top to bottom, the risk assessment matrices go from the least restrictive category to the most restrictive category (the matrix could have been arranged in opposite order). These risk assessment matrices are only illustrations of aids that can be useful in prioritizing and in determining remedial actions that should be considered and taken. There are many different risk assessment matrices available, such as ANSI/AIHA Z10, *American National Standard for Occupational Safety and Health Management Systems*, which displays a different presentation than either of the risk assessment matrices shown in Exhibit F.1 and Exhibit F.2.

Likelihood of occurrence in period	Severity of the injury (consequences)				
	Slight	Minor	Medium	Critical	Catastrophic
Unlikely	L	L	L	M	M
Seldom	L	L	M	M	H
Occasional	L	M	M	H	E
Likely	M	M	H	E	E
Definite	M	H	E	E	E

EXHIBIT F.1

Risk assessment matrix using letter designations.

Likelihood of occurrence in period	Severity of the injury (consequences)				
	Slight	Minor	Medium	Critical	Catastrophic
Unlikely	1	2	3	4	5
Seldom	2	4	6	8	10
Occasional	3	6	9	12	15
Likely	4	8	12	16	20
Definite	5	10	15	20	25

EXHIBIT F.2

Risk assessment matrix using risk numbers.

Notes:
1. Extreme equal 25 through 15.
2. High equals 12 through 9.
3. Moderate equals 8 through 4.
4. Low equals 3 through 1.

F.3 Risk Reduction.

Although it can never be entirely eliminated, risk can be significantly reduced through the application of the hierarchy of health and safety management controls, which is detailed in the commentary on Informative Annex P. Whenever the residual risk is unacceptable, additional safety measures must be taken to reduce the risk to an acceptable level. One of the higher levels of controls, engineering controls, and all three of the lower levels of controls — warnings, administrative controls, and PPE — are discussed in this section.

F.3.1 Protective Measures. Once the risk prior to the application of protective measures has been estimated, all practicable efforts must be made to reduce the risk of injury or damage to health. Careful consideration of failure modes is an important part of risk reduction. Care should be taken to ensure that both technical and behavioral failures, which could result in ineffective risk reduction, are taken into account during the risk reduction stage of the risk assessment.

Situations in which hazard elimination cannot be attained typically require a balanced approach in order to reduce the likelihood of injury or damage to health. For example, the effective control of access to an electrical system requires the use of barriers, awareness placards, safe operating instructions, qualification and training, and PPE personnel protective equipment as required by this standard, as well as initial and refresher or periodic training for all affected personnel in the area. Engineering controls alone are not sufficient to reduce the remaining risk to a tolerable level. Typically, all five areas of risk reduction must be implemented to achieve the desired result.

Consideration of all five of the items in F.3.1.1 through F.3.1.5 is required to establish an adequate risk reduction strategy.

Section F.3 identifies four areas of risk reduction, and F.3.1 stresses that a combination of controls may be the most effective method. For instance, where the higher order controls such as elimination or substitution do not reduce the risk to a satisfactory level, a combination of the lower order controls, such as warning and PPE, may be required, in addition to a higher level control to mitigate the risk.

The following is a list of some of the possible failure modes that could contribute to a shock or arc blast event:

1. Loose connections
2. Loose object in enclosure, such as a screw, bolt, tool, or component
3. Equipment, component, or part at the end of its rated life (i.e., deteriorated due to time, number of operations under certain conditions, and so forth)
4. Improperly grounded equipment
5. Vermin in the equipment
6. Defective component or part
7. Moisture or other contaminant in the equipment
8. Improper size or type of component

F.3.1.1 Engineering Controls. Engineering controls can have a substantial impact on risk. They should, where practicable, be considered and analyzed. Typically, engineering controls take the form of barriers and other safeguarding devices as described in *NFPA 70, National Electrical Code*; IEC 60204-1 ed 5.1 Consol. with am 1, *Safety of Machinery — Electrical Equipment of Machines — Part 1: General Requirements*; and NFPA 79, *Electrical Standard for Industrial Machinery*.

F.3.1.2 Awareness Devices. Awareness means can be used to complement the effects of engineering controls with regard to risk reduction. They should be chosen based on the design configuration for each specific application and their potential effectiveness during foreseen interaction. Each design and configuration can require unique awareness devices in order to have the desired impact on risk. Typically, awareness means take the form of signs, visual alarms, audible alarms, and so forth.

F.3.1.3 Procedures. Procedures and instructions that are required for individual(s) to safely interact with the electrical system should be identified. The procedures and instructions should include descriptions of the hazards, the possible hazardous events, hazardous situations, and the protective measures that need to be implemented. Procedures and instructions should also be used to communicate foreseeable misuse of the system that could contribute to an increased level of risk. Typically, formal procedures are provided in written form; however, in some cases, verbal instruction can be provided. Care should be taken in the latter case to ensure that the verbal instructions will have the desired impact on risk.

F.3.1.4 Training. Training, with regard to the proper interaction and for foreseeable inappropriate interaction with the electrical system, must be completed. The intent of the training is to ensure that all affected personnel are able to understand when and how hazardous situations can arise and how to best reduce the risk associated with those situations. Typically, training for individuals interacting with electrical systems will include technical information regarding hazards, hazardous situations, or both as well as information related to potential failure modes that could affect risk. This type of training generally will be provided by a trainer who has an in-depth understanding of electrical system design, as well as experience in the field of adult education. Less technical training content could be appropriate in situations in which only awareness of electrical hazards is needed to ensure that unqualified personnel do not interact with the electrical system.

F.3.1.5 Personal Protective Equipment (PPE). The electrical system must be analyzed in order to determine the appropriate PPE. Once the appropriate PPE has been determined, personnel must maintain and use it as required in order to ensure that residual risk remains at the desired level.

F.4 Risk Evaluation.

F.4.1 Risk Evaluation. Once the appropriate protective measures described in F.3.1 have been applied, the effect of those measures on the elements of risk *(see Figure F.2.1)* should be taken into account. Each type of protective measure could affect one or more of the elements that contribute to risk. The effects on risk or on the individual elements of risk, should be considered in the final risk estimation. The cumulative effect of the final combination of protective measures can then be used to estimate the residual risk. Paragraphs F.4.1.1 through F.4.1.5 provide a general, nonexhaustive outline that can be used as a guide to the final estimation of risk.

F.4.1.1 Design — Elimination or Substitution by Design.

    (a) Elimination of the hazard — impacts both severity of injury or damage to health and likelihood of injury or damage to health

Failure mode(s) examples:

(1) Component(s) failure
(2) Application of an incorrect construction or manufacturing specification
(3) Incorrect calculation (that is, potential energy, toxicity, strength, durability)
(4) Inadequate procurement control
(5) Incorrect or insufficient maintenance, or both

    (b) Substitution — can affect severity of injury or damage to health, frequency of exposure to the hazard under consideration, or the likelihood of avoiding or limiting injury or damage to health, depending on which method of substitution or combination thereof is applied.

Failure mode(s) examples:

(1) Unexpected or unanticipated interaction
(2) Excessive production pressure
(3) Inadequate procurement control

F.4.1.2 Design — Use of Engineering Controls.

(a) Greatest impact on the likelihood of a hazardous event(s) under certain circumstances
(b) No impact on severity of injury or damage to health

Failure mode(s) examples:

(1) Incorrect application of construction or manufacturing specification
(2) Unanticipated tasks
(3) Incentive to circumvent or reduce effectiveness
(4) Excessive production pressure
(5) Protective system failure

F.4.1.3 Use of Systems that Increase Awareness of Potential Hazards.

(a) Potential impact on avoiding or limiting injury or damage to health
(b) Potential impact on inadvertent exposure
(c) Minimal or no impact on severity of injury or damage to health

Failure mode(s) examples:

(1) Too many warning signs

(2) Depreciation of effect over time

(3) Lack of understanding

F.4.1.4 Organization and Application of a Safe System of Work.

F.4.1.4.1 Personnel training.

(a) Greatest impact on avoiding or limiting injury or damage to health

(b) Minimal, if any, impact on severity of injury or damage to health

(c) Possible impact on the likelihood of a hazardous event(s) under certain circumstances

Failure mode(s) examples:

(1) Training not understood

(2) Identified hazards not clearly communicated

(3) Depreciation of effect over time

(4) Training material not current

(5) Training not consistent with instructions

(6) Training material not inclusive of detail regarding how to perform work

F.4.1.4.2 Access restrictions.

(a) Greatest impact on exposure

(b) No impact on severity of injury or damage to health

Failure mode(s) examples:

(1) Work permit system does not exist

(2) Competency complacency

(3) Insufficient monitoring, control, or corrective actions, or combination thereof

F.4.1.4.3 Safe work procedures.

(a) Greatest impact on avoiding or limiting injury or damage to health

(b) Minimal, if any, impact on severity of injury or damage to health

(c) Possible impact on the likelihood of a hazardous event(s) under certain circumstances

Failure mode(s) examples:

(1) Inconsistent with the current culture

(2) Procedures not current or accessible

(3) Does not consider all tasks, hazards, hazardous situations, or combination thereof

(4) Insufficient monitoring, control, corrective actions, or combination thereof

(5) Instructions not consistent with training content

(6) Content too general (e.g., "Don't touch the live parts; Be careful.")

F.4.1.4.4 Policies and instructions.

(a) Greatest impact on exposure

(b) Possible impact on the likelihood of a hazardous event(s) under certain circumstances

(c) Minimal or no impact on severity of injury or damage to health

Failure mode(s) examples:

(1) Policies and instructions inconsistent
(2) Instructions not clearly communicated or accessible
(3) Insufficient monitoring, control, or corrective actions, or combination thereof
(4) Allows personnel to make the decision to work live without adequate justification

F.4.1.5 Personal Protective Equipment (PPE).

(a) Greatest impact on avoiding or limiting injury or damage to health
(b) Potential impact on inadvertent exposure
(c) Minimal impact on severity of injury or damage to health
(d) No impact on the likelihood of a hazardous event(s)

Failure mode(s) examples:

(1) Reason for use not understood
(2) Creates barriers to effective completion of the work
(3) PPE specification inappropriate for the considered hazards
(4) Production pressure does not afford time to use or maintain
(5) Worker forgets to use when needed
(6) Excessive discomfort
(7) Perceived invulnerability
(8) Insufficient monitoring, control, corrective actions, or combination thereof

The preceding hazard identification and risk assessment procedures may be condensed into nine basic steps, as defined in Annex E of ANSI/AIHA Z10:

1. Identify the electrical work task to be performed, the electrical system to be analyzed, or the electrical process to be analyzed. Break the task down where needed to assist in analysis.
2. Identify the electrical hazards associated with the task, system, or process involved.
3. Define the possible failure modes that result in exposure to electrical hazards and the potential resultant realization harm from the electrical hazards. "How could an objectionable occurrence ensue for a task and each of its possible associated electrical hazards?"
4. Estimate the frequency and duration of exposure to the electrical hazards.
5. Assess the severity of the potential injury from the electrical hazard(s).
6. Determine the likelihood of the occurrence of the hazardous event(s). Normally, the likelihood of occurrence is associated with a time interval, such as hourly, daily, weekly, monthly, yearly, etc.
7. Define the level of risk for the associated electrical hazard(s). The level of risk is determined as the product of the likelihood of an occurrence times the potential severity of the injury.
8. Determine if the level of risk is connected with the potential hazard(s) associated with the task, system, or process is acceptable or not.
9. If the level of risk is not acceptable, identify the additional measure or corrective actions to be taken.

F.5 Risk Reduction Verification.

F.5.1 Verification. Once the assessment has been completed and protective measures have been determined, it is imperative to ensure that the protective measures are implemented prior

Hazard (situation)	Risk Reduction Strategy	Confirmation (in place) Yes / No
Human factors (mistakes)	Training and instructions include details regarding hazardous situations that could arise.	
Human factors (willful disregard)	Policies and supervision are in place in order to ensure that instructions are followed.	
Unqualified person performing electrical work	Work permit system is in place to control personnel activities.	
Inappropriate overcurrent protection	Instructions include details regarding the selection or replacement of fuses and/or circuit breakers.	
Short circuits between test leads	Training and instructions include details regarding care and inspection of testing equipment.	
Meter malfunctions	Training and instructions include details regarding care and inspection of testing equipment.	
Meter misapplication	Training and instructions include details regarding use of testing equipment.	
Qualified person performing electrical work that exceeds individual's qualification	Work permit system is in place to control personnel activities.	

FIGURE F.5.2   Sample Auditing Form.

to initiating the electrical work. While this procedure might not result in a reduction of the PPE required, it could improve the understanding of the properties of the hazards associated with a task to a greater extent and thus allow for improvement in the implementation of the protective measures that have been selected.

F.5.2 Auditing. For each activity that has been assessed, it might be necessary to audit the risk reduction strategy that is applicable. If an audit is required, the auditing process should take place prior to commencement of work on electrical systems. An example of a nonexhaustive audit is shown in Figure F.5.2. Each audit process might need to be specific to the properties of the electrical system, to the task to be performed, or to both.

LIKELIHOOD OF OCCURRENCE OF ARC FLASH/BLAST EVENT AND CIRCUIT BREAKER OPERATION

Experience has shown that the likelihood of an arc flash/blast event increases as people interact with electrical equipment and that some types of interactions with electrical equipment increase the likelihood of an arc flash/blast event happening more than others. Turning on a circuit breaker (CB) may increase the likelihood of an arc flash/blast event more than turning off a CB, for instance. CBs housed in electrical equipment enclosures that have been properly installed, applied within their ratings, and properly maintained have a lower likelihood of causing an arc flash/blast event than those that have not. Furthermore, such CBs have a greater likelihood of clearing (interrupting and extinguishing) a fault within the time period predicted by its time–current characteristic (TCC) curve. Additionally, it is more likely that the actual incident energy level will be approximately the calculated level. It is known that incident energy level available affects the seriousness of the harm — the greater the incident energy level, the more harm can be expected.

The risk of injury largely depends on the amount of energy available to the breaker, how old it is, how well it is maintained, and the task that is to be performed, among other factors. For example, there will be little risk when simply operating (turning on and off) a well-maintained breaker in a house with a 240-volt service and 10,000 amperes available. In contrast, a commercial building with an equally well-maintained breaker typically will have 40,000 amperes, which poses greater risk. In addition, switching a breaker on can carry more risk than turning a breaker off.

The collective experience of the Technical Committee on Electrical Safety in the Workplace is that normal operation of enclosed electrical equipment operating at 600 volts or less that has been properly installed and maintained by qualified persons is not likely to expose the employee to an electrical hazard. Therefore, the likelihood of occurrence of an arc flash event from this electrical equipment can be considered as unlikely.

EXAMPLE: CIRCUIT BREAKER OPERATION WHILE ENERGIZED

Background Facts

Using a risk register and a risk assessment matrix, an electrical safety committee is analyzing two task scenarios to determine if additional safety controls are required. The first task scenario is the operation (turning on and off) of a 1600-A circuit breaker (CB) at a main 480Y/277-V, non-arc-rated switchboard in a building where the short-circuit current is calculated to be 35,000 A. The second task scenario is the operation (turning on and off) of a 20-A HID CB in a 480Y/277-V lighting panelboard where the short-circuit current is 25,000 A. In both cases, the electrical equipment is assumed to have been properly applied (installed, selected, and maintained properly), the equipment is within its rated life, a moist or damp environment is not involved, the equipment is dead front (enclosed), and the CBs are operated while the equipment is in a dead front configuration.

From Table 130.7(C)(15)(A)(b), at 18 in. it can be assumed that the switchboard has an incident energy level greater than 25 cal/cm^2 and less than or equal to 40 cal/cm^2, and that the lighting panelboard has an incident energy level greater than 4 cal/cm^2 and less than or equal to 8 cal/cm^2. Furthermore, the 1,600-A switchboard CB is operated twice a year when maintenance is scheduled for the facility, and the 20-A HID CB is operated twice a day to turn on and turn off the lighting. Should an arc flash incident occur, it is assumed that at 8 cal/cm^2 or less the integrity of the lighting panelboard will not be breached. Of the possible electrical hazards, only a risk assessment for the arc flash/blast hazard will be performed.

Analysis Using Risk Register

Based on the background facts, a copy of the risk register identified as Commentary Table F.1 will be filled out. Scenario 1 illustrates the risk evaluation of the 1600-A CB in the switchboard, and Scenario 2 illustrates the 20-A CB in the panelboard.

Scenario 1: The switchboard is not arc rated and the incident energy level is between 25 and 40 cal/cm^2, therefore the severity determined from Table F.2.3 is 6. The CB is operated twice a year, so the frequency and duration of exposure from Table F.2.4.1 is determined to be 3. The likelihood of a hazardous event occurring is considered to be negligible due to the proper application of the CB and switchboard, and the value is determined to be 1 from Table F.2.4.2. Finally, since it is not anticipated that in every case the switchboard enclosure would be breached should an arc flash event occur while operating the CB, the likelihood of avoiding or limiting injury is rare, and the value is determined to be 3 from Table F.2.4.3.

Scenario 2: Based on a level of incident energy greater than 4 cal/cm^2 and equal to or less than 8 cal/cm^2, the lighting panelboard enclosure integrity is assumed not to be breached if an arc flash incident occurs while switching the CB, and the severity of injury is determined to be 1 from Table F.2.3. Based on the CB being operated twice per day, the frequency and duration of exposure from Table F.2.4.1 is determined to be 5. For the same reason as for the 1600-A switchboard CB, the likelihood is determined to be 1 from Table F.2.4.2. Finally, since it is assumed that the enclosure integrity remains intact, the likelihood of avoiding or limiting injury is considered to be probable and is determined to be 1 from Table F.2.4.3.

The completed risk register is displayed as Commentary Table F.3. After evaluation, the electrical safety committee has determined that a risk score higher than 10 requires consideration of additional safety controls. Therefore, based on their analysis as depicted in the table, the committee has determined that the use of PPE is required when operating the 1600-A SWB CB but not required when operating the 20-A HID lighting panelboard CBs.

COMMENTARY TABLE F.3 *Risk Register for Example Scenario 1 (1600-A SWB CB) and Scenario 2 (20-A HID Lighting Panelboard CB)*

| Scenario No. | Hazard | Severity Se (Table F.2.3) | Probability of Occurrence of Harm, Po = (Fr + Pr + Av) | | | | Risk Score (R) |
			Fr (Table F.2.4.1)	Pr (Table F.2.4.2)	Av (Table F.2.4.3)	Total	Se × Po
1	Arc flash/ blast	6	3	1	3	6	36
2	Arc flash/ blast	1	5	1	1	7	7

Analysis Using Risk Assessment Matrix

The electrical safety committee revised Table F.2 to include incident energy levels to assist in using Commentary Table F.3, based on their assumption that incident energy levels greater than 4 cal/cm² and less than or equal to 8 cal/cm² will not breach the integrity of the enclosure. Further, they have determined that the likelihood of occurrence of an arc flash/blast event based on their equipment being properly selected, installed, and maintained is unlikely. The risk assessment matrix for these scenarios is shown in Exhibit F.3.

Since the incident energy level in Scenario 1 is greater than 8 cal/cm² and less than 40 cal/cm² and the likelihood is considered unlikely, the risk is color-coded yellow and is assumed to be moderate. Therefore, additional controls as indicated in Commentary

EXHIBIT F.3

Risk assessment matrix for example Scenario 1 (1600-A SWB CB) and Scenario 2 (20-A HID lighting panelboard CB)

| Likelihood of occurrence in period | Severity of the injury (consequences) | | | | |
	Slight	Minor	Medium	Critical	Catastrophic
cal/cm²	< 1.2	≥ 1.2 to ≤ 8		> 8 to ≤ 40	> 40
Unlikely	1	2	3	4	5
Seldom	2	4	6	8	10
Occasional	3	6	9	12	15
Likely	4	8	12	16	20
Definite	5	10	15	20	25

Notes:
1. Extreme equals 25 through 15.
2. High equals 12 through 9.
3. Moderate equals 8 through 4.
4. Low equals 3 through 1.

Table F.2 should be considered by management. For Scenario 2, the risk assessment level is determined to be low, since the incident energy level is greater than 4 cal/cm² and less than or equal to 8 cal/cm² and the electrical safety committee recommends continued monitoring. They further recommend that the existing control remain in place as detailed in Commentary Table F.2.

Summary

The example should not be relied upon for any particular installation. It is simply meant to demonstrate the thinking process required to perform hazard assessment and risk analysis based on the principles included in Informative Annex F.

Sample Lockout/Tagout Procedure

Summary of Annex G Changes

Informative Annex G: Changed text to be consistent with the redefined hazard and risk terminology. Replaced the phrase *use of tags and warning signs* with *alerting techniques*, and replaced *voltage-detecting instruments* with *test instruments* to clarify and promote the consistent use of terminology associated with hazard and risk.

G.13: Deleted item (8) *Individual employee control of energy.*

This informative annex is not a part of the requirements of this NFPA document but is included for informational purposes only.

Informative Annex G illustrates how a lockout/tagout procedure might be published. An employer could "fill in the blanks" and publish the resulting procedure for his or her organization. See Supplement 3 for another possible safety procedure format.

Lockout is the preferred method of controlling personnel exposure to electrical energy hazards. Tagout is an alternative method that is available to employers. To assist employers in developing a procedure that meets the requirement of 120.2 of *NFPA 70E*, the sample procedure that follows is provided for use in lockout and tagout programs. This procedure can be used for a simple lockout/tagout, or as part of a complex lockout/tagout. A more comprehensive plan will need to be developed, documented, and used for the complex lockout/tagout.

LOCKOUT /TAGOUT PROCEDURE FOR [COMPANY NAME]
OR
TAGOUT PROCEDURE FOR [COMPANY NAME]

1.0 Purpose.

This procedure establishes the minimum requirements for lockout/tagout of electrical energy sources. It is to be used to ensure that conductors and circuit parts are disconnected from sources of electrical energy, locked (tagged), and tested before work begins where employees could be exposed to dangerous conditions. Sources of stored energy, such as capacitors or springs, shall be relieved of their energy, and a mechanism shall be engaged to prevent the reaccumulation of energy.

2.0 Responsibility.

All employees shall be instructed in the safety significance of the lockout/tagout procedure. All new or transferred employees and all other persons whose work operations are or might be in the area shall be instructed in the purpose and use of this procedure. *[Name(s) of the person(s) or the job title(s) of the employee(s) with responsibility]* shall ensure that appropriate personnel receive instructions on their roles and responsibilities. All persons installing a lockout/tagout device shall sign their names and the date on the tag *[or state how the name of the individual or person in charge will be available]*.

3.0 Preparation for Lockout/Tagout.

3.1 Review current diagrammatic drawings (or their equivalent), tags, labels, and signs to identify and locate all disconnecting means to determine that power is interrupted by a physical break and not de-energized by a circuit interlock. Make a list of disconnecting means to be locked (tagged).

3.2 Review disconnecting means to determine adequacy of their interrupting ability. Determine if it will be possible to verify a visible open point, or if other precautions will be necessary.

3.3 Review other work activity to identify where and how other personnel might be exposed to electrical hazards. Review other energy sources in the physical area to determine employee exposure to those sources of other types of energy. Establish energy control methods for control of other hazardous energy sources in the area.

3.4 Provide an adequately rated test instrument to test each phase conductor or circuit part to verify that they are de-energized *(see Section 11.3)*. Provide a method to determine that the test instrument is operating satisfactorily.

3.5 Where the possibility of induced voltages or stored electrical energy exists, call for grounding the phase conductors or circuit parts before touching them. Where it could be reasonably anticipated that contact with other exposed energized conductors or circuit parts is possible, call for applying ground connecting devices.

4.0 Simple Lockout/Tagout.

The simple lockout/tagout procedure will involve 1.0 through 3.0, 5.0 through 9.0, and 11.0 through 13.0.

5.0 Sequence of Lockout/Tagout System Procedures.

5.1 The employees shall be notified that a lockout/tagout system is going to be implemented and the reason for it. The qualified employee implementing the lockout/tagout shall know the disconnecting means location for all sources of electrical energy and the location of all sources of stored energy. The qualified person shall be knowledgeable of hazards associated with electrical energy.

5.2 If the electrical supply is energized, the qualified person shall de-energize and disconnect the electric supply and relieve all stored energy.

5.3 Wherever possible, the blades of disconnecting devices should be visually verified to be fully opened, or draw-out type circuit breakers should be verified to be completely withdrawn to the fully disconnected position.

5.4 Lockout/tagout all disconnecting means with lockout/tagout devices.

Informational Note: For tagout, one additional safety measure must be employed, such as opening, blocking, or removing an additional circuit element.

5.5 Attempt to operate the disconnecting means to determine that operation is prohibited.

5.6 A test instrument shall be used. *(See 11.3.)* Inspect the instrument for visible damage. Do not proceed if there is an indication of damage to the instrument until an undamaged device is available.

5.7 Verify proper instrument operation and then test for absence of voltage.

5.8 Verify proper instrument operation after testing for absence of voltage.

5.9 Where required, install a grounding equipment/conductor device on the phase conductors or circuit parts, to eliminate induced voltage or stored energy, before touching them. Where it has been determined that contact with other exposed energized conductors or circuit parts is possible, apply ground connecting devices rated for the available fault duty.

5.10 The equipment, electrical source, or both are now locked out (tagged out).

6.0 Restoring the Equipment, Electrical Supply, or Both to Normal Condition.

6.1 After the job or task is complete, visually verify that the job or task is complete.

6.2 Remove all tools, equipment, and unused materials and perform appropriate housekeeping.

6.3 Remove all grounding equipment/conductors/devices.

6.4 Notify all personnel involved with the job or task that the lockout/tagout is complete, that the electrical supply is being restored, and that they are to remain clear of the equipment and electrical supply.

6.5 Perform any quality control tests or checks on the repaired or replaced equipment, electrical supply, or both.

6.6 Remove lockout/tagout devices. The person who installed the devices is to remove them.

6.7 Notify the owner of the equipment, electrical supply, or both, that the equipment, electrical supply, or both are ready to be returned to normal operation.

6.8 Return the disconnecting means to their normal condition.

7.0 Procedure Involving More Than One Person.

For a simple lockout/tagout and where more than one person is involved in the job or task, each person shall install his or her own personal lockout/tagout device.

8.0 Procedure Involving More Than One Shift.

When the lockout/tagout extends for more than one day, it shall be verified that the lockout/tagout is still in place at the beginning of the next day. When the lockout/tagout is continued on successive shifts, the lockout/tagout is considered to be a complex lockout/tagout.

For a complex lockout/tagout, the person in charge shall identify the method for transfer of the lockout/tagout and of communication with all employees.

9.0 Complex Lockout/Tagout.

A complex lockout/tagout plan is required where one or more of the following exist:

(1) Multiple energy sources (more than one)
(2) Multiple crews
(3) Multiple crafts
(4) Multiple locations
(5) Multiple employers
(6) Unique disconnecting means
(7) Complex or particular switching sequences
(8) Lockout/tagout for more than one shift; that is, new shift workers

9.1 All complex lockout/tagout procedures shall require a written plan of execution. The plan shall include the requirements in 1.0 through 3.0, 5.0, 6.0, and 8.0 through 12.0.

9.2 A person in charge shall be involved with a complex lockout/tagout procedure. The person in charge shall be at the procedure location.

9.3 The person in charge shall develop a written plan of execution and communicate that plan to all persons engaged in the job or task. The person in charge shall be held accountable for safe execution of the complex lockout/tagout plan. The complex lockout/tagout plan must address all the concerns of employees who might be exposed, and they must understand how electrical energy is controlled. The person in charge shall ensure that each person understands the electrical hazards to which they are exposed and the safety-related work practices they are to use.

9.4 All complex lockout/tagout plans identify the method to account for all persons who might be exposed to electrical hazards in the course of the lockout/tagout.

One of the following methods is to be used:

(1) Each individual shall install his or her own personal lockout or tagout device.
(2) The person in charge shall lock his/her key in a lock box.
(3) The person in charge shall maintain a sign-in/sign-out log for all personnel entering the area.
(4) Another equally effective methodology shall be used.

9.5 The person in charge can install locks/tags or direct their installation on behalf of other employees.

9.6 The person in charge can remove locks/tags or direct their removal on behalf of other employees, only after all personnel are accounted for and ensured to be clear of potential electrical hazards.

9.7 Where the complex lockout/tagout is continued on successive shifts, the person in charge shall identify the method for transfer of the lockout and the method of communication with all employees.

10.0 Discipline.

10.1 Knowingly violating this procedure will result in [*state disciplinary actions that will be taken*].

10.2 Knowingly operating a disconnecting means with an installed lockout device (tagout device) will result in [*state disciplinary actions to be taken*].

11.0 Equipment.

11.1 Locks shall be [*state type and model of selected locks*].

11.2 Tags shall be [*state type and model to be used*].

11.3 The test instrument(s) to be used shall be [*state type and model*].

12.0 Review.

This procedure was last reviewed on [date] and is scheduled to be reviewed again on [date] (not more than 1 year from the last review).

13.0 Lockout/Tagout Training.

Recommended training can include, but is not limited to, the following:

(1) Recognition of lockout/tagout devices
(2) Installation of lockout/tagout devices
(3) Duty of employer in writing procedures
(4) Duty of employee in executing procedures
(5) Duty of person in charge
(6) Authorized and unauthorized removal of locks/tags
(7) Enforcement of execution of lockout/tagout procedures

(8) Simple lockout/tagout
(9) Complex lockout/tagout
(10) Use of single-line and diagrammatic drawings to identify sources of energy
(11) Alerting techniques
(12) Release of stored energy
(13) Personnel accounting methods
(14) Temporary protective grounding equipment needs and requirements
(15) Safe use of test instruments

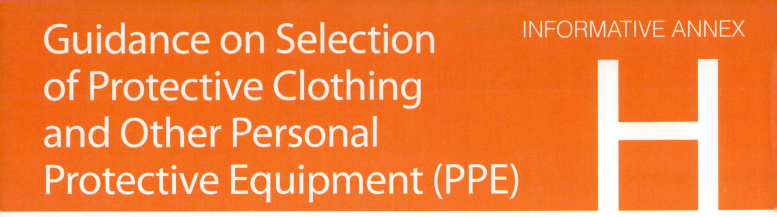

Guidance on Selection of Protective Clothing and Other Personal Protective Equipment (PPE)

Summary of Annex H Changes

H.2: Replaced the term *hazard/risk category* with *arc flash PPE category*, *voltage-rated gloves* and *voltage-rated tools* with *rubber insulating gloves*, *assumed* with *estimated available*, and *arc flash hazard analysis* with *arc flash risk assessment*. Revised Table H.2 notes to be consistent with terminology used in the rest of the standard and to reflect changes to the new hazard/risk terminology and the new terminology in the PPE Category Tables. Table designations have been revised in the text for correlation.

H.3: Replaced the term *hazard analysis* with *risk assessment of electrical hazards*, *personal protective equipment (PPE)* with *PPE*, and *incident exposure* with *incident energy exposure*, to clarify and promote the consistent use of terminology associated with hazard and risk. Editorial changes were made to the table references.

This informative annex is not a part of the requirements of this NFPA document but is included for informational purposes only.

Informative Annex H provides guidance on the selection of clothing and personal protective equipment (PPE) regardless of whether the arc flash PPE category method or the incident energy method is used to perform an arc flash risk assessment. This informative annex also provides guidance for the selection of shock protective equipment.

H.1 Arc-Rated Clothing and Other Personal Protective Equipment (PPE) for Use with Arc Flash PPE Categories.

Table 130.7(C)(15)(A)(a), Table 130.7(C)(15)(A)(b), Table 130.7(C)(15)(B), and Table 130.7(C)(16) provide guidance for the selection and use of PPE when using arc flash PPE categories.

Tables 130.7(C)(15)(A)(b), 130.7(C)(15)(B), and 130.7(C)(16) are only applicable when using the arc flash PPE category method to perform an arc flash risk assessment for alternating current (ac) or direct current (dc) systems. Arc flash PPE categories may be applicable when using the incident energy method to perform an arch flash risk assessment. When performing an incident energy analysis, the arc flash labels created may include an arc flash PPE category.

H.2 Simplified Two-Category Clothing Approach for Use with Table 130.7(C)(15)(A)(a), Table 130.7(C)(15)(A)(b), Table 130.7(C)(15)(B), and Table 130.7(C)(16).

The use of Table H.2 is a simplified approach to provide minimum PPE for electrical workers within facilities with large and diverse electrical systems. The clothing listed in Table H.2 fulfills the minimum arc-rated clothing requirements of Table 130.7(C)(15)(A)(a), Table 130.7(C)(15)(A)(b), Table 130.7(C)(15)(B), and Table 130.7(C)(16). The clothing systems listed in this table should be used with the other PPE appropriate for the arc flash PPE category [see Table 130.7(C)(16)]. The notes to Table 130.7(C)(15)(A)(a), Table 130.7(C)(15)(A)(b), and Table 130.7(C)(15)(B), must apply as shown in those tables.

After an employer develops and publishes a procedure that describes how thermal protection is determined based on the use of the arc flash PPE categories method, the employer must develop a system that enables the procedure requirements to be efficiently and effectively administered. Section H.2 illustrates one method that enables a PPE program to be efficiently and effectively administered where the arc flash PPE category method has been selected for performing an arc flash risk assessment.

Table H.2 Simplified Two-Category, Arc-Rated Clothing System

Clothing[a]	Applicable Tasks
Everyday Work Clothing Arc-rated long-sleeve shirt with arc-rated pants (minimum arc rating of 8) *or* Arc-rated coveralls (minimum arc rating of 8)	All arc flash PPE category 1 and arc flash PPE category 2 tasks listed in Table 130.7(C)(15)(A)(a), Table 130.7(C)(15)(A)(b), and Table 130.7(C)(15)(B)[b]
Arc Flash Suit A total clothing system consisting of arc-rated shirt and pants and/or arc-rated coveralls and/or arc flash coat and pants (clothing system minimum arc rating of 40)	All arc flash PPE category 3 and arc flash PPE category 4 tasks listed in Table 130.7(C)(15)(A)(a), Table 130.7(C)(15)(A)(b), and Table 130.7(C)(15)(B)[b]

[a]Note that other PPE listed in Table 130.7(C)(16), which include arc-rated face shields or arc flash suit hoods, arc-rated hard hat liners, safety glasses or safety goggles, hard hats, hearing protection, heavy-duty leather gloves, rubber insulating gloves, and leather protectors, could be required. The arc rating for a garment is expressed in cal/cm².
[b]The estimated available short-circuit current capacities and fault clearing times or arcing durations are listed in the text of Table 130.7(C)(15)(A)(b) and Table 130.7(C)(15)(B). Various tasks are listed in Table 130.7(C)(15)(A)(a). For tasks not listed or for power systems with greater than the estimated available short-circuit capacity or with longer than the assumed fault clearing times or arcing durations, an arc flash risk assessment is required in accordance with 130.5.

H.3 Arc-Rated Clothing and Other Personal Protective Equipment (PPE) for Use with Risk Assessment of Electrical Hazards.

Table H.3(a) provides a summary of specific sections within the *NFPA 70E* standard describing PPE for electrical hazards. Table H.3(b) provides guidance on the selection of arc-rated and other PPE for users who determine the incident energy exposure (in cal/cm²).

Even when using the arc flash PPE category method, the sections identified in Table H.3(a) should be reviewed to ensure that appropriate PPE is being used to protect employees from the electrical hazards identified for the task involved. Further guidance is provided in

Table H.3(b), where an incident energy analysis has been used to perform the arc flash risk assessment.

> Tables H.3(a) and (b) were revised by a tentative interim amendment (TIA).

TABLE H.3(a)   *Summary of Specific Sections Describing PPE for Electrical Hazards*

Shock Hazard PPE	Applicable Section(s)
Rubber insulating gloves and leather protectors (unless the requirements of ASTM F 496 are met)	130.7(C)(7)(a)
Rubber insulating sleeves (as needed)	130.7(C)(7)(a)
Class G or E hard hat (as needed)	130.7(C)(3)
Safety glasses or goggles (as needed)	130.7(C)(4)
Dielectric overshoes (as needed)	130.7(C)(8)
Arc Flash Hazard PPE	
Incident energy exposures up to 1.2 cal/cm²	
Clothing: nonmelting or untreated natural fiber long-sleeve shirt and long pants or coverall	130.7(C)(1); 130.7(C)(9)(d)
Gloves: heavy-duty leather	130.7(C)(7)(b); 130.7(C)(10)(d)
Hard hat: class G or E	130.7(C)(3)
Face shield: covers the face, neck, and chin (as needed)	130.7(C)(3)
Safety glasses or goggles	130.7(C)(4); 130.7(C)(10)(c)
Hearing protection	130.7(C)(5)
Footwear: heavy-duty leather (as needed)	130.7(C)(10)(e)
Incident Energy Exposures ≥ 1.2 cal/cm²	
Clothing: arc-rated clothing system with an arc rating appropriate to the anticipated incident energy exposure	130.7(C)(1); 130.7(C)(2); 130.7(C)(6); 130.7(C)(9)(d)
Clothing underlayers (when used): arc-rated or nonmelting untreated natural fiber	130.7(C)(9)(c); 130.7(C)(11); 130.7(C)(12)
Gloves:	130.7(C)(7)(b); 130.7(C)(10)(d)
Exposures ≥ 1.2 cal/cm² and ≤ 8 cal/cm²: heavy-duty leather gloves	
Exposures > 8 cal/cm²: rubber insulating gloves with their leather protectors; or arc-rated gloves	
Hard hat: class G or E	130.7(C)(1); 130.7(C)(3)
Face shield:	130.7(C)(1); 130.7(C)(3); 130.7(C)(10)(a); 130.7(C)(10)(b); 130.7(C)(10)(c)
Exposures ≥1.2 cal/cm² 12 cal/cm²: arc-rated face shield that covers the face, neck, and chin and an arc-rated balaclava or an arc-rated arc flash suit hood	
Exposures> 12 cal/cm²: arc-rated arc flash suit hood	
Safety glasses or goggles	130.7(C)(4); 130.7(C)(10)(c)
Hearing protection	130.7(C)(5)
Footwear:	130.7(C)(10)(e)
Exposures ≤4 cal/cm²: heavy-duty leather footwear (as needed)	
Exposures > 4 cal/cm²: heavy-duty leather footwear	

TABLE H.3(b)   Guidance on Selection of Arc-Rated Clothing and Other PPE for Use When Incident Energy Exposure Is Determined

Incident Energy Exposure	Protective Clothing and PPE
≤ 1.2 cal/cm² Protective clothing, nonmelting (in accordance with ASTM F 1506) or untreated natural fiber	Shirt (long sleeve) and pants (long) or coverall
Other PPE	Face shield for projectile protection (AN) Safety glasses or safety goggles (SR) Hearing protection Heavy-duty leather gloves or rubber insulating gloves with leather protectors (AN)
> 1.2 to 12 cal/cm² Arc-rated clothing and equipment with an arc rating equal to or greater than the determined incident energy *(See Note 3.)*	Arc-rated long-sleeve shirt and arc-rated pants or arc-rated coverall or arc flash suit (SR) *(See Note 3.)* Arc-rated face shield and arc-rated balaclava or arc flash suit hood (SR) *(See Note 1.)* Arc-rated jacket, parka, or rainwear (AN)
Other PPE	Hard hat Arc-rated hard hat liner (AN) Safety glasses or safety goggles (SR) Hearing protection Heavy-duty leather gloves or rubber insulating gloves with leather protectors (SR) *(See Note 4.)* Leather footwear
> 12 cal/cm² Arc-rated clothing and equipment with an arc rating equal to or greater than the determined incident energy *(See Note 3.)*	Arc-rated long-sleeve shirt and arc-rated pants or arc-rated coverall and/or arc flash suit (SR) Arc-rated arc flash suit hood Arc-rated gloves Arc-rated jacket, parka, or rainwear (AN)
Other PPE	Hard hat Arc-rated hard hat liner (AN) Safety glasses or safety goggles (SR) Hearing protection Arc-rated gloves or rubber insulating gloves with leather protectors (SR) *(See Note 4.)* Leather footwear

AN: As needed [in addition to the protective clothing and PPE required by 130.5(C)(1)].

SR: Selection of one in group is required by 130.5(C)(1).

Notes:

(1) Face shields with a wrap-around guarding to protect the face, chin, forehead, ears, and neck area are required by 130.7(C)(10)(c). For full head and neck protection, use a balaclava or an arc flash hood.

(2) All items not designated "AN" are required by 130.7(C).

(3) Arc ratings can be for a single layer, such as an arc-rated shirt and pants or a coverall, or for an arc flash suit or a multi-layer system consisting of a combination of arc-rated shirt and pants, coverall, and arc flash suit.

(4) Rubber insulating gloves with leather protectors provide arc flash protection in addition to shock protection. Higher class rubber insulating gloves with leather protectors, due to their increased material thickness, provide increased arc flash protection.

•

Job Briefing and Planning Checklist

This informative annex is not a part of the requirements of this NFPA document but is included for informational purposes only.

Informative Annex I illustrates various subjects that should be discussed when a job briefing is held. Other subjects might need to be discussed. The purpose of the checklist in Figure I.1 is to help facilitate the conversation.

I.1 Job Briefing and Planning Checklist.

Figure I.1 illustrates considerations for a job briefing and planning checklist.

Identify

- ❏ Hazards
- ❏ Voltage levels involved
- ❏ Skills required
- ❏ Any "foreign" (secondary source) voltage source
- ❏ Any unusual work conditions
- ❏ Number of people needed to do the job

- ❏ Shock protection boundaries
- ❏ Available incident energy
- ❏ Potential for arc flash (Conduct an arc flash hazard analysis.)
- ❏ Arc flash boundary

Ask

- ❏ Can the equipment be de-energized?
- ❏ Are backfeeds of the circuits to be worked on possible?

- ❏ Is a standby person required?

Check

- ❏ Job plans
- ❏ Single-line diagrams and vendor prints
- ❏ Status board
- ❏ Information on plant and vendor resources is up to date

- ❏ Safety procedures
- ❏ Vendor information
- ❏ Individuals are familiar with the facility

Know

- ❏ What the job is
- ❏ Who else needs to know — Communicate!

- ❏ Who is in charge

Think

- ❏ About the unexpected event . . . What if?
- ❏ Lock — Tag — Test — Try
- ❏ Test for voltage — FIRST
- ❏ Use the right tools and equipment, including PPE

- ❏ Install and remove temporary protective grounding equipment
- ❏ Install barriers and barricades
- ❏ What else . . . ?

Prepare for an emergency

- ❏ Is the standby person CPR trained?
- ❏ Is the required emergency equipment available? Where is it?
- ❏ Where is the nearest telephone?
- ❏ Where is the fire alarm?
- ❏ Is confined space rescue available?

- ❏ What is the exact work location?
- ❏ How is the equipment shut off in an emergency?
- ❏ Are the emergency telephone numbers known?
- ❏ Where is the fire extinguisher?
- ❏ Are radio communications available?

FIGURE I.1 *Sample Job Briefing and Planning Checklist*

Energized Electrical Work Permit

This informative annex is not a part of the requirements of this NFPA document but is included for informational purposes only.

Informative Annex J illustrates an energized electrical work permit. The format of this work permit example is not fixed by requirement, although many employers have used this template successfully. The basic purpose of a work permit is to ensure that people in responsible positions are involved in the decision whether or not to accept the increased risk associated with working on energized electrical conductors or circuit parts. An additional benefit of the work permit is that its review might initiate a decision to perform the work de-energized.

J.1 Energized Electrical Work Permit Sample.

Figure J.1 illustrates considerations for an energized electrical work permit.

J.2 Energized Electrical Work Permit.

Figure J.2 illustrates items to consider when determining the need for an energized electrical work permit.

ENERGIZED ELECTRICAL WORK PERMIT

PART I: TO BE COMPLETED BY THE REQUESTER:

Job/Work Order Number _____

(1) Description of circuit/equipment/job location: _____
_____

(2) Description of work to be done: _____
_____

(3) Justification of why the circuit/equipment cannot be de-energized or the work deferred until the next scheduled outage:
_____
_____

_____   _____
Requester/Title   Date

PART II: TO BE COMPLETED BY THE ELECTRICALLY QUALIFIED PERSONS *DOING* THE WORK:

Check when complete

(1) Detailed job description procedure to be used in performing the above detailed work: ☐
_____
_____

(2) Description of the safe work practices to be employed: _____ ☐
_____

(3) Results of the shock risk assessment: _____
   (a) Voltage to which personnel will be exposed   ☐
   (b) Limited approach boundary   ☐
   (c) Restricted approach boundary   ☐
   (d) Necessary shock, personal, and other protective equipment to safely perform assigned task   ☐

(4) Results of the arc flash risk assessment: _____
   (a) Available incident energy at the working distance or arc flash PPE category   ☐
   (b) Necessary arc flash personal and other protective equipment to safely perform the assigned task   ☐
   (c) Arc flash boundary   ☐

(5) Means employed to restrict the access of unqualified persons from the work area: _____ ☐
_____

(6) Evidence of completion of a job briefing, including discussion of any job-related hazards: _____ ☐
_____

(7) Do you agree the above-described work can be done safely? ☐ Yes ☐ No (If *no*, return to requester.)

_____   _____
Electrically Qualified Person(s)   Date

_____   _____
Electrically Qualified Person(s)   Date

PART III: APPROVAL(S) TO PERFORM THE WORK WHILE ELECTRICALLY ENERGIZED:

_____   _____
Manufacturing Manager   Maintenance/Engineering Manager

_____   _____
Safety Manager   Electrically Knowledgeable Person

_____   _____
General Manager   Date

Note: Once the work is complete, forward this form to the site Safety Department for review and retention.

© 2014 National Fire Protection Association   NFPA 70E

FIGURE J.1 *Sample Permit for Energized Electrical Work.*

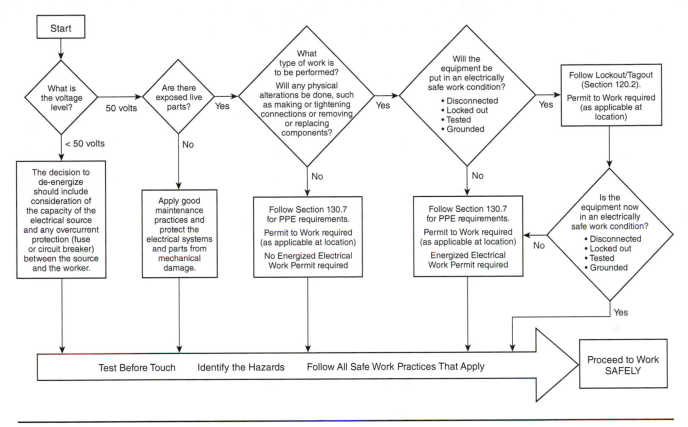

FIGURE J.2 *Energized Electrical Work Permit Flow Chart.*

General Categories of Electrical Hazards

Summary of Annex K Changes

K.4: Replaced *1600 km/hr* with *1120 km/hr* to correct a mathematical error from the previous edition.

This informative annex is not a part of the requirements of this NFPA document but is included for informational purposes only.

Informative Annex K provides an abbreviated discussion of known electrical hazards. Critical information is provided for an employee to use in support of an improved electrical safety program.

K.1 General Categories.

There are three general categories of electrical hazards: electrical shock, arc flash, and arc blast.

K.2 Electric Shock.

Approximately 30,000 nonfatal electrical shock accidents occur each year. The National Safety Council estimates that about 1000 fatalities each year are due to electrocution, more than half of them while servicing energized systems of less than 600 volts.

Electrocution is the fourth leading cause of industrial fatalities, after traffic, homicide, and construction accidents. The current required to light a 7½-watt, 120-volt lamp, if passed across the chest, is enough to cause a fatality. The most damaging paths through the body are through the lungs, heart, and brain.

Protection against electric shock exposure and accidents was the original mission of *NFPA 70E®, Standard for Electrical Safety in the Workplace®*. Establishing an electrically safe work condition is the desired approach to mitigating electric shock exposure, but where tasks are permitted on energized electrical conductors and circuit parts, the requirements in Article 130 provide protection strategies that allow an employee to safely work on energized electrical equipment. Although 130.2(A)(3) establishes 50 volts and higher as the threshold at which shock protection for personnel is required, voltages lower than 50 volts can be hazardous under certain conditions. Wet, damp, and submersed conditions lower the body resistance to electric current, and voltages lower than 50 volts in these conditions are potentially dangerous. The lower voltage levels may not necessarily pose an injurious electric shock hazard, but the effects of electric current on muscular control can be dangerous to an employee who might be partially or completely immersed in a body of water. Swimming pool maintenance is an activity that could pose this threat to an employee.

According to the Bureau of Labor Statistics (BLS), contact with electricity was the seventh leading cause of total occupational fatalities from 2003 to 2010. During this period there were 42,882 fatalities from all causes, and 1,738 (about 4 percent) of those were due to contact with electric current. Some of the statistics from this time period are shown in Commentary Table K.1. The percentages of electrical fatalities that occurred in this period, by type of activity, are shown in Commentary Table K.2.

COMMENTARY TABLE K.1 *Number of Deaths from Contact with Electricity, 2003 to 2010, by Employment*

Employment	Number of Fatalities
Electricians	300
Construction laborers	146
Electrical power line installers and repairers	132
Tree trimmers and pruners	79
Industrial machinery installation, repair, and maintenance employees	71
Heating, air conditioning, and refrigeration mechanics and installers	55
Driver/sales employees and truck drivers	50
Material moving employees	44

COMMENTARY TABLE K.2 *Percentage of Electrical Deaths, 2003 to 2010, by Activity*

Percent of Electrical Fatality (percentage)	Activity
45	Contact with overhead power lines
28	Contact with wiring transformers or other electrical components
18	Contact with electric current of machine, tool, appliance, or light fixture
4	Struck by lightning

In 2010, contact with transformers or other electrical components amounted to around 30 percent of the fatalities, and contact with overhead power lines amounted to around 46 percent of the fatalities.

From the period of 1992 to 2010, the BLS documented 5,096 fatal electrical injuries and estimated 66,748 electrical injuries. Contact with overhead power lines accounted for 44 percent of electrical fatalities but only 3 percent of nonfatal electrical injuries. Contact with wiring, transformers, or other electrical components accounted for 27 percent of fatal injuries and 35 percent of nonfatal injuries. Additional data and information regarding electrical injury statistics can be found in the resource library of the Electrical Safety Foundation International (ESFI) at www.esfi.org.

K.3 Arc Flash.

When an electric current passes through air between ungrounded conductors or between ungrounded conductors and grounded conductors, the temperatures can reach 35,000°F. Exposure to these extreme temperatures both burns the skin directly and causes ignition of clothing, which adds to the burn injury. The majority of hospital admissions due to electrical accidents are from arc flash burns, not from shocks. Each year more than 2000 people are admitted to burn centers with severe arc flash burns. Arc flashes can and do kill at distances of 3 m (10 ft).

K.4 Arc Blast.

The tremendous temperatures of the arc cause the explosive expansion of both the surrounding air and the metal in the arc path. For example, copper expands by a factor of 67,000 times when it turns from a solid to a vapor. The danger associated with this expansion is one of high pressures, sound, and shrapnel. The high pressures can easily exceed hundreds or even thousands of pounds per square foot, knocking workers off ladders, rupturing eardrums, and collapsing lungs. The sounds associated with these pressures can exceed 160 dB. Finally, material and molten metal are expelled away from the arc at speeds exceeding 1120 km/hr (700 mph), fast enough for shrapnel to completely penetrate the human body.

Protection against the arc flash hazards described in Section K.3 is to help prevent employees from receiving incurable burns. The advice given in Informational Note No. 3 to 130.7(A) recognizes the extremely dangerous concussive forces, sound, and shrapnel that occur in arc blast events where the incident energy exceeds 40 cal/cm^2, even though arc-rated garments are available with greater ratings. Arc-rated garments with higher ratings are not intended to provide arc blast protection.

Arc terminal temperature is estimated to be in excess of 35,000°F (19,430°C). The plasma of vaporizing metal has a temperature of 23,000°F (12,760°C). An atomic bomb after 0.3 seconds reaches only 12,600°F (6980°C), and the surface of the sun is only 10,000°F (5540°C). The superheating of the air during an arcing incident creates an acoustic wave similar to the generation of thunder by lightning.

Ralph H. Lee, in his paper "Pressures Developed by Arcs" *(IEEE Transactions on Industry Applications*, Volume 1A-23, No. 4, pp. 760–764, July/August 1987), provides information regarding arc blast pressures, including the following. The acoustic and pressure waves from arcs are developed from the expansion of boiling metal and the superheating of air by the arc passing through it. Vaporizing copper expands 67,000 times in volume, whereas water expands 1670 times while becoming steam. This expansion accounts for the expulsion of molten metal droplets from the arc, which can be thrust up to distances around 10 ft (3 m). This pressure generates ionized vapor (plasma) outward from the arc for distances that are proportional to the arc power. Fifty-three kilowatts of power will vaporize 0.05 in.^{3} (0.328 cm^3) of copper into 3350 in.^{3} (54,907 cm^3) of vapor, and 1 in.^{3} (16.39 cm^3) vaporizes into 1.44 yd^3 (1.098 m^3).

Immediately, the hot vapor forming an arc starts to cool, but while hot it combines with oxygen in the air to become an oxide of the metal vapor. It solidifies as it cools and becomes minute particles that appear as smoke — copper and iron are black, and aluminum is grey. These particles are poisonous if inhaled, are quite hot, and will cling to any surface they touch.

The pressure from a 100 kA arc can reach around 400 lb/ft^2 (1950 kg/m^2) at a distance of 3.3 ft (1 m). This pressure is about ten times the value of wind resistance that walls are generally built to withstand; it could destroy a conventional wall at a distance of approximately 40 ft (12 m). A 25 kA arc at a distance of 2 ft (0.6 m) can produce a pressure of around 160 lb/ft^2 (780 kg/m^2). The average man's front body projects around 3 ft^2 (0.28 m^2) of body area, while the upper body projects about 2.2 ft^2 (0.2 m^2) of body area. The 25 kA arc at 2 ft (0.6 m) is sufficient to place a total pressure of about 480 lb (3×160) on the front of the average man's body. Arc blasts have propelled large objects — such as personnel, switchboard doors, and bus bars — several feet at high rates of speed. A 50 kA arc fault can produce enough energy to drive a 160 lb (73 kg) person standing 2 ft (0.6 m) from the arc at nearly 110 mph (50 m/sec). Arc blasts can provide enough pressure to collapse lungs and rupture eardrums if hearing protection is not used.

The following equation from information contained in Ralph H. Lee's paper cited earlier on pressures developed by arcs can be used to estimate the arc blast pressure:

$$P = (11.5 \times I_a)/D^{0.9}$$

where:
P = pressure (lb/ft^2)
I_a = arcing current (kA)
D = distance from the center of the arc (ft)

However, it should be noted that this equation has not been generally adopted and additional testing is required.

Typical Application of Safeguards in the Cell Line Working Zone

Summary of Annex L Changes

Informative Annex L: Replaced the phrase *hazardous condition(s)* with *electrical hazard(s)* throughout annex.

L.1: Deleted the phrase *the hazardous electrical condition will be removed* from items (5) and (7) to promote the consistent use of terminology associated with hazard and risk.

This informative annex is not a part of the requirements of this NFPA document but is included for informational purposes only.

Informative Annex L discusses and illustrates the application of typical safeguards in a process area associated with a dc electrical system that cannot be de-energized.

L.1 Application of Safeguards.

This informative annex permits a typical application of safeguards in electrolytic areas where electrical hazards exist. Take, for example, an employee working on an energized cell. The employee uses manual contact to make adjustments and repairs. Consequently, the exposed energized cell and grounded metal floor could present an electrical hazard. Safeguards for this employee can be provided in the following ways:

(1) Protective boots can be worn that isolate the employee's feet from the floor and that provide a safeguard from the electrical hazard.
(2) Protective gloves can be worn that isolate the employee's hands from the energized cell and that provide a safeguard.
(3) If the work task causes severe deterioration, wear, or damage to personal protective equipment (PPE), the employee might have to wear both protective gloves and boots.
(4) A permanent or temporary insulating surface can be provided for the employee to stand on to provide a safeguard.
(5) The design of the installation can be modified to provide a conductive surface for the employee to stand on. If the conductive surface is bonded to the cell, a safeguard will be provided by voltage equalization.

(6) Safe work practices can provide safeguards. If protective boots are worn, the employee should not make long reaches over energized (or grounded) surfaces such that his or her elbow bypasses the safeguard. If such movements are required, protective sleeves, protective mats, or special tools should be used. Training on the nature of electrical hazards and proper use and condition of safeguards is, in itself, a safeguard.

(7) The energized cell can be temporarily bonded to ground.

L.2 Electrical Power Receptacles.

Power supply circuits and receptacles in the cell line area for portable electric equipment should meet the requirements of 668.21 of *NFPA 70, National Electrical Code*. However, it is recommended that receptacles for portable electric equipment not be installed in electrolytic cell areas and that only pneumatic-powered portable tools and equipment be used.

Layering of Protective Clothing and Total System Arc Rating

This informative annex is not a part of the requirements of this NFPA document but is included for informational purposes only.

Informative Annex M discusses how layering of protective clothing can impact the overall rating of the layered protection. The total arc rating of a layered clothing system is to be determined by testing the multilayer system as it would be worn in the field. The total system arc rating cannot be determined simply by adding together the arc ratings of the individual layers. Exhibit M.1 shows an arc-rated PPE system of clothing.

Exhibit M.2 is a manufacturer's table indicating the typical number of arc-rated clothing layers to attain an arc thermal performance value (ATPV).

When layers of clothing are worn, some air space is captured between the layers. Air is a good thermal insulator and modifies the protective nature of the sum of the protective clothing. The resulting protection offered by the layers of clothing can be greater or less than the sum of the protection afforded by the individual layers. Although this annex discusses layering of protective clothing, consultation with the clothing manufacturer is recommended.

M.1 Layering of Protective Clothing.

M.1.1 Layering of arc-rated clothing is an effective approach to achieving the required arc rating. The use of all arc-rated clothing layers will result in achieving the required arc rating with the lowest number of layers and lowest clothing system weight. Garments that are not arc-rated should not be used to increase the arc rating of a garment or of a clothing system.

M.1.2 The total system of protective clothing can be selected to take credit for the protection provided by all the layers of clothing that are worn. For example, to achieve an arc rating of 40 cal/cm^2, an arc flash suit with an arc rating of 40 cal/cm^2 could be worn over a cotton shirt and cotton pants. Alternatively, an arc flash suit with a 25 cal/cm^2 arc rating could be worn over an arc-rated shirt and arc-rated pants with an arc rating of 8 cal/cm^2 to achieve a total system arc rating of 40 cal/cm^2. This latter approach provides the required arc rating at a lower weight and with fewer total layers of fabric and, consequently, would provide the required protection with a higher level of worker comfort.

M.2 Layering Using Arc-Rated Clothing over Natural Fiber Clothing Underlayers.

M.2.1 Under some exposure conditions, natural fiber underlayers can ignite even when they are worn under arc-rated clothing.

M.2.2 If the arc flash exposure is sufficient to break open all the arc-rated clothing outerlayer or underlayers, the natural fiber underlayer can ignite and cause more severe burn injuries to

EXHIBIT M.1

An arc-rated PPE system of clothing. (Courtesy of Salisbury by Honeywell)

Salisbury pro-wear color code	Hazard risk category (HRC)	Clothing description (Typical number of clothing layers given in parentheses)	Minimum arc thermal performance exposure value (ATPV) rating of PPE
Navy blue	1	FR long-sleeve shirt and FR pants or FR coverall plus arc-rated face shield or switching hood (1)	4 cal/cm²
Navy blue	2	FR long-sleeve shirt and FR pants or FR coverall plus switching hood (or face shield with balaclava) (1 or 2)	8 cal/cm²
Royal blue	3	FR long-sleeve shirt and FR pants or FR coverall and FR jacket and FR pants or total FR clothing system with hood (2 or 3)	25 cal/cm²
Gray/Khaki	4	FR long-sleeve shirt and FR pants or FR coverall and FR jacket and FR pants or total FR clothing system with hood (2 or 3)	40 cal/cm²

an expanded area of the body. This is due to the natural fiber underlayers burning onto areas of the worker's body that were not exposed by the arc flash event. This can occur when the natural fiber underlayer continues to burn underneath arc-rated clothing layers even in areas in which the arc-rated clothing layer or layers are not broken open due to a "chimney effect."

M.3 Total System Arc Rating.

M.3.1 The total system arc rating is the arc rating obtained when all clothing layers worn by a worker are tested as a multilayer test sample. An example of a clothing system is an arc-rated coverall worn over an arc-rated shirt and arc-rated pants in which all of the garments are constructed from the same arc-rated fabric. For this two-layer arc-rated clothing system, the arc rating would typically be more than three times higher than the arc ratings of the individual layers; that is, if the arc ratings of the arc-rated coverall, shirt, and pants were all in the range of 5 cal/cm² to 6 cal/cm², the total two-layer system arc rating would be over 20 cal/cm².

M.3.2 It is important to understand that the total system arc rating cannot be determined by adding the arc ratings of the individual layers. In a few cases, it has been observed that the total system arc rating actually decreased when another arc-rated layer of a specific type was added to the system as the outermost layer. The only way to determine the total system arc rating is to conduct a multilayer arc test on the combination of all of the layers assembled as they would be worn.

The arc rating of a layered system of PPE is not simply a matter of adding together the ratings of the individual pieces; most manufacturers provide data on the arc ratings of individual parts and of the whole system as layer systems require testing. Exhibit M.3 is an example of how a manufacturer might supply such data.

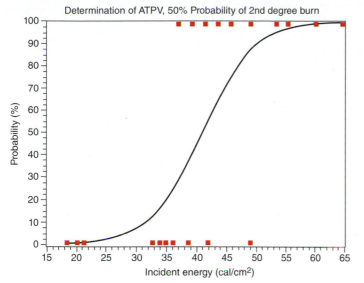

Determination of ATPV, 50% Probability of 2nd degree burn

ATPV = 40.9 cal/cm²

Probability of burn	Ei
5%	28.5
10%	31.7
20%	35.1
30%	37.4
40%	39.2
50%	**40.9**
60%	42.7
70%	44.5
80%	46.8
90%	50.2

Pts = 21
Pts above stoll = 10
Pts break-open = 3
Pts always >stoll = 5
Pts always <stoll = 8
Pts within 20% = 12
Pts in mix zone = 8

EXHIBIT M.3

Curve showing the probability of injury based on arc rating of materials in a layered PPE system. (Courtesy of ArcWear.com)

Fabric description:

Two layers, style 85917 - protera 1808.0 oz/yd² 271 g/m² 2 × 1 LH twill, 65% modacrylic 35% N317, navy 10057Q, AAD 8.0 oz/yd² 271 g/m² over style S961 indura ultra soft 11 oz/yd 373 g/m² duck, 88% cotton 12% nylon, brown, AAD 11.6 oz/yd² 393 g/m², ArcWear# 1102P86

Example Industrial Procedures and Policies for Working Near Overhead Electrical Lines and Equipment

This informative annex is not a part of the requirements of this NFPA document but is included for informational purposes only.

Informative Annex N illustrates electrical safety measures that might be appropriate for working near overhead lines. Although the content of this informative annex is not enforceable, the measures contained here are effective. See Supplement 3 for another possible safety procedure format.

N.1 Introduction.

This informative annex is an example of an industrial procedure for working near overhead electrical systems. Areas covered include operations that could expose employees or equipment to contact with overhead electrical systems.

When working near electrical lines or equipment, avoid direct or indirect contact. Direct contact is contact with any part of the body. Indirect contact is when part of the body touches or is in dangerous proximity to any object in contact with energized electrical equipment. The following two assumptions should always be made:

(1) Lines are "live" (energized).
(2) Lines are operating at high voltage (over 1000 volts).

As the voltage increases, the minimum working clearances increase. Through arc-over, injuries or fatalities could occur, even if actual contact with high-voltage lines or equipment is not made. Potential for arc-over increases as the voltage increases.

N.2 Overhead Power Line Policy (OPP).

This informative annex applies to all overhead conductors, regardless of voltage, and requires the following:

(1) That employees not place themselves in close proximity to overhead power lines. "Close proximity" is within a distance of 3 m (10 ft) for systems up to 50 kV, and should be increased 100 mm (4 in.) for every 10 kV above 50 kV.
(2) That employees be informed of the hazards and precautions when working near overhead lines.
(3) That warning decals be posted on cranes and similar equipment regarding the minimum clearance of 3 m (10 ft).
(4) That a "spotter" be designated when equipment is working near overhead lines. This person's responsibility is to observe safe working clearances around all overhead lines and to direct the operator accordingly.

(5) That warning cones be used as visible indicators of the 3 m (10 ft) safety zone when working near overhead power lines.

Informational Note: "Working near," for the purpose of this informative annex, is defined as working within a distance from any overhead power line that is less than the combined height or length of the lifting device plus the associated load length and the required minimum clearance distance [as stated in N.2(1)]. Required clearance is expressed as follows: Required clearance = lift equipment height or length + load length + at least 3 m (10 ft)

(6) That the local responsible person be notified at least 24 hours before any work begins to allow time to identify voltages and clearances or to place the line in an electrically safe work condition.

N.3 Policy.

All employees and contractors shall conform to the OPP. The first line of defense in preventing electrical contact accidents is to remain outside the limited approach boundary. Because most company and contractor employees are not qualified to determine the system voltage level, a qualified person shall be called to establish voltages and minimum clearances and take appropriate action to make the work zone safe.

N.4 Procedures.

N.4.1 General. Prior to the start of all operations where potential contact with overhead electrical systems is possible, the person in charge shall identify overhead lines or equipment, reference their location with respect to prominent physical features, or physically mark the area directly in front of the overhead lines with safety cones, survey tape, or other means. Electrical line location shall be discussed at a pre-work safety meeting of all employees on the job (through a job briefing). All company employees and contractors shall attend this meeting and require their employees to conform to electrical safety standards. New or transferred employees shall be informed of electrical hazards and proper procedures during orientations.

On construction projects, the contractor shall identify and reference all potential electrical hazards and document such actions with the on-site employers. The location of overhead electrical lines and equipment shall be conspicuously marked by the person in charge. New employees shall be informed of electrical hazards and of proper precautions and procedures.

Where there is potential for contact with overhead electrical systems, local area management shall be called to decide whether to place the line in an electrically safe work condition or to otherwise protect the line against accidental contact. Where there is a suspicion of lines with low clearance [height under 6 m (20 ft)], the local on-site electrical supervisor shall be notified to verify and take appropriate action.

All electrical contact incidents, including "near misses," shall be reported to the local area health and safety specialist.

N.4.2 Look Up and Live Flags. In order to prevent accidental contacts of overhead lines, all aerial lifts, cranes, boom trucks, service rigs, and similar equipment shall use look up and live flags. The flags are visual indicators that the equipment is currently being used or has been returned to its "stowed or cradled" position. The flags shall be yellow with black lettering and shall state in bold lettering "LOOK UP AND LIVE."

The procedure for the use of the flag follows.

(1) When the boom or lift is in its stowed or cradled position, the flag shall be located on the load hook or boom end.

(2) Prior to operation of the boom or lift, the operator of the equipment shall assess the work area to determine the location of all overhead lines and communicate this information to all crews on site. Once completed, the operator shall remove the flag from the load hook

or boom and transfer the flag to the steering wheel of the vehicle. Once the flag is placed on the steering wheel, the operator can begin to operate the equipment.

(3) After successfully completing the work activity and returning the equipment to its stowed or cradled position, the operator shall return the flag to the load hook.

(4) The operator of the equipment is responsible for the placement of the look up and live flag.

N.4.3 High Risk Tasks.

N.4.3.1 Heavy Mobile Equipment. Prior to the start of each workday, a high-visibility marker (orange safety cones or other devices) shall be temporarily placed on the ground to mark the location of overhead wires. The supervisors shall discuss electrical safety with appropriate crew members at on-site tailgate safety talks. When working in the proximity of overhead lines, a spotter shall be positioned in a conspicuous location to direct movement and observe for contact with the overhead wires. The spotter, equipment operator, and all other employees working on the job location shall be alert for overhead wires and remain at least 3 m (10 ft) from the mobile equipment.

All mobile equipment shall display a warning decal regarding electrical contact. Independent truck drivers delivering materials to field locations shall be cautioned about overhead electrical lines before beginning work, and a properly trained on-site or contractor employee shall assist in the loading or off-loading operation. Trucks that have emptied their material shall not leave the work location until the boom, lift, or box is down and is safely secured.

N.4.3.2 Aerial Lifts, Cranes, and Boom Devices. Where there is potential for near operation or contact with overhead lines or equipment, work shall not begin until a safety meeting is conducted and appropriate steps are taken to identify, mark, and warn against accidental contact. The supervisor will review operations daily to ensure compliance.

Where the operator's visibility is impaired, a spotter shall guide the operator. Hand signals shall be used and clearly understood between the operator and spotter. When visual contact is impaired, the spotter and operator shall be in radio contact. Aerial lifts, cranes, and boom devices shall have appropriate warning decals and shall use warning cones or similar devices to indicate the location of overhead lines and identify the 3 m (10 ft) minimum safe working boundary.

N.4.3.3 Tree Work. Wires shall be treated as live and operating at high voltage until verified as otherwise by the local area on-site employer. The local maintenance organization or an approved electrical contractor shall remove branches touching wires before work begins. Limbs and branches shall not be dropped onto overhead wires. If limbs or branches fall across electrical wires, all work shall stop immediately and the local area maintenance organization is to be called. When climbing or working in trees, pruners shall try to position themselves so that the trunk or limbs are between their bodies and electrical wires. If possible, pruners shall not work with their backs toward electrical wires. An insulated bucket truck is the preferred method of pruning when climbing poses a greater threat of electrical contact. Personal protective equipment (PPE) shall be used while working on or near lines.

N.4.4 Underground Electrical Lines and Equipment. Before excavation starts and where there exists reasonable possibility of contacting electrical or utility lines or equipment, the local area supervision (or USA DIG organization, when appropriate) shall be called and a request is to be made for identifying/marking the line location(s).

When USA DIG is called, their representatives will need the following:

(1) Minimum of two working days' notice prior to start of work, name of county, name of city, name and number of street or highway marker, and nearest intersection

(2) Type of work

(3) Date and time work is to begin

(4) Caller's name, contractor/department name and address

(5) Telephone number for contact

(6) Special instructions

Utilities that do not belong to USA DIG must be contacted separately. USA DIG might not have a complete list of utility owners. Utilities that are discovered shall be marked before work begins. Supervisors shall periodically refer their location to all workers, including new employees, subject to exposure.

N.4.5 Vehicles with Loads in Excess of 4.25 m (14 ft) in Height. This policy requires that all vehicles with loads in excess of 4.25 m (14 ft) in height use specific procedures to maintain safe working clearances when in transit below overhead lines.

The specific procedures for moving loads in excess of 4.25 m (14 ft) in height or via routes with lower clearance heights are as follows:

(1) Prior to movement of any load in excess of 4.25 m (14 ft) in height, the local health and safety department, along with the local person in charge, shall be notified of the equipment move.

(2) An on-site electrician, electrical construction representative, or qualified electrical contractor should check the intended route to the next location before relocation.

(3) The new site is to be checked for overhead lines and clearances.

(4) Power lines and communication lines shall be noted, and extreme care used when traveling beneath the lines.

(5) The company moving the load or equipment will provide a driver responsible for measuring each load and ensuring each load is secured and transported in a safe manner.

(6) An on-site electrician, electrical construction representative, or qualified electrical contractor shall escort the first load to the new location, ensuring safe clearances, and a service company representative shall be responsible for subsequent loads to follow the same safe route.

If proper working clearances cannot be maintained, the job must be shut down until a safe route can be established or the necessary repairs or relocations have been completed to ensure that a safe working clearance has been achieved.

All work requiring movement of loads in excess of 4.25 m (14 ft) in height are required to begin only after a general work permit has been completed detailing all pertinent information about the move.

N.4.6 Emergency Response. If an overhead line falls or is contacted, the following precautions should be taken:

(1) Keep everyone at least 3 m (10 ft) away.

(2) Use flagging to protect motorists, spectators, and other individuals from fallen or low wires.

(3) Call the local area electrical department or electric utility immediately.

(4) Place barriers around the area.

(5) Do not attempt to move the wire(s).

(6) Do not touch anything that is touching the wire(s).

(7) Be alert to water or other conductors present.

(8) Crews shall have emergency numbers readily available. These numbers shall include local area electrical department, utility, police/fire, and medical assistance.

(9) If an individual becomes energized, DO NOT TOUCH the individual or anything in contact with the person. Call for emergency medical assistance and call the local utility immediately. If the individual is no longer in contact with the energized conductors, CPR, rescue breathing, or first aid should be administered immediately, but only by a trained person. It is safe to touch the victim once contact is broken or the source is known to be de-energized.

(10) Wires that contact vehicles or equipment will cause arcing, smoke, and possibly fire. Occupants should remain in the cab and wait for the local area electrical department or utility. If it becomes necessary to exit the vehicle, leap with both feet as far away from the vehicle as possible, without touching the equipment. Jumping free of the vehicle is the last resort.

(11) If operating the equipment and an overhead wire is contacted, stop the equipment immediately and, if safe to do so, jump free and clear of the equipment. Maintain your balance, keep your feet together and either shuffle or bunny hop away from the vehicle another 3 m (10 ft) or more. Do not return to the vehicle or allow anyone else for any reason to return to the vehicle until the local utility has removed the power line from the vehicle and has confirmed that the vehicle is no longer in contact with the overhead lines.

Safety-Related Design Requirements

Summary of Annex O Changes

O.1.2: Replaced the phrase *eliminate or reduce exposure risks* with *eliminate hazards or reduce risk,* for consistency with other safety-related standards that address hazard, risk, and risk assessment. When a hazard is eliminated, the risk associated with that hazard is also eliminated. However, the elimination of all hazards may not be feasible and some risk may remain. Risk is reduced by applying the hierarchy of risk control methods.

O.2.1: Deleted the reference to 130.3(B)(1) and replaced the phrase *electrical hazard analysis* with *that electrical hazards risk assessments are performed,* to align with the revisions to Article 130. Restructured text into a list format.

O.2.2: Restructured text into a list format to clarify and promote the consistent use of terminology associated with hazard and risk.

O.2.3: Changed title from *Arc Energy Reduction to Incident Energy Reduction Methods.* Replaced the phrase *potentially hazardous work* with *work.* Added description of the functionality of an energy-reducing maintenance switch. Added methods with definitions or descriptive text to provide a list of what can assist the user in reducing incident energy levels.

This informative annex is not a part of the requirements of this NFPA document but is included for informational purposes only.

The design of a facility, equipment, or circuit could determine if a work task can be performed safely. In a large measure, the facility and circuit design determines whether or how a worker is or might be exposed to an electrical hazard when performing tasks necessary to troubleshoot, repair, or maintain a facility.

The circuit and equipment design determines the amount of incident energy that might be available at various points in the system and whether an electrically safe work condition can be created for sectors of the circuit. The location of components and isolating or insulating barriers determines whether or how an employee might be exposed to shock or electrocution.

O.1 Introduction.

This informative annex addresses the responsibilities of the facility owner or manager or the employer having responsibility for facility ownership or operations management to perform a risk assessment during the design of electrical systems and installations.

O.1.1 This informative annex covers employee safety-related design concepts for electrical equipment and installations in workplaces covered by the scope of this standard. This informative annex discusses design considerations that have impact on the application of the safety-related work practices only.

O.1.2 This informative annex does not discuss specific design requirements. The facility owner or manager or the employer should choose design options that eliminate hazards or reduce risk and enhance the effectiveness of safety-related work practices.

O.2 General Design Considerations.

O.2.1 Employers, facility owners, and managers who have responsibility for facilities and installations having electrical energy as a potential hazard to employees and other personnel should ensure that electrical hazards risk assessments are performed during the design of electrical systems and installations.

O.2.2 Design option decisions should facilitate the ability to eliminate hazards or reduce risk by doing the following:

(1) Reducing the likelihood of exposure
(2) Reducing the magnitude or severity of exposure
(3) Enabling achievement of an electrically safe work condition

O.2.3 Incident Energy Reduction Methods. The following methods have proved to be effective in reducing incident energy:

(1) Zone-selective interlocking. A method that allows two or more circuit breakers to communicate with each other so that a short circuit or ground fault will be cleared by the breaker closest to the fault with no intentional delay. Clearing the fault in the shortest time aids in reducing the incident energy.
(2) Differential relaying. The concept of this protection method is that current flowing into protected equipment must equal the current out of the equipment. If these two currents are not equal, a fault must exist within the equipment, and the relaying can be set to operate for a fast interruption. Differential relaying uses current transformers located on the line and load sides of the protected equipment and fast acting relay.
(3) Energy-reducing maintenance switching with a local status indicator. An energy-reducing maintenance switch allows a worker to set a circuit breaker trip unit to operate faster while the worker is working within an arc flash boundary, as defined in NFPA 70E, and then to set the circuit breaker back to a normal setting after the work is complete.

O.2.4 Other Methods.

(1) Energy-reducing active arc flash mitigation system. This system can reduce the arcing duration by creating a low impedance current path, located within a controlled compartment, to cause the arcing fault to transfer to the new current path, while the upstream breaker clears the circuit. The system works without compromising existing selective coordination in the electrical distribution system.

(2) Arc flash relay. An arc flash relay typically uses light sensors to detect the light produced by an arc flash event. Once a certain level of light is detected the relay will issue a trip signal to an upstream overcurrent device.

(3) High-resistance grounding. A great majority of electrical faults are of the phase-to-ground type. High-resistance grounding will insert an impedance in the ground return path and will typically limit the fault current to 10 amperes and below (at 5 kV nominal or below), leaving insufficient fault energy and thereby helping reduce the arc flash hazard level. High-resistance grounding will not affect arc flash energy for line-to-line or line-to-line-to-line arcs.

(4) Current-limiting devices. Current-limiting protective devices reduce incident energy by clearing the fault faster and by reducing the current seen at the arc source. The energy reduction becomes effective for current above the current-limiting threshold of the current-limiting fuse or current limiting circuit breaker.

There are two IEEE projects the reader should be aware of regarding safety-related design requirements:

- IEEE P1683 — *Draft Guide for Motor Control Centers Rated up to and Including 600 Vac or 1000 Vdc with Recommendations Intended to Help Reduce Electrical Hazards.* This proposed guide would provide functional design and factory verification test recommendations for motor control centers (MCC) intended to reduce the likelihood of shock and arc flash injuries. The proposed guide would apply to single- and three-phase, 50- and 60-Hz ac and dc MCCs rated not more than 600 V ac or 1000 V dc. The recommendations are to facilitate performance of defined maintenance tasks on energized equipment and augment the existing requirements of the applicable MCC safety standards for MCCs, such as NEMA ICS 18-2001, *Motor Control Centers*, and ANSI/UL 845-2005, *Motor Control Centers.* The proposed guide would identify field practices and interface relationships between the specifier, manufacturer, installer, and user for safety concerns.

- IEEE P1814 — *Recommended Practice for Electrical System Design Techniques to Improve Electrical Safety.* This proposed recommended practice would address system and equipment design techniques and equipment selection to improve electrical safety. The techniques would supplement the minimum requirements of installation codes and equipment standards.

In addition, the reader should also be aware of the following documents:

- ANSI/IEEE C37.20.7-2007, *IEEE Guide for Testing Metal-Enclosed Switchgear Rated Up to 38 kV for Internal Arcing Faults.* It includes a procedure for testing and evaluating the performance of metal-enclosed switchgear for internal arcing faults. Furthermore, it includes a method of identifying the capabilities of this equipment. It also discusses service conditions, installation, and application of equipment.

- ANSI/IEEE C37.20.7-2007/Cor. 1-2010, *IEEE Guide for Testing Metal-Enclosed Switchgear Rated up to 38 kV for Internal Arcing Faults, Corrigendum 1.* This corrigendum corrects technical errors found in IEEE C37.20.7-2007 during use in laboratories concerning current values and arc initiation in low-voltage testing and supply frequency for equipment.

Aligning Implementation of This Standard with Occupational Health and Safety Management Standards

P

This informative annex is not a part of the requirements of this NFPA document but is included for informational purposes only.

ANSI/AIHA Z10, *American National Standard for Occupational Health and Safety Management Systems*, defines the minimum basic requirements for a company's or an organization's overall occupational health and safety management system. The standard addresses management leadership, employee participation, planning, implementation, evaluation, corrective action, and management review. It covers basic activities such as incident investigation, inspections, and training. The standard includes required interrelated processes suitable for continual improvement, and it aligns with the Deming Cycle. The Deming Cycle, Plan – Do – Check – Act, or PDCA, is an iterative four-step management process also known as the Deming circle or wheel, or Shewhart Cycle or PDSA (plan – do – study – act). The Deming Cycle is shown in Exhibit P.1. The management system in ANSI/AIHA Z10 is planned to allow for continual improvement of health and safety management.

PDCA is a successive cycle that starts off from a known base and tests small potential effects on the process, then gradually leads to larger and more targeted change based on a continuous process basis. Under the Plan phase, the objectives and processes necessary to deliver results in accordance with the expected output or goal are established. Under the Do phase, the new process is implemented on a small scale to test possible effects. Under the Check phase, the new process is measured and the results of the Do phase are compared against the expected results to determine any differences. Finally, under the Act phase, the differences are analyzed to determine their cause. Under this phase, the changes to be made and where they apply to cause improvements are determined. When a cycle through these four steps does not result in an improvement, the original base is returned to and the process is tried again until there is a plan that involves improvement.

A fundamental principle of PDCA is reiteration. Reiterating the PDCA cycle works to bring a company, organization, group, or production area closer to its goals. PDCA needs to be repeated in ever-tightening circles that represent increasing knowledge of the system, ultimately ending on or near the fundamental goals of that system. If a small misstep is made, it is easy to repeat the previous iteration. Small steps from a solid base provide feedback to justify hypotheses and increase understanding of the system being studied.

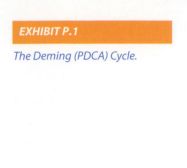

EXHIBIT P.1

The Deming (PDCA) Cycle.

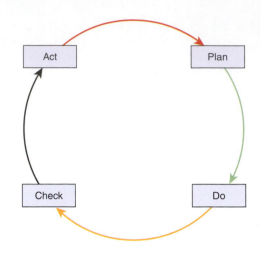

The power of this concept lies in its simplicity. The simple cycle can be continually reapplied to the process in question for improvement.

P.1 General.

Injuries from electrical energy are a significant cause of occupational fatalities in the workplace in the United States. This standard specifies requirements unique to the hazards of electrical energy. By itself, however, this standard does not constitute a comprehensive and effective electrical safety program. The most effective application of the requirements of this standard can be achieved within the framework of a recognized health and safety management system standard. ANSI/AIHA Z10, *American National Standard for Occupational Health and Safety Management Systems*, provides comprehensive guidance on the elements of an effective health and safety management system and is one recognized standard. ANSI/AIHA Z10 is harmonized with other internationally recognized standards, including CAN/CSA Z1000, *Occupational Health and Safety Management; ANSI/*ISO 14001, *Environmental Management Systems - Requirements with Guidance for Use*; and **BS** OSHAS 18001, *Occupational Health and Safety Management Systems.* Some companies and other organizations have proprietary health and safety management systems that are aligned with the key elements of ANSI/AIHA Z10.

The most effective design and implementation of an electrical safety program can be achieved through a joint effort involving electrical subject matter experts and safety professionals knowledgeable about safety management systems.

Such collaboration can help ensure that proven safety management principles and practices applicable to any hazard in the workplace are appropriately incorporated into the electrical safety program.

This informative annex provides guidance on implementing this standard within the framework of ANSI/AIHA Z10 and other recognized or proprietary comprehensive occupational health and safety management system standards.

According to ANSI/AIHA Z10, a *hazard* is defined as "[a] condition, set of circumstances, or inherent property that can cause injury, illness, or death". *Incident* is defined as "[a]n event in which a work-related injury or illness (regardless of severity) or fatality occurred or could have occurred (commonly referred to as a 'close call' or 'near miss')." *Risk* is defined as "[a]n estimate of the combination of the likelihood of an occurrence of a hazardous event or exposure(s), and the severity of injury or illness that may be caused by the event or exposures."

ANSI/AIHA Z10 outlines provisions for the use of a hierarchy of controls — health and safety controls. From an electrical safety perspective, we are specifically interested in electrical safety controls. The hierarchy of health and safety controls offers a methodical approach to reducing the risk associated with a hazard. ANSI/AIHA Z10 lists the following methods of reducing risk, from most effective to least effective: elimination, substitution, engineering controls, warnings, administrative controls, and personal protective equipment (PPE). A control that can be used after the fact is mitigation. Exhibit P.2 provides a graphical representation of the hierarchy of health and safety controls.

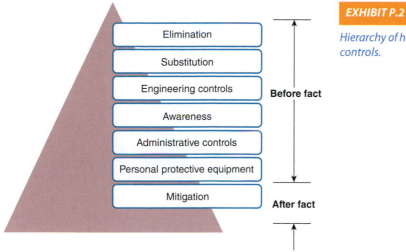

EXHIBIT P.2

Hierarchy of health and safety controls.

The highest level of practical control should be used. In many cases, it may be found that a combination of controls is the most effective method. This is frequently the case where a higher order control, such as substitution, is not practical or does not reduce the risk to an acceptable level. In this case a combination of the lower order controls, such as warnings and PPE, may be required when performing tasks such as testing equipment or creating an electrically safe work condition.

The use of arc-rated switchgear may be considered substitution or an engineering control. Administrative controls include such items as job planning, training, work procedures and practices, temporary barricades, and the like. Mitigation, such as an incident response plan, is one additional control indicated in Exhibit P.2. Mitigation is applied after an event or incident to minimize the damage; it is not one of the controls identified in ANSI/AIHA Z10.

PART 2

Supplements

The four supplements included in Part Two of the *Handbook for Electrical Safety in the Workplace* provide additional information as well as supporting reference material to assist users and enforcers of *NFPA 70E®*, *Standard for Electrical Safety in the Workplace®*. The supplements are not part of *NFPA 70E* or the commentary presented in Part One of this book.

National Electrical Code
Requirements Associated with Safety-Related Work Practices

Editor's Note: *This supplement consists of a list of requirements from the 2014 edition of NFPA 70®, National Electrical Code®, that directly impact the implementation of safety-related work practices from NFPA 70E®, Standard for Electrical Safety in the Workplace®.*

NEC Requirements for Disconnecting Means

110.25	Lockable Disconnecting Means	230.93	Protection of Specific Circuits
225.30	Number of Supplies	230.94	Relative Location of Overcurrent Device and Other Service Equipment
225.31	Disconnecting Means	230.200	General
225.32	Location	230.204	Isolating Switches
225.33	Maximum Number of Disconnects	230.205	Disconnecting Means
225.34	Grouping of Disconnects	230.206	Overcurrent Devices as Disconnecting Means
225.35	Access to Occupants	240.24	Location in or on Premises
225.40	Access to Overcurrent Protective Devices	240.40	Disconnecting Means for Fuses
225.51	Isolating Switches	240.41	Arcing or Suddenly Moving Parts
225.52	Disconnecting Means	404.4	Damp or Wet Locations
230.2	Number of Services	404.6	Position and Connection of Switches
230.3	One Building or Other Structure Not to Be Supplied Through Another	404.7	Indicating
230.70	General	404.8	Accessibility and Grouping
230.71	Maximum Number of Disconnects	404.9	Provisions for General-Use Snap Switches
230.72	Grouping of Disconnects	404.11	Circuit Breakers as Switches
230.74	Simultaneous Opening of Poles	404.13	Knife Switches
230.75	Disconnection of Grounded Conductor	422.30	General
230.76	Manually or Power Operable	422.31	Disconnection of Permanently Connected Appliances
230.77	Indicating	422.35	Switch and Circuit Breaker to Be Indicating
230.82	Equipment Connected to the Supply Side of Service Disconnect	424.19	Disconnecting Means
230.91	Location	424.21	Switch and Circuit Breaker to Be Indicating
230.92	Locked Service Overcurrent Devices	424.65	Location of Disconnecting Means

NEC Requirements for Disconnecting Means (*contd*)

424.66	Installation
424.86	Markings
426.50	Disconnecting Means
426.51	Controllers
427.55	Disconnecting Means
430.74	Electrical Arrangement of Control Circuits
430.75	Disconnection
430.95	Service Equipment
430.102	Location
430.103	Operation
430.111	Switch or Circuit Breaker as Both Controller and Disconnecting Means
430.112	Motors Served by Single Disconnecting Means
430.113	Energy from More Than One Source
430.227	Disconnecting Means
430.232	Where Required
440.13	Cord-Connected Equipment
440.14	Location
440.63	Disconnecting Means
455.8	Disconnecting Means
455.9	Connection of Single-Phase Loads
455.10	Terminal Housings
455.20	Disconnecting Means
455.21	Start-Up
455.22	Power Interruption
460.28	Means for Discharge
490.21	Circuit-Interrupting Devices
490.22	Isolating Means
490.38	Door Stops and Cover Plates
490.39	Gas Discharge from Interrupting Devices
490.40	Visual Inspection Windows
490.41	Location of Industrial Control Equipment
490.45	Circuit Breakers — Interlocks
490.46	Circuit Breaker Locking
490.48	Substation Design, Documentation, and Required Diagram
490.51	General
514.11	Circuit Disconnects

514.13	Provisions for Maintenance and Service of Dispensing Equipment
600.6	Disconnects
610.32	Disconnecting Means for Cranes and Monorail Hoists
620.51	Disconnecting Means
620.52	Power from More Than One Source
620.53	Car Light, Receptacle(s), and Ventilation Disconnecting Means
620.54	Heating and Air-Conditioning Disconnecting Means
620.55	Utilization Equipment Disconnecting Means
620.71	Guarding Equipment
625.18	Interlock
625.42	Disconnecting Means
625.46	Loss of Primary Source
630.13	Disconnecting Means
630.33	Disconnecting Means
645.10	Disconnecting Means
645.11	Uninterruptible Power Supplies (UPSs)
665.7	Remote Control
665.12	Disconnecting Means
665.19	Component Interconnection
668.13	Disconnecting Means
670.4	Supply Conductors and Overcurrent Protection
680.12	Maintenance Disconnecting Means
690.13	Building or Other Structure Supplied by a Photovoltaic System
690.15	Disconnection of Photovoltaic Equipment
690.16	Fuses
690.17	Disconnect Type
690.18	Installation and Service of an Array
690.71	Installation
694.18	Stand-Alone Systems
694.20	All Conductors
694.22	Additional Provisions
694.24	Disconnection of Wind Electric System Equipment
694.26	Fuses

694.28	Installation and Service of a Wind Turbine
694.52	Power Systems Employing Energy Storage
694.54	Identification of Power Sources
694.56	Instructions for Disabling Turbine
695.4	Continuity of Power
700.16	Emergency Illumination

700.26	Accessibility
705.12	Point of Connection
705.20	Disconnecting Means, Sources
705.21	Disconnecting Means, Equipment
705.22	Disconnect Device

NEC Requirements for Working Space and Clearances

110.26	Spaces About Electrical Equipment
110.27	Guarding of Live Parts
110.30	General
110.31	Enclosure for Electrical Installations
110.32	Work Space About Equipment
110.33	Entrance to Enclosures and Access to Working Space
110.34	Work Space and Guarding
110.70	General
110.71	Strength
110.72	Cabling Work Space
110.73	Equipment Work Space
110.74	Conductor Installation.
110.75	Access to Manholes
110.76	Access to Vaults and Tunnels
110.77	Ventilation
110.78	Guarding
110.79	Fixed Ladders
225.18	Clearance for Overhead Conductors and Cables
225.19	Clearances from Buildings for Conductors of Not over 1000 Volts, Nominal
225.25	Location of Outdoor Lamps
225.26	Vegetation as Support
225.60	Clearances over Roadways, Walkways, Rail, Water, and Open Land
225.61	Clearances over Buildings and Other Structures
230.9	Clearances on Buildings
230.24	Clearances
230.26	Point of Attachment
230.62	Service Equipment — Enclosed or Guarded
408.5	Clearance for Conductor Entering Bus Enclosures
408.17	Location Relative to Easily Ignitible Material
408.18	Clearances

408.20	Location of Switchboards and Switchgear
408.56	Minimum Spacings
410.5	Live Parts
410.6	Listing Required
410.8	Inspection
410.10	Luminaires in Specific Locations
410.11	Luminaires Near Combustible Material
410.12	Luminaires over Combustible Material
410.14	Luminaires in Show Windows
430.231	General
430.232	Where Required
430.233	Guards for Attendants
445.14	Protection of Live Parts
445.15	Guards for Attendants
450.8	Guarding
460.2	Enclosing and Guarding
480.9	Battery Locations
490.24	Minimum Space Separation
490.32	Guarding of High-Voltage Energized Parts Within a Compartment
490.33	Guarding of Energized Parts Operating at 1000 Volts, Nominal, or Less Within Compartments
490.35	Accessibility of Energized Parts
490.42	Interlocks — Interrupter Switches
490.43	Stored Energy for Opening
590.7	Guarding
610.57	Clearance
620.4	Live Parts Enclosed
620.5	Working Clearances
665.22	Access to Internal Equipment
668.10	Cell Line Working Zone
680.8	Overhead Conductor Clearances
690.34	Access to Boxes

NEC Requirements for Field Markings

110.15	High-Leg Marking		490.53	Enclosures
110.16	Arc-Flash Hazard Warning		490.55	Power Cable Connections to Mobile Machines
110.21	Marking		669.8	Disconnecting Means
110.22	Identification of Disconnecting Means		690.4	General Requirements
110.24	Available Fault Current		690.10	Stand-Alone Systems
210.5	Identification for Branch Circuits		690.35	Ungrounded Photovoltaic Power Systems
215.12	Identification for Feeders		690.54	Interactive System Point of Interconnection
225.37	Identification		694.50	Interactive System Point of Interconnection
230.56	Service Conductor with the Higher Voltage to Ground		700.7	Signs
408.4	Field Identification Required		701.7	Signs
408.58	Panelboard Marking		702.7	Signs
409.110	Marking		705.10	Directory
490.48	Substation Design, Documentation, and Required Diagram			

NEC Requirements for Overcurrent Protection

110.9	Interrupting Rating		240.83	Marking
110.10	Circuit Impedance, Short-Circuit Current Ratings, and Other Characteristics		240.86	Series Ratings
			240.87	Arc Energy Reduction
210.13	Ground-Fault Protection of Equipment		240.90	General
210.20	Overcurrent Protection		240.91	Protection of Conductors
215.10	Ground-Fault Protection of Equipment		240.92	Location in Circuit
240.4	Protection of Conductors		490.52	Overcurrent Protection
240.8	Fuses or Circuit Breakers in Parallel		620.62	Selective Coordination
240.12	Electrical System Coordination		700.26	Accessibility
240.60	General		700.27	Ground-Fault Protection of Equipment
240.61	Classification		700.28	Selective Coordination
240.80	Method of Operation		701.25	Accessibility
240.81	Indicating		701.26	Ground-Fault Protection of Equipment
240.82	Nontamperable		700.27	Selective Coordination

NEC Requirements for Equipment for General Use

110.2	Approval		250.116	Nonelectrical Equipment
110.3	Examination, Identification, Installation, and Use of Equipment		250.188	Grounding of Systems Supplying Portable or Mobile Equipment
110.11	Deteriorating Agents		406.4	General Installation Requirements
250.34	Portable and Vehicle-Mounted Generators		406.8	Noninterchangeability
250.94	Bonding for Other Systems		406.9	Receptacles in Damp or Wet Locations

406.10	Grounding-Type Receptacles, Adapters, Cord Connectors, and Attachment Plugs
410.30	Supports
410.36	Means of Support
410.82	Portable Luminaires
410.84	Cord Bushings
410.96	Lampholders in Wet or Damp Locations
410.97	Lampholders Near Combustible Material
410.145	Exposure to Damage

410.146	Marking
422.4	Live Parts
426.30	Personnel Protection
427.25	Personnel Protection
430.14	Location of Motors
430.16	Exposure to Dust Accumulations
450.48	Storage in Vaults
600.10	Portable or Mobile Signs

NEC Requirements for Wiring Methods

250.114	Equipment Connected by Cord and Plug

NEC Requirements for Flexible Cords and Cables

400.7	Uses Permitted
400.8	Uses Not Permitted
400.9	Splices

400.10	Pull at Joints and Terminals
400.14	Protection from Damage
400.36	Splices and Terminations

NEC Circuit Requirements

210.4	Multiwire Branch Circuits
210.7	Multiple Branch Circuits
250.4	General Requirements for Grounding and Bonding

250.6	Objectionable Current
250.138	Cord-and-Plug-Connected Equipment
422.51	Vending Machines
422.52	Electric Drinking Fountains

Other *NEC* Requirements Related to Safe Work Practices

90.1	Purpose
90.4	Enforcement
90.7	Examination of Equipment for Safety
110.18	Arcing Parts
110.23	Current Transformers
210.8	Ground-Fault Circuit-Interrupter Protection for Personnel
215.9	Ground-Fault Circuit-Interrupter Protection for Personnel
225.56	Inspections and Tests
230.10	Vegetation as Support

460.6	Discharge of Stored Energy
590.1	Scope
590.2	All Wiring Installations
590.3	Time Constraints
590.4	General
590.5	Listing of Decorative Lighting
590.6	Ground-Fault Protection for Personnel
610.55	Limit Switch
700.3	Tests and Maintenance
701.3	Tests and Maintenance

Electrical Preventive Maintenance Programs

NFPA 70B provides detailed information on the maintenance of electrical equipment. Chapter 4 explains the benefits of an electrical preventative maintenance program, and Chapters 5 and 6 provide specific information on how to set up a maintenance program. Scheduled maintenance tends to be more systematic and orderly, while breakdown maintenance is often performed under stressful conditions that could tempt workers to take dangerous safety shortcuts.

Businesses can no longer tolerate equipment failures as a signal that maintenance of equipment is necessary. Just-in-time (JIT) manufacturing and Six Sigma programs dictate an effective preventive maintenance program that ensures continuity of operations and prevents equipment breakdowns. Breakdowns affect quality and production schedules but they also disrupt customers' supply chains. JIT falls apart when suppliers cannot perform.

CHAPTER 4   Why an Effective Electrical Preventive Maintenance (EPM) Program Pays Dividends

4.1 Why EPM?

4.1.1 Electrical equipment deterioration is normal, and equipment failure is inevitable. However, equipment failure can be delayed through appropriate EPM. As soon as new equipment is installed, a process of normal deterioration begins. Unchecked, the deterioration process can cause malfunction or an electrical failure. Deterioration can be accelerated by factors such as a hostile environment, overload, or severe duty cycle. An effective EPM program identifies and recognizes these factors and provides measures for coping with them.

4.1.2 In addition to normal deterioration, other potential causes of equipment degradation can be detected and corrected through EPM. Among these are load changes or additions, circuit alterations, improperly set or improperly selected protective devices, and changing voltage conditions.

4.1.3 Without an EPM program, management assumes a greatly increased risk of a serious electrical failure and its consequences.

4.2 Value and Benefits of a Properly Administered EPM Program.

4.2.1 A well-administered EPM program reduces accidents, saves lives, and minimizes costly breakdowns and unplanned shutdowns of production equipment. Impending troubles can be identified — and solutions applied — before they become major problems requiring more expensive, time-consuming solutions.

4.2.2 Benefits of an effective EPM program fall into two general categories. Direct, measurable economic benefits are derived from reduced cost of repairs and reduced equipment downtime. Less measurable but very real benefits result from improved safety. To understand fully how personnel and equipment safety are served by an EPM program, the mechanics of the program — inspection, testing, and repair procedures — should be understood. Such an understanding explains other intangible benefits such as improved employee morale, better workmanship and increased productivity, reduced absenteeism, reduced interruption of production, and improved insurance considerations. Improved morale comes with employee awareness of a conscious management effort to promote safety by reducing the likelihood of electrical injuries or fatalities, electrical explosions, and fires. Reduced personnel injuries and property loss claims can help keep insurance premiums at favorable rates.

4.2.3 Some of the benefits that result from improved safety are difficult to measure. However, direct and measurable economic benefits can be documented by equipment repair cost and equipment downtime records after an EPM program has been implemented.

4.2.4 Dependability can be engineered and built into equipment, but effective maintenance is required to keep it dependable. Experience shows that equipment is reduced when it is covered by an EPM program. In many cases, the investment in EPM is small compared with the cost of accidents, equipment repair, and the production losses associated with unexpected outages.

4.2.5 Careful planning is the key to the economic success of an EPM program. With proper planning, maintenance costs can be held to a practical minimum, while production is maintained at a practical maximum.

4.2.6 An EPM program requires the support of top management, because top management provides the funds that are required to initiate and maintain the program. The maintenance of industrial electrical equipment is essentially a matter of business economics. Maintenance costs can be placed in either of two basic categories: preventive maintenance or breakdown repairs. The money spent for preventive maintenance will be reflected as less money required for breakdown repairs. An effective EPM program holds the sum of these two expenditures to a minimum.

4.2.7 An EPM program is a form of protection against accidents, lost production, and loss of profit. An EPM program enables management to place a monetary value on the cost of such protection. An effective EPM program satisfies an important part of management's responsibility for keeping costs down and production up.

CHAPTER 5 What Is an Effective Electrical Preventive Maintenance (EPM) Program?

5.1 Introduction. An effective electrical preventive maintenance (EPM) program should enhance safety and also reduce equipment failure to a minimum consistent with good economic judgment.

5.2 Essential Elements of an EPM Program. An EPM program should consist of the following essential elements:

(1) Responsible and qualified personnel
(2) Regularly scheduled inspection, testing, and servicing of equipment
(3) Survey and analysis of electrical equipment and systems to determine maintenance requirements and priorities
(4) Programmed routine inspections and suitable tests

(5) Accurate analysis of inspection and test reports so that proper corrective measures can be prescribed

(6) Performance of necessary work

(7) Concise but complete records

NFPA 70B contains detailed information regarding each of these elements. A survey and analysis of essential equipment will evaluate its operating parameters and current condition to determine any necessary repair as well as the nature and frequency of required inspection and testing. Analysis of the inspection reports requires follow-through with necessary repairs, replacement, and adjustment for an effective electrical preventive maintenance (EPM) program. A facility might not have qualified personnel to perform the maintenance duties and qualified contractors may be necessary to follow the EPM program. Programmed inspection and testing may require coordination between maintenance and production to schedule outages of production equipment under the maintenance program. Records should be kept not only of the condition and repair of equipment but also of the training and skills of the employee conducting the maintenance.

5.3 Planning an EPM Program. The following factors should be considered in the planning of an EPM program.

(1) *Personnel Safety:* Will an equipment failure endanger or threaten the safety of any personnel? What can be done to ensure personnel safety?

(2) *Equipment Loss:* Is installed equipment — both electrical and mechanical — complex or so unique that required repairs would be unusually expensive?

(3) *Production Economics:* Will breakdown repairs or replacement of failed equipment require extensive downtime? How many production dollars will be lost in the event of an equipment failure? Which equipment is most vital to production?

A priority plan of essential equipment will determine how, where, and when each piece fits into the maintenance program. An effective EPM program needs to consider not only equipment loss but the economics of lost production and the impact of an equipment failure on employee safety.

Effective EPM programs begin with good design. In the design of new facilities, a conscious effort to ensure optimum maintainability is recommended. For example, auxiliary power sources can make it easier to schedule and perform maintenance work with minimum interruption of production.

CHAPTER 6 Planning and Developing an Electrical Preventive Maintenance (EPM) Program

6.1 Introduction.

6.1.1 The purpose of an EPM program is to reduce hazard to life and property resulting from the failure or malfunction of electrical systems and equipment. This chapter explains the planning and development considerations that can be used to establish such a program.

6.1.2 The following four basic steps should be taken in the planning and development of an EPM program:

(1) Compile a listing of all equipment and systems.

(2) Determine which equipment and systems are most critical.

(3) Develop a system for monitoring.

(4) Determine the internal and/or external personnel needed to implement and maintain the EPM program.

6.1.5 The work center of each maintenance work group should be conveniently located. This work center should contain the following:

(1) Copies of all the inspection and testing procedures for that zone
(2) Copies of previous reports
(3) Single-line diagrams
(4) Schematic diagrams
(5) Records of complete nameplate data
(6) Vendors' catalogs
(7) Facility stores' catalogs
(8) Supplies of report forms

6.2 Survey of Electrical Installation.

6.2.1 Data Collection.

6.2.1.1 The first step in organizing a survey should be to examine available resources. Will the available personnel permit the survey of an entire system, process, or building, or should it be divided into segments?

6.2.1.2 Where the project will be divided into segments, a priority should be assigned to each segment. Segments found to be related should be identified before the actual work commences.

6.2.1.3 The third step should be the assembling of all documentation. This might necessitate a search of desks, cabinets, computers, and such, and might also require that manufacturers be contacted, to replace lost documents. All of the documents should be centralized, controlled, and maintained. The documentation should include recommended practices and procedures for some or all of the following:

(1) Installation
(2) Disassembly/assembly (interconnections)
(3) Wiring diagrams, schematics, bills of materials
(4) Operation (set-up and adjustment)
(5) Maintenance (including parts list and recommended spares)
(6) Software program (if applicable)
(7) Troubleshooting

6.2.2 Diagrams and Data. The availability of up-to-date, accurate, and complete diagrams is the foundation of a successful EPM program. The diagrams discussed in 6.2.2.1 through 6.2.2.8.2 are some of those in common use.

Single line, circuit-routing, and schematic diagrams are just a few of the diagrams discussed in 6.2.2. These diagrams assist maintenance personnel by providing information such as equipment ratings, the physical location of conductors and equipment, and access points for raceways or pull boxes. The manufacturer's service manuals and instructions are also important for the diagrams and recommended practices and procedures. Another important aspect of the EPM program is that equipment installation changes be highlighted and noted or revised in an appropriate manner to keep the documentation current.

6.2.3 System Diagrams. System diagrams should be provided to complete the data being assembled. The importance of the system determines the extent of information shown. The information can be shown on the most appropriate type of diagram but should include the same basic information, source and type of power, conductor and raceway information, and switching and protective devices with their physical locations. It is vital to show where the system might interface with another system, such as with emergency power; hydraulic, pneumatic, or mechanical systems; security and fire-alarm systems; and monitoring and control systems. Some of the more common of these are described in 6.2.3.1 through 6.2.3.3.

System diagrams for lighting; heating, ventilation, and air-conditioning; and control and monitoring are also important. Lighting diagrams may show lighting that is used for night security personnel or the location of emergency lighting. Diagrams for ventilation systems should show the interface with other systems, such as for smoke removal. Control and monitoring systems are often complicated, and the diagram should describe how these systems function. The referenced sections of NFPA 70B provide further information on these system diagrams.

6.2.5 Test and Maintenance Equipment.

6.2.5.1 All maintenance work requires the use of proper tools and equipment to properly perform the task to be done. In addition to their ordinary tools, maintenance personnel (such as carpenters, pipe fitters, and machinists) use special tools or equipment based on the nature of the work to be performed. The electrician is no exception, but for EPM, special-use tools should be readily available. The size of the facility, the nature of its operations, and the extent of its maintenance, repair, and test facilities are all factors that determine the use frequency of the equipment. Economics seldom justify purchasing an infrequently used, expensive tool when it can be rented. However, a corporation having a number of facilities in the area might well justify common ownership of the same device for joint use, making it quickly available at any time to any facility. Typical examples might be high-current test equipment, infrared thermography equipment, or a ground-fault locator.

6.2.5.2 Because a certain amount of mechanical maintenance is often a part of the EPM program being conducted on associated equipment, the electrical maintenance personnel should have ready access to such items as the following:

(1) Assorted lubrication tools and equipment
(2) Various types and sizes of wrenches
(3) Nonmetallic hammers and blocks to protect against injury to machined surfaces
(4) Feeler gauges to function as inside- and outside-diameter measuring gauges
(5) Instruments for measuring torque, tension, compression, vibration, and speed
(6) Standard and special mirrors with light sources for visual inspection
(7) Industrial-type portable blowers and vacuums having insulated nozzles for removal of dust and foreign matter
(8) Nontoxic, nonflammable cleaning solvents
(9) Clean, lint-free wiping cloths

6.2.5.3 The use of well-maintained safety equipment is essential and should be mandatory for work on energized electrical conductors or circuit parts. Prior to performing maintenance on energized electrical conductors or circuit parts, *NFPA 70E, Standard for Electrical Safety in the Workplace,* should be used to identify the degree of personal protective equipment (PPE) required. Some of the more important equipment that should be provided includes the following:

(1) Heavy leather gloves

(2) Insulating gloves, mats, blankets, baskets, boots, jackets, and coats

(3) Insulated hand tools such as screwdrivers and pliers

(4) Nonmetallic hard hats with suitable arc-rated face protection

(5) Poles with hooks and hot sticks to safely open isolating switches

Paragraphs 6.2.5.4 through 6.2.5.9 of NFPA 70B further describe various test and maintenance equipment.

6.3 Identification of Critical Equipment.

6.3.1 Equipment (electric or otherwise) should be considered critical if its failure to operate normally and under complete control will cause a serious threat to people, property, or the product. Electric power, like process steam, water, and so forth, might be essential to the operation of a machine, but unless loss of one or more of these supplies causes the machine to become hazardous to people, property, or production, that machine might not be critical. The combined knowledge and experience of several people might be needed to make this determination. In a small facility, the facility engineer or master mechanic working with the operating superintendent should be able to make this determination.

Section 6.3 describes some systems or segments of systems that may be critical to the operation of a facility. The equipment's relation to the entire operation and the effect of its loss on safety and production should be considered. An entire system may be critical but due to its size or complexity may be segmented.

6.4 Establishment of a Systematic Program.
The purpose of any inspection and testing program is to establish the condition of equipment to determine what work should be done and to verify that it will continue to function until the next scheduled servicing occurs. Inspection and testing are best done in conjunction with routine maintenance. In this way, many minor items that require no special tools, training, or equipment can be corrected as they are found. The inspection and testing program is probably the most important function of a maintenance department in that it establishes what should be done to keep the system in service to perform the function for which it is required.

Several aspects of an installation can have an impact on the frequency of inspection and testing. Section 6.4 points out that the atmosphere or environment, as well as the load the equipment is operating under, should be taken into consideration for the EPM program. For example, dust may impede a motor's ability to dissipate heat, or a motor designed for continuous operation but installed with an intermittent load could overheat the motor windings. Section 6.4 also expands on considerations for inspection frequencies, such as for equipment operating continuously or considered critical.

6.5 Methods and Procedures.

6.5.1 General.

6.5.1.1 If a system is to operate without failure, not only should the discrete components of the system be maintained, but the connections between these components also should be covered by a thorough set of methods and procedures. Overlooking this important link in the system causes many facilities to suffer high losses every year.

6.5.1.2 Other areas where the maintenance department should develop its own procedures are shutdown safeguards, interlocks, and alarms. Although the individual pieces of equipment can have testing and calibrating procedures furnished by the manufacturer, the equipment application is probably unique, so the system should have an inspection and testing procedure developed.

Section 6.5 breaks methods and procedures into five categories; forms and reports, planning, analysis of safety procedures, records, and emergency procedures. Forms can come in many varieties, and Section 6.5 lists items that should be included on a form. Some examples of forms for inspection, testing, and repair are provided in Annex H of NFPA 70B. The planning stage is where the scheduling of the inspections can be set to reduce production downtime. Many details involved in setting safety procedures are beyond the scope of NFPA 70B. Nonetheless, general considerations are covered in Section 6.5. Records are important to provide a method of evaluating the results of the EPM program, including the cost not only of the program but of the estimated cost of business interruption. Lastly, Section 6.5 recognizes the need for maintenance personnel to be prepared for emergencies.

6.8 Outsourcing of Electrical Equipment Maintenance.

6.8.1 General. This section describes the process for a facility to request the services of qualified contractors to perform maintenance on electrical equipment.

6.8.2 Contract Elements. Elements in a contract for outsourcing electrical maintenance service are to include, but are not limited to, the following:

(1) Define project scope of work, what is included and not included, along with equipment specifications on any new or replacement parts, and the time period(s) in which the activities are to be performed.

(2) Determine if it is a performance-based or detailed (step-by-step) specification.

(3) Determine which safety and maintenance codes and standards are to be followed, including appropriate permits.

(4) Determine methodology for pricing: lump sum or unit price.

(5) Determine the qualifications of potential contractors and develop and maintain a list of such.

(6) Obtain appropriate liability, insurance coverage, and warranty information.

(7) Assemble the appropriate up-to-date and accurate facility and equipment specific documents, such as, but not limited to, the following:

    (a) Facility one line diagrams

    (b) Facility layout drawings showing location of substations and major facility electrical equipment

    (c) Facility equipment list (if facility drawings show equipment, facility drawings may be used in lieu of specific equipment lists)

    (d) Equipment manufacturers' requirements (these include equipment service manuals, equipment drawings, etc.)

    (e) Arc flash hazard analysis, short circuit analysis, and time-current coordination studies

(8) Conduct a pre-bid/negotiation walk through with the potential contractor.

(9) Conduct a post-work walk-through to verify proper completion of scope of work, and review written report from the contractor on findings and recommendations, as applicable.

Often facilities do not have qualified personnel to conduct inspections, tests, or repairs of all or some of the equipment installed. *NFPA 70E* provides guidance on host and contractor relationships from the worker safety aspect. NFPA 70B addresses items that should be included in a service contract for outsourced equipment maintenance.

Typical Safety Procedure (Procedure for Selection, Inspection, and Care of Rubber Insulating Gloves and Leather Protectors)

Editor's Note: In order to assist readers of this handbook develop safety procedures, this supplement describes a typical safety procedure regarding a common subject in an electrical safety program — rubber insulating gloves and leather protectors. This procedure illustrates the type of information and format that can be used to develop electrical safety procedures. The readers of this handbook need to develop a method for identifying their procedures so that each electrical safety procedure is uniquely identified.

Title: Procedure for Selection, Inspection, and Care of Rubber Insulating Gloves and Leather Protectors	Doc. ID No.: ESP.100.000.003.01
	Page: 001 of 010
Eff. Date: 06/24/2013	Revision No. : 001 Revision Date: 06/06/2013	Supersedes: N/A
Reviewer: MK Rev. Date: 06/07/2013	Approved by: MDF Approval Date: 06/07/2013	Covers: All Personnel Within Restricted Approach Boundary

1.0 **Purpose:** The purpose of this procedure is to detail the selection, inspection, and care requirements for rubber insulating gloves and leather protectors.

2.0 **Scope:** This procedure covers all rubber insulating gloves regardless of Class and Type.

3.0 **Revision History:** This is the first revision and is a general upgrade.

4.0 **Background:** This procedure is required to meet Occupational Safety and Health Administration (OSHA) regulations and the requirements contained in *NFPA 70E*®.

5.0 **Definitions:**

5.1 Class. A rubber insulating glove *class* defines the maximum ac rms and dc use values in volts and the ac rms and dc proof test voltage values to be used in volts. There are six voltage classes as follows:

  a. 00, 0, 1, 2, 3, and 4, with cuff labels colored beige, red, white, yellow, green, and orange, respectively.

5.2 Work Instruction. A work instruction tells the reader the necessary steps and the order in which the steps are to be taken to safely perform the required task, and is action oriented.

  a. Each step is stated simply in a clear and concise manner that is easy to read and understand.
  b. Work instructions can be prepared using an outline type format, a flow chart format, a playscript format, or a question and answer format.
  c. The format selected is to be the one that is best suited for the situation.

5.3 Type. A rubber insulating glove *type* defines whether the material is ozone (natural) resistant – Type I, or non-ozone (synthetic) resistant – Type II.

6.0 **Work Instructions:**

6.1 Cleaning:

  a. General:

    1) Cleaning should be done under good lighting conditions.

    2) Cleaning can uncover damage that may be hidden under soiled surfaces as contamination may mask color or hide defects.

    3) Cleaning removes contaminants (chemicals) that are less obvious or cannot be seen but that can degrade rubber such as fertilizers, herbicides, and pesticide residues.

    4) Contaminants may be conductive, especially when wet.

    5) Regular cleaning is good practice.

Title: Procedure for Selection, Inspection, and Care of Rubber Insulating Gloves and Leather Protectors	Doc. ID No.: ESP.100.000.003.01 Page: 002 of 010	
Eff. Date: 06/24/2013	Revision No. : 001 Revision Date: 06/06/2013	Supersedes: N/A
Reviewer: MK Rev. Date: 06/07/2013	Approved by: MDF Approval Date: 06/07/2013	Covers: All Personnel Within Restricted Approach Boundary

   b. Hands Prior to Inspection, Handling, or Use:

     1) Prior to wearing or handling rubber insulating gloves or leather protectors, hands are to be cleaned using a non-petroleum based hand cleaner or towelettes made specifically for workers who wear rubber gloves.

     2) The cleaner is to be capable of dissolving and removing grease, oil, ink, tar, pipe dope, creosote, paint, etc., without harming natural or synthetic rubber.

     3) The cleaner is to be capable of cleaning hands with or without water.

     4) Workers should be aware that the cleaner may contain skin conditioners or perfume.

     5) Only approved hand cleaner is to be used.

     6) Approved hand cleaner for those using rubber insulating gloves or leather protectors is to be obtained from inventory control.

   c. Rubber Insulating Gloves:

     1) Prior to inspection or use, as necessary, rubber insulating gloves should be cleaned to ensure that only clean gloves are used.

     2) Rubber insulating gloves are generally to be cleaned by washing with a mild soap and water combination.

     3) The label area may be cleaned using soapy water or denatured alcohol.

     4) After washing, the gloves are to be thoroughly rinsed with water.

     5) Finally, the gloves are to be air-dried in an area away from direct sunlight and other sources (ozone) and where the temperature is less than 120°F (49°C).

     6) The cleaning is to be done, as necessary, prior to use to ensure that only clean gloves are used.

     7) The mild soap to be used is to be a concentrated detergent with a special grease release formula that removes oils, grease, and dirt from both natural and synthetic rubber.

     8) Only an approved cleaner for use on rubber insulating gloves obtained from inventory control is to be used.

6.2 Use of Sunscreen:

   a. Only approved sunscreen is to be used.

   b. If sunscreen is used by those handling or using rubber insulating gloves and leather protectors, it must be of a non-oily type that leaves no residue or slippery hands.

   c. It must be of a type made specifically for use with rubber protectors and rubber insulating gloves (e.g., it must be safe for use with leather protectors and rubber insulating gloves).

   d. Approved sunscreen for those using or handling rubber insulating gloves and leather protectors is to be obtained from inventory control.

6.3 Determine that the rubber insulating gloves have an adequate voltage rating (class) for the nominal system voltage.

   a. This is done by examining the glove label and the *NFPA 70E* equipment label.

   b. Verify that the glove class is suitable for the maximum possible rms voltage in accordance with the ANSI C84.1 "B" range, based on nominal system voltage designated on the *NFPA 70E* equipment label and the location of use within the system:

Title: Procedure for Selection, Inspection, and Care of Rubber Insulating Gloves and Leather Protectors	**Doc. ID No.:** ESP.100.000.003.01 **Page:** 003 of 010	
Eff. Date: 06/24/2013	**Revision No. :** 001 **Revision Date:** 06/06/2013	**Supersedes:** N/A
Reviewer: MK **Rev. Date:** 06/07/2013	**Approved by:** MDF **Approval Date:** 06/07/2013	**Covers:** All Personnel Within Restricted Approach Boundary

          1) 508 V for a 480 V nominal system at the service location

     c) Refer to the Electrical Safety Procedure (ESP) on *Matching Rubber Insulating Goods to Nominal System Voltage at Point of Use* for further information.

6.4 Use only properly sized gloves:

    a. Glove size is measured by using a soft tape measure.

    b. Measure the dominant hand around the palm, not including the thumb, at its widest point in inches and the distance from the base of the palm to the tip of the index finger in inches of the dominant hand.

    c. The glove size is the longer of the above two measurement taken to the closest ½ inch above the measured value (e.g., 8" = 8, 8¼" = 8.5). Gloves come in sizes from 7 to 11 inches depending on the class, type, and length specified.

    d. If cotton glove liners are used, the glove size is to be increased one half-size (e.g., for 8.5, use 9).

6.5 Each rubber insulating glove is labeled in a color denoting the "class" and the maximum voltage the glove can be used for as indicated in the following table:

Class	**Label Color[1]**	**Proof Test Voltage**		**Max. Use Voltage**	
N/A	**N/A**	**ac (rms)**	**dc**	**ac (rms)[2]**	**dc[2]**
00	Beige	2,500	10,000	500	750
0	Red	5,000	20,000	1,000	1,500
1	White	10,000	40,000	7,500	11,250
2	Yellow	20,000	50,000	17,000	25,500
3	Green	30,000	60,000	26,500	39,750
4	Orange	40,000	70,000	36,000	54,000

Notes:

1. Rubber insulating gloves are to have a color-coded label that meets the requirements contained in ASTM D120.
2. Maximum use voltage when worn with leather protectors.

6.6 Verify that the rubber insulating gloves are the right type (category) for the use location (e.g., ozone resistant or non-ozone resistant):

    a. Type I – natural rubber – non-ozone resistant – indoor use

    b. Type II – synthetic – ozone resistant – outdoor or indoor use

    c. Where Type I gloves are used in outdoor locations, the amount of time they are exposed to sunlight (ozone) is to be minimized.

Title: Procedure for Selection, Inspection, and Care of Rubber Insulating Gloves and Leather Protectors		Doc. ID No.: ESP.100.000.003.01 Page: 004 of 010
Eff. Date: 06/24/2013	Revision No. : 001 Revision Date: 06/06/2013	Supersedes: N/A
Reviewer: MK Rev. Date: 06/07/2013	Approved by: MDF Approval Date: 06/07/2013	Covers: All Personnel Within Restricted Approach Boundary

6.7 The following table and notes indicate a breakdown of rubber insulating gloves that are generally available:

Breakdown of Generally Available Rubber Insulating Gloves[1]

Class	Length in. (mm)[2]	Type[3]	Cuff Type[4]	Color[5,6]	Sizes[7,8]
00	11 (279), 14 (356)	I	SC	R, B	7, 8, 8H, 9, 9H, 10, 10H, 11, 12
00	11 (279), 14 (356)	II	SC	BL, BLO	7, 8, 8H, 9, 9H, 10, 10H, 11, 12
0	11 (279), 14 (356)	I	SC	R, B, Y, BY	7, 8, 8H, 9, 9H, 10, 10H, 11, 12
0	11 (279), 14 (356)	II	SC	BL, BLO	7, 8, 8H, 9, 9H, 10, 10H, 11, 12
1	14 (356), 16 (406), 18 (457)	I	SC, BC, CC	B, YB, RB	7, 8, 8H, 9, 9H, 10, 10H, 11, 12
2	14 (356), 16 (406), 18 (457)	I	SC, BC, CC	B, YB, RB	7, 8, 8H, 9, 9H, 10, 10H, 11, 12
3	14 (356), 16 (406), 18 (457)	I	SC, BC, CC	B, YB, RB	7, 8, 8H, 9, 9H, 10, 10H, 11, 12
4	14 (356), 16 (406), 18 (457)	I	SC, BC, CC	B, YB, RB	9, 9H, 10, 10H, 11, 12

Notes:

1. This chart is intended to provide guidance on the types of rubber insulating gloves that are generally available. The various manufacturers and suppliers of rubber insulated gloves and leather protectors should be consulted to determine the types and characteristics of gloves that are readily available.
2. Contour cuff (CC) gloves are available only on 18 in. (457 mm) gloves.
3. Type I equals natural rubber, which is non-ozone resistant; Type II equals synthetic rubber, which is ozone resistant.
4. Rubber insulating gloves come in one of three types of cuffs styles — straight cuff, contour cuff, or bell cuff:
   a. SC equals straight cuff, which is the default cuff style.
   b. CC equals a contour cuff, which is angled to prevent bunching or binding at the elbow when bent, and is only available on 18 in. gloves.
   c. BC equals a bell cuff, which accommodates heavier winter clothing and allows for greater air flow in warmer weather.

Title: Procedure for Selection, Inspection, and Care of Rubber Insulating Gloves and Leather Protectors	Doc. ID No.: ESP.100.000.003.01
	Page: 005 of 010
Eff. Date: 06/24/2013	Revision No. : 001 Revision Date: 06/06/2013	Supersedes: N/A
Reviewer: MK Rev. Date: 06/07/2013	Approved by: MDF Approval Date: 06/07/2013	Covers: All Personnel Within Restricted Approach Boundary

Notes (continued):

5. Color code is as follows:

   a. R equals Red.
   b. B equals Black.
   c. Y equals Yellow.
   d. O equals Orange.
   e. RB equals Red on the inside and Black on the outside.
   f. YB equals Yellow on the inside and Black on the outside.
   e. BLO equals Blue inner color and Orange outer color.

6. The contrast between the outer color and the inner color makes inspection for cuts and tears easier when the glove is inflated or stretched.
7. Bell cuff (BC) gloves are not available in sizes 7, 8, or 8H.
8. The H in the size designation stands for a half (½) size.

6.8 Verify that the voltage insulation certification is still current:

   a. Check the marked test date and issue date or expiration date, as appropriate, to confirm that glove voltage insulation certification is still current.
   b. Rubber insulating gloves must be issued from inventory by inventory control within 1 year of the date they are voltage tested or they must be retested, and they must be retested no later than 6 months after issue.
   c. The gloves must be returned to inventory control to be sent for retesting at the time their replacement is obtained.

6.9 Ensure that hand jewelry, or other objects which could damage the gloves, are removed prior to inspection or use of the gloves.

6.10 Perform a visual inspection prior to each use, paying particular attention to the body of the glove, between the fingers, and each finger, inspecting both the inside surface and the outside surface for:

   a. Holes
   b. Rips
   c. Tears
   d. Cuts
   e. Cracks
   f. Punctures
   g. Snags
   h. Ozone cutting, checking, cracking, breaks, or pitting
   i. UV checking
   j. Embedded material, such as wire or metal shavings, that could cause punctures
   k. Textural changes, such as the following:

Title: Procedure for Selection, Inspection, and Care of Rubber Insulating Gloves and Leather Protectors		Doc. ID No.: ESP.100.000.003.01
		Page: 006 of 010
Eff. Date: 06/24/2013	Revision No. : 001 Revision Date: 06/06/2013	Supersedes: N/A
Reviewer: MK Rev. Date: 06/07/2013	Approved by: MDF Approval Date: 06/07/2013	Covers: All Personnel Within Restricted Approach Boundary

      1) Swelling
      2) Softening
      3) Hardening
      4) Stickiness
      5) Inelasticity

  l. Oil contamination

  m. Other such defects that may damage the insulating properties

      Note: Surface irregularities may be present on all rubber goods because of imperfections on forms or molds or because of inherent difficulties in the manufacturing process. They may appear as indentations, protuberances or imbedded foreign materials that are acceptable under the following conditions:

      1) The indentation or protuberance blends into a smooth slope when the material is stretched.
      2) Foreign material remains in place when the insulating material is folded and stretches with the insulating material surrounding it.

      If defects are found that cannot be rectified, cut a finger off of each glove and turn into inventory control for proper disposal.

6.11 Perform an air leak test to test for punctures for each glove that may not be caught by the visual inspection by one of the following means or methods:

  a. Inflating gloves makes checking for cuts, tears, or ozone easier.
  b. The roll test:

      1) Hold the glove downward and grasp the cuff.
      2) Roll the glove cuff tightly, forcing air into the palm and finger area.
      3) Apply pressure to the various areas of the glove.
      4) Hold the glove close to the face and ear to listen for or feel leaking air from holes while the glove is inflated and pressure applied.

  c. Check for air leaks by means of a glove inflator:

      1) Inflate the glove following the instructions included with the glove inflator.
      2) Hold the glove close to the face and ear to listen for or feel leaking air from holes in the inflated glove.

  d. Do not over inflate:

      1) Type I – natural rubber – maximum of 1.5 times normal
      2) Type II – synthetic rubber – maximum of 1.25 times normal

Title: Procedure for Selection, Inspection, and Care of Rubber Insulating Gloves and Leather Protectors	Doc. ID No.: ESP.100.000.003.01 Page: 007 of 010	
Eff. Date: 06/24/2013	Revision No. : 001 Revision Date: 06/06/2013	Supersedes: N/A
Reviewer: MK Rev. Date: 06/07/2013	Approved by: MDF Approval Date: 06/07/2013	Covers: All Personnel Within Restricted Approach Boundary

6.12 Use leather protectors over rubber insulating gloves:

   a. Leather protectors must be sized for the rubber insulating gloves (e.g., size 10 for size 10).
   b. Perform a visual inspection of the leather protectors, checking for

   1) Cuts
   2) Tears
   3) Rips
   4) Abrasions
   5) Holes
   6) Contaminants, such as oil or petroleum products, or
   7) Other such damage
   8) If damaged discard by cutting off at least one glove finger on each glove and turn into in inventory control
   9) Leather protectors are not to be used as work gloves and work gloves are not to be used as leather protectors

6.13 Verify that the distance between the end of the cuff of the rubber insulating glove and the cuff of the leather protector is not less than the distance required by ASTM F496, as shown in the following table:

Class	Distance (inches)
00	½
0	½
1	1
2	2
3	3
4	4

6.14 Use of Glove Liners and Glove Dust:

   a. Only approved glove liners of an appropriate material, such as cotton, cotton with Lycra, or wool blends with substantial wool content, can be used for perspiration control or to provide additional warmth.
   b. Glove liners come only in one size (e.g., one size fits all).
   c. Glove liners can have a knit or straight cuff.
   d. Where glove liners are used, glove size is required to be increased one half (½) size.
   e. Only approved glove dust may be used to control perspiration.
   f. Do not use talc, baby powder, or similar products as they may contain acids and dyes that can cause the rubber insulating gloves to deteriorate.
   g. The glove dust is to be a type suggested or approved by the manufacturer of the gloves.
   h. Only approved glove liners are to be used, and they are to be obtained from inventory control.
   i. Only approved glove dust is to be used, and it may be obtained from inventory control.

Title: Procedure for Selection, Inspection, and Care of Rubber Insulating Gloves and Leather Protectors	Doc. ID No.: ESP.100.000.003.01
	Page: 008 of 010
Eff. Date: 06/24/2013	Revision No. : 001 Revision Date: 06/06/2013	Supersedes: N/A
Reviewer: MK Rev. Date: 06/07/2013	Approved by: MDF Approval Date: 06/07/2013	Covers: All Personnel Within Restricted Approach Boundary

    j. The request to inventory control is to indicate whether the liners are to be used for perspiration control or warmth. The request may be made for more than one pair of glove liners.

    k. Damaged or soiled glove liners are not be used.

    l. Damaged glove liners must have a finger cut off and be returned to inventory control for replacement.

    m. The cleaner is to be diluted in accordance with the manufacturer's instructions prior to use and applied by means of a clean rag or sponge obtained from inventory control.

    n. Only cleaners tested for compatibility with the type of rubber compound employed in accordance with ASTM D471 and ASTM F496 are to be used.

6.15 Properly store:

    a. Store in a canvas storage bag with ventilation holes in bottom for proper ventilation.

    b. Do not store more than a single pair of rubber insulating gloves or leather protectors in a single bag that has a storage compartment for only one glove.

        1) Where the rubber insulating glove manufacturer permits, leather protectors may be stored over rubber insulating gloves inside a single glove bag storage compartment.

    c. Bags with multiple storage compartments designed to store more than one pair of gloves may be used to store both the rubber insulating gloves and the leather protectors.

    d. Store rubber insulating gloves with the cuffs down and the fingers up.

    e. Ensure that gloves lay flat in the bag.

    f. Store only in properly sized canvas storage bag(s).

    g. Do not store inside out, as storing inside out strains the rubber severely and promotes early ozone cutting.

    h. Do not allow gloves to be stored in a folded condition, as folds and creases strain rubber and cause it to crack from ozone prematurely.

    i. Do not store near sources of heat; store in cool, dry areas only.

6.16 Follow the manufacturer's instructions regarding inspection, storage, and use.

6.17 Training:

    a. Employees are to receive training on the proper use, care, storage, and replacement of rubber insulating gloves prior to issue of their first pair. This training is to include at least the following:

        1) How to determine proper size

        2) How to determine proper voltage class

        3) How to determine proper glove category

        4) How to determine the certification is current

        5) How to properly inspect prior to use

        6) How to clean properly

        7) How to store properly

        8) How to take damaged gloves out of service properly

        9) How to request a new or replacement pair of rubber insulating gloves

Title: Procedure for Selection, Inspection, and Care of Rubber Insulating Gloves and Leather Protectors	**Doc. ID No.:** ESP.100.000.003.01
	Page: 009 of 010
Eff. Date: 06/24/2013	**Revision No. :** 001 **Revision Date:** 06/06/2013	**Supersedes:** N/A
Reviewer: MK **Rev. Date:** 06/07/2013	**Approved by:** MDF **Approval Date:** 06/07/2013	**Covers:** All Personnel Within Restricted Approach Boundary

    b. After training, the employee must demonstrate to the satisfaction of the trainer that the employee is competent in the above procedures.

    c. The training, the content of the training, and demonstrated ability are to be documented for each employee.

    d. This training is to be done on at least a yearly basis or any time it is determined that an employee is failing to comply with this procedure.

6.18 Rubber insulating gloves, leather protectors, and glove liners must conform to company specifications.

7.0 Responsibilities:

    a. Supervisors are responsible for seeing that this procedure is followed and enforced.

    b. Trainers are responsible for seeing:

        1) That workers are properly trained

        2) That workers show demonstrated ability in the selection, use, and care of rubber insulating gloves and leather protectors

        3) That training is properly documented

8.0 References:

    a. Company specifications:

        1) Specifications for rubber insulating gloves

        2) Specifications for leather protectors

        3) Specifications for cotton glove liners

        4) Specifications for canvas carrying bags for rubber insulating gloves

        5) Specifications for cleaning products for rubber insulating goods

    b. Governmental standards:

        1) Occupational Safety and Health Administration (OSHA):

            a). 29 CFR 1910.137, *Occupational Safety and Health Standards, Personal Protective Equipment, Electrical Protective Devices*

    c. Industry standards:

        1) American Society of Testing and Materials (ASTM):

            a) ASTM D120, *Standard Specification for Rubber Insulating Gloves*

            b) ASTM F496, *Standard Specification for In-Service Care of Insulating Gloves and Sleeves*

            c) ASTM F696, *Standard Specification for Leather Protectors for Insulating Gloves and Mittens*

            d) ASTM F2675, *Test Method For Determining Arc Ratings of Hand Protective Products Developed and Used for Electrical Arc Flash Protection*

Title: Procedure for Selection, Inspection, and Care of Rubber Insulating Gloves and Leather Protectors		Doc. ID No.: ESP.100.000.003.01
		Page: 010 of 010
Eff. Date: 06/24/2013	Revision No. : 001	Supersedes: N/A
	Revision Date: 06/06/2013	
Reviewer: MK	Approved by: MDF	Covers: All Personnel Within Restricted Approach Boundary
Rev. Date: 06/07/2013	Approval Date: 06/07/2013	

      2) National Fire Protection Association (NFPA):

          a) *NFPA 70E, Standard for Electrical Safety in the Workplace*

    d. Company electrical safety procedures:

      1) ESP-100.000.001.01, *Matching Rubber Insulating Goods to Nominal System Voltage at Point of Use*

      2) ESP-100.000.002.01, *Purchasing Procedures for Rubber Insulating Gloves and Leather Protectors*

      3) ESP-100.000.004.01, *Limitations on Use of Rubber Insulating Gloves without Leather Protectors*

      4) ESP-100.000.005.01, *Issue and Replacement of Electrical Protective Equipment*

      5) ESP-100.000.006.01, *After Issue Storage of Rubber Insulating Goods*

      6) ESP-100.000.007.01, *Inventory Storage of Electrical Protective Equipment*

      7) ESP-100.000.008.01, *Cleaning of Rubber Insulating Goods*

      8) ESP-100.000.009.01, *Proper Disposal of Electrical Protective Equipment* c. Industry Standards:

      9) ESP-100.000.010.01, *Electrical Safety Training on the Issue, Inspection, Use, Cleaning, Replacement, and Proper Disposal of Rubber Insulating Gloves and Leather Protectors*

      10) ESP-100.000.011.01, *Procedure for Having Rubber Insulating Goods Retested*

      11) ESP-100.000.012.01, *Repair Procedure for Allowable Repairs to Rubber Insulating Gloves*

9.0 **Attachments:**

    a. 29 CFR 1910.137, *Occupational Safety and Health Standards, Personal Protective Equipment, Electrical Protective Devices*

Steve and Dela Lenz: One Family's Experience with an Arc-Flash Incident

Editor's Note*: Although Steve was working on the exterior of an enclosure and performing a task he had done repeatedly without incident, the fact that he could not see what was behind the enclosure wall elevated the risk of exposure to an arc flash event. What happened to Steve emphasizes the importance of performing a thorough hazard identification and risk assessment procedure. Although Steve was performing a seemingly innocuous task and could not have known that an arc flash would ensue, the level of incident energy that would result from a screw coming in contact with an energized conductor warranted performing this task with the equipment in an electrically safe work condition. If an analysis concludes it is infeasible to an employee should wear PPE that is rated for the available incident energy and protects all parts of the body that will be within the arc flash boundary.*

On a cold December morning in 2010, Dela Lenz was driving to her Bible study group. It was her turn to bring snacks, and that's what she was thinking about as she approached the center of town and her cell phone rang. It was 9 a.m. Steve Lenz, her husband of 20 years and father of their two children, was working a few towns over in an unoccupied building. "I've been burned bad," she recalls hearing. "I burnt my face off and I'm on my way to the hospital." Steve was surprised he could even speak on the phone call. His lips felt like they were frozen.

Steve's job that morning was to install a monitoring box on the outside end of a switchgear, a job he had done more than a dozen times. After reviewing the project with the engineer and deciding on the best place to install the monitor, Steve stayed behind to complete the job. The part of the gear he was to install was an entrance cabinet that carried parallel 500 kcmil 480-volt feeder wires but with no live exposed parts. The wires were

EXHIBIT S4.1

Steve on the day of the accident, immediately after admission to the hospital. (Courtesy of the Lenz family)

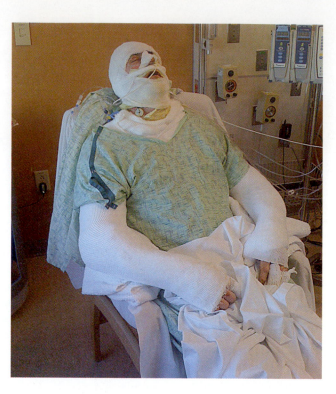

EXHIBIT S4.2

Steve 5 days after the accident. (Courtesy of the Lenz family)

not secured into the gear, but there was more than four inches of clearance on the end of the gear he would be working on. Steve used a self-tapping screw that would penetrate the panel by less than half an inch and then stop.

Steve considered the installation a "zero-risk job" — he would be on the outside of the switchgear and the screws would fall at least 3½ inches from the live wires. He drilled the top left, then the top right, the bottom right, and finally the bottom left. As the last screw came to a stop against the box flange, he heard a loud boom inside the feeder cabinet and a buzzing sound, which were followed by a spray of metal and fire. Instinctively, he ducked. Backing up to avoid the flames, he hit a wall. Then, as he turned the corner, he realized he was on fire.

After frantically ripping off his t-shirt and slapping out the flames, he looked down at his badly burned hands and felt an agonizing pain on his face. As he stood there, wondering what had just happened, a warm feeling on the left side of his neck crept upward. When he looked down at his torso, he realized he was still on fire. He started whipping at the flames and finally put them out.

The 29-year veteran electrician had just experienced an arc flash, an electric current that passes through air when insulation or isolation between electrified conductors is no longer sufficient to withstand the applied voltage. The flash is immediate and can cause severe injury, including burns. Each year, more than 2,000 people are treated in burn centers with severe arc-flash injuries. The arc flash that Steve experienced was unique in that he wasn't working in the panel: the arc flash had come through the side of the switchgear and the monitor box.

Badly burned, Steve looked around the smoke-filled room and realized there was nothing he could do. The cabinet door was still closed, so he grabbed his phone and called the engineer he had met with that morning to make sure the building hadn't lost power. Steve explained that he had been in an electrical explosion and asked the engineer to secure the building. Then, leaving his tools behind, he went to his van, searched his GPS for the closest hospital, and called his wife.

After urging Steve without success to wait for an ambulance, Dela headed in his direction. Steve had told her he was driving himself to the hospital and that he would call her when he got there. Dela knew it would take her more than an hour to reach him. Feeling sheer panic, she somehow focused during the drive on what needed to be done. She remembered that Steve's mother was substitute teaching at the children's school that day, so she called the school, told her mother-in-law that Steve had been in an accident, and asked her to take the kids, then 12 and 15, out of class and explain to them what had happened to their dad.

Word traveled fast. Dela's phone didn't stop ringing. She couldn't answer all the calls. Steve was the provider of her family, the love of her life, and she had never imagined life without him. "I didn't know what to expect," says Dela. "Everything goes through your mind — is he going to live, is he going to be an electrician, are we going to go bankrupt?"

With his keys dangling from the end of his finger, Steve walked into the hospital. Doctors and nurses rushed Steve onto a gurney and covered him in yellow ice packs to stop the burning. As Dela and Steve would learn later, in an arc-flash event the skin keeps burning for 72 hours. Covered in ice packs, Steve started shaking violently. A warm blanket was pulled up to his chin and he blacked out. The next thing he felt was someone prying off his wedding ring. He awoke and offered to go to his van and get a tool to help, not realizing that more than 30 percent of his body was burned, then blacked out again. His next vision was the spinning of blades in slow motion as three men hoisted him into a helicopter to transport him to a hospital with a specialized burn unit.

Having received a call from a nurse who was treating Steve, Dela headed straight to the second hospital. The next thing Steve remembers is waking up in a hospital bed wrapped in gauze.

Steve's mother brought the children to the hospital later that day. The man they had always viewed as invincible was unrecognizable. Gauze encircled his whole upper body. Only his swollen, blister-ridden eyes, burnt lips, and a small patch of neck were visible. His son went to touch one of his dad's wrapped hands, not knowing if Steve could acknowledge or even feel the touch. His daughter, strong-willed, shut down. Clad in gloves and gowns to prevent the spread of germs in the Intensive Care Unit (ICU), they cried.

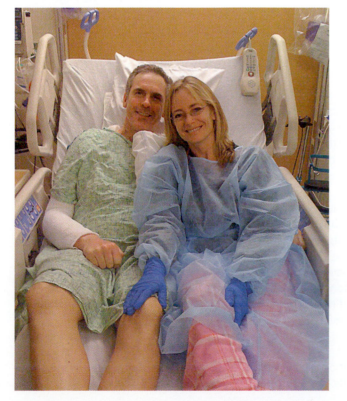

EXHIBIT S4.3

Twenty-one days after the accident, Steve and Dela watch a movie that Steve doesn't remember. (Courtesy of the Lenz family)

"It's so painful to see your kids hurting," says Dela. "It was scary, and we didn't know what was going to happen."

The first week in the hospital, Dela couldn't remember to eat. It would come to mind only when she started feeling light-headed. Eventually, she realized she had to take care of herself to stay strong for Steve. Steve was Dela's first priority, but everything else wouldn't just take care of itself. Her kids still needed their mother; her grandmother still needed help during the week; and everyone needed to be updated on Steve's condition. For the first time, Dela and Steve needed to ask for help instead of giving it.

Steve and Dela's children would spend the next month with their grandparents and aunt and uncle, while Dela slept on a small sofa in the ICU. Out of 27 days, she spent only a few nights away from Steve's side — one night right after the accident to get the kids settled and another to stay in a nearby hotel room with them so they could be close to Steve over Christmas.

The kids visited often, but the distance between them and their father was more than just miles. In the required gown and gloves, their son would sit in the farthest chair and stare out the door and down the hallway. Their daughter had few words.

"I was the guy who coaches baseball and basketball," says Steve. "Suddenly, I wasn't their indestructible dad; I was something different."

Dela remembers almost every moment in the hospital, while Steve has to piece together the bits he can recall — a result of many intense hallucinations brought on by the four narcotics, two nerve drugs, and amnesia medications it took to alleviate some of the pain. In his mind, the clock was always at 9 a.m. in his small hospital room, which would meld into a farmhouse, then a hotel room, then a janitor's closet as he grasped for reality. During one episode, he thought that hospital staff was harvesting his organs. Sure, there were the funny moments — like the singing pickle he envisioned — but mostly it was a fight to come back to reality. Dela was his anchor. "When she was there I knew what I was seeing was real; when she wasn't I assumed I was hallucinating," says Steve.

Some memories were all too real. The trips to "the cleaning room" were the most painful. Nearly every day, burn techs would take Steve to a room and place him in a large metal trough. Then they would strip him down and scrub the dead skin off his arms and chest. Dela couldn't stand to see the pain in Steve's face. She tried to stay a couple of times, but when he started verbalizing the pain she had to leave. After the agonizing 2-hour process, Steve would be wiped out for another couple of hours.

After a while, every time the burn techs walked into Steve's room, his heart monitor would go off as his blood pressure rose. "I remember it was like a game to me," Steve recalls. "It was like being tortured, and I wasn't going to make a sound."

As days passed, the hallucinations and detachment from reality were making Steve feel crazy, and the pain just got worse with each surgery.

A few days after arriving at the hospital, Steve had his first surgery. The surgeons took all the dead skin off his chest and sides and covered the areas with cadaver skin to protect him until he was ready to have grafts from his own body. A silver spongy material was then stapled directly to his skin.

By the end of the third surgery, skin from his lower right leg and both his thighs covered his arms, chest and collarbone. A large brace kept his arm from bending. Immediately following the surgery, Steve could hardly move. The pain was nearly unbearable despite the massive amount of pain medication. Doctors say that the process of taking healthy "donor skin" from the body is even more painful than the burns themselves.

Dela became Steve's personal nurse. In an ICU gown and gloves 24 hours a day, she rarely left his side. She washed his burnt lips throughout the day, applied antibiotic ointment, fed him ice, helped him get up and around when possible, and called for help when he needed it. Steve maintains that his wife was the one who got punished through the whole ordeal.

Steve would go in and out of sleep during the day and then sleep at night, but that's when Dela had to contact family and friends and check on the kids to make sure what used to be her normal responsibilities were being covered. The nights got late, leaving Dela little sleep by the time the nurses started their rounds in the morning.

"There was one day when I thought 'I have to get out of this room — I just have to get out of here; I'm going to go nuts,'" says Dela.

Steve and Dela were eager to get home, but before Steve could be released he needed to eat on his own and gain strength. His body was in what doctors call hyperdrive, which means that as his body was trying to reverse the effects of the burn, it was eating away at its own protein and muscle. His daily nourishment consisted of bags of what Steve referred to as "goo" squirted through a tube that had been placed in his nose. Those were his meals. Over the course of a month, Steve lost 20 pounds, a significant amount for someone his size.

"I remember lying there one night, and I said to my wife, 'Why doesn't God just pull the plug — why doesn't he just finish?' because I realized I can't go back to who I was and I didn't see an end in sight," says Steve. "I felt like I was stuck in perpetual hell, and I couldn't get out."

Steve begged Dela not to sign the release for a fourth surgery. At that point, he was willing to risk serious infection to avoid the pain of taking more skin from his legs. But he needed the fourth surgery, and Dela signed the forms. This time they took large amounts of skin from the back of his left thigh and calf and grafted it to the sides of his upper body.

"If you've never been burnt, you don't understand," says Steve. "It is so quick, it is so hard, and it lasts so long."

Going Home

Three days after the fourth surgery, the doctors said Steve could go home. It was good news, but going home proved especially hard for Dela.

Steve was sent home with nine different prescriptions, but administering his medications wasn't something Dela had to do while Steve was in the hospital. Some were once a day, some twice a day, some three times a day; one was as needed. She learned the hard way that timing was essential for some of the meds. And then, on his first day back, Steve fell while Dela was changing his bandages. Steve wasn't hurt, but it was a sign of things to come.

On his second day home, Steve fell again. This time he blacked out in the shower and crashed through the shower door, landing in a pile of glass. The heat from the hot water, combined with his blood pressure medication and the fact that he had half the amount of blood in his body compared to a healthy person, had caused the blackout. Dela wrapped his new wounds and got him to the hospital. After hours in the emergency room and numerous stitches, Steve was back home the next day.

"I was overwhelmed," admits Dela. Having him at home was harder than she had thought it would be. In addition to caring for Steve and finding out what works and what doesn't outside the hospital, Dela had the food shopping, cooking, laundry, and all the other daily chores to manage again. She also had two kids who needed to continue their regular activities.

Since Steve had been home, his pain medication had changed a number of times as the doctors searched for the right combination. He finally ended up with a fentanyl transdermal patch, a narcotic used in chronic pain management. The patch, which had to be changed every 3 days, made the pain disappear, but Steve began noticing that after 2½ days he would start to fidget and become cold, sweaty, and irritable. "I was miserable," says Steve. Dela, along with his doctor, thought it was anxiety, but Steve started noticing that when he replaced the patch the symptoms vanished and he felt like himself again.

"I wish someone had told me that if you've been on narcotics for more than 3 weeks that you're addicted," says Steve, who never drank, did drugs, or let any substance control him.

After his self-diagnosed dependency on narcotics was confirmed by his doctor, Steve ripped off the patch and then gradually weaned himself off all medication. After 2 weeks,

Steve was no longer dependent on pain medication, but he was in constant pain. Of the withdrawal, Steve says, "It was a crappy couple of weeks, but I guess compared to the month of December, it wasn't so bad."

Doctors estimated it would be 6 to 8 months before Steve was back to work, but by mid-March, just 10 weeks out of the hospital, Steve was working again. He was in pain, he was tired, and Dela had to go everywhere with him, but he was back.

Lessons Learned

Steve considers himself lucky. People die from arc flash accidents; many others are disfigured and live with chronic pain. Eight months and $300,000 worth of medical bills later, Steve has full mobility and faint scars, but his life is not the same. As an electrician, he is more cautious. He still works alone, but he takes precautions he never considered before the incident. If he is going to be in a panel or drilling into a switchgear, he covers everything with insulating rubber sheeting to protect himself from shock and burns. He has an arc-flash outfit, which includes a face shield, gloves, and an inexpensive shirt made of protective material. He also carries the knowledge that if he had been wearing the outfit, he would have walked away that December morning without a scratch.

"If I could change one thing, I would have worn the stupid $35 shirt," says Steve. "I would have put on the gloves; I would have put on the shield."

Steve comes from a family of electricians, received extensive training, and worked for a number of large companies before the accident, but he had never had specific arc-flash training. He learned the hard way and respects what some companies, like the one his brother works for, are doing to raise awareness and to prevent accidents through advanced arc-flash training and arc-flash analysis that provide employees guidance on specific jobs.

"There is equipment, there are classes, there are warnings," says Steve. "It's just not in place in every application yet."

Steve and Dela's family members, who were always close, are even closer now. Steve is back coaching baseball and recently traveled for his daughter's basketball tournament, but he will never forget the looks on his kids' faces while he was in the hospital or the trauma his wife went through.

"One tiny mistake, one oversight, a fraction of a second, a dropped tool, somebody else's mistake — it doesn't matter — it's going to catch you," says Steve. "We all think that we are indestructible, and I learned that I'm not indestructible: Hopefully I won't have to prove that to my kids again."

Index

A

Accessible (as applied to equipment)
  Definition, Art. 100
Accessible (as applied to wiring methods)
  Definition, Art. 100
Accessible, readily (readily accessible)
  Battery enclosures, 320.3(A)(2)(a)
  Definition, Art. 100
Aerial lifts, 130.8(F)(1)
Alarms, battery operation, 320.3(A)(4)
Approach distances; *see also* Boundary
  Limits of, Annex C
  Preparation for, C.1
  Qualified persons, 130.4(C), 130.4(D), C.1.1, C.1.2
  Unqualified persons, 130.4(C), 130.8(E), C.1.1
Approved (definition), Art. 100
Arc blast, K.4
Arc flash boundary; *see* Boundary, arc flash
Arc flash hazard, 130.2(2), 320.3(5), K.3
  Analysis, 310.5(C)
  Definition, Art. 100
  Protection from; *see* Arc flash protective equipment
Arc flash protective equipment, 130.5(C), 130.5(D), 130.7
  Qualified persons, use by, C.1.2.1, C.1.2.3
  Unqualified persons, use by, C.1.1
Arc flash risk assessment, 130.2(B)(2), 130.3(A), 130.5
  Electrolytic cell line working zones, 310.5(C)(1)
Arc flash suit, 130.7(C)(10)(a), 130.7(C)(13), 130.7(C)(16)
  Definition, Art. 100
Arc rating
  Definition, Art. 100
  Total system arc rating, protective clothing, M.3
Arc-resistant switchgear, 130.7(C)(15)
Attachment plug (plug cap) (plug), 110.4(B)(2)(b),
    110.4(B)(3)(c), 110.4(B)(5)
  Definition, Art. 100
  Maintenance, 245.1
Attendants, to warn and protect employees, 130.7(E)(3)
Authority having jurisdiction (definition), Art. 100
Authorized personnel
  Battery rooms or areas restricted to, 320.3(A)(2)(a)
  Definition, 320.2
Automatic (definition), Art. 100

B

Balaclava (sock hood definition), Art. 100
Barricades, 130.7(E)(2)
  Definition, Art. 100
Barriers
  Definition, Art. 100
  Electrolytic cells, safe work practices, 310.5(D)(3)
  Physical or mechanical, 130.6(F), 130.7(D)(1)(i)
  Rotating equipment, 230.2
Batteries
  Abnormal battery conditions, alarms for, 320.3(A)(4)
  Cell flame arresters, 320.3(D)
  Definition, 320.2
  Direct-current ground-fault detection, 320.3(C)(1)
  Electrolyte hazards, 320.3(B)
  Maintenance requirements, safety-related, Art. 240, 320.3(C)
  Operation, 320.3(C)
  Personal protective equipment (PPE), use of, 320.3(A)(5)
  Safety requirements, Art. 320
  Testing, 320.3(C)
  Tools and equipment, use of, 320.3(C)(2)
  Valve-regulated lead acid cell (definition), 320.2
  Vented cell (definition), 320.2
  Ventilation, 240.1, 320.3(D)
  VRLA (valve-regulated lead acid cell) (definition), 320.2
Battery effect (definition), 310.2
Battery enclosures, 320.3(A)(2)
Battery rooms
  Definition, 320.2
  Requirements, 320.3(A)(2)
Blind reaching, electrical safety program, 130.6(B)
Body wash apparatus, 240.2
Bonded (bonding)
  Definition, Art. 100
  Maintenance of, 205.6
Bonding conductor or jumper (definition), Art. 100
Boundary; *see also* Approach distances
  Approach boundaries to energized conductors or circuit,
    130.2(B)(2), 130.4, C.1.1, C.1.2
  Arc flash, 130.2(B)(2), 130.5(1), 130.5(B), C.1.1, C.1.2
    Calculations, Annex D
    Definition, Art. 100
    Protective equipment, use of, 130.7(C)(16), C.1.1, C.1.2.1, C.1.2.3

Limited approach, 130.2(1), 130.2(B)(2), 130.2(B)(3), 130.4(C), 130.7(D)(1), 130.8, C.1.1, C.1.2.2; *see also* Approach distances
  Definition, Art. 100
Restricted approach, 130.4(D), C.1.1, C.1.2.3; *see also* Approach distances
  Definition, Art. 100
Shock protection, 130.2(B)(2), 130.4
Branch circuit (definition), Art. 100
Building (definition), Art. 100

C
Cabinets (definition), Art. 100
Cable
  Flexible; *see* Flexible cords and cables
  Maintenance of, 205.13
Cable trays, maintenance, 215.3
Cell
  Definition, 320.2
  Electrolytic; *see* Electrolytic cell
  Valve-regulated lead acid (definition), 320.2
  Vented (definition), 320.2
Cell line; *see* Electrolytic cell line
Chemical hazard, 320.3(5)
Circuit breakers
  Definition, Art. 100
  Low-voltage circuit breakers, calculations for incident energy and arc flash protection boundary for, D.4.7
  Molded-case, 225.2
  Reclosing circuits after operation, 130.6(L)
  Routine opening and closing of circuits, 130.6(L)
  Safety-related maintenance requirements, Art. 225
  Testing, 225.3
Circuit protective device operation, reclosing circuit after, 130.6(M); *see also* Circuit breakers; Disconnecting means; Fuses; Overcurrent protection
Circuits
  De-energized; *see* De-energized
  Energized, working on or near parts that are or might become; *see* Working on energized electrical conductors or circuit parts
  Identification, maintenance of, 205.12
  Impedance, 120.3(D)
  Protection and control, 220.2
  Reclosing after protective device operation, 130.6(M)
  Routine opening and closing of, 130.6(L)
Clear spaces, 130.6(H), 205.9
Combustible dust, 130.6(J)
Competent person, 350.4
  Definition, 350.2
Conductive (definition), Art. 100
Conductive work locations, 110.4(B)(4)
Conductors
  Bare (definition), Art. 100
  Covered (definition), Art. 100
  De-energized; *see* De-energized
  Energized; *see* Working on energized electrical conductors or circuit parts
  Grounding conductors, equipment, 110.4(B)(2)(a)
    Definition, Art. 100
  Grounding electrode conductors (definition), Art. 100
  Identification; *see* Identified/identification
  Insulated

  Definition, Art. 100
  Integrity of insulation, maintenance of, 210.4
  Maintenance of, 205.13, 210.3
Contractors, relationship with, 110.3
Controllers (definition), Art. 100
Cord- and plug-connected equipment, 110.4(B)
  Connecting attachment plugs, 110.4(B)(5)
  Grounding-type equipment, 110.4(B)(2)
  Handling, 110.4(B)(1)
  Safety-related maintenance requirements, Art. 245
  Visual inspection, 110.4(B)(3)
Cords, flexible; *see* Flexible cords and cables
Covers, 215.1
Cranes, 310.5(D)(9)
Current-limiting overcurrent protective device (definition), Art. 100
Cutout
  Definition, Art. 100
  Portable cutout-type switches, 310.5(D)(8)
Cutting, 130.10

D
De-energized, 130.7(A); *see also* Electrically safe work condition
  Conductors or circuit parts that have lockout/tagout devices applied, 120.2
  Definition, Art. 100
  Process to de-energize equipment, 120.2(F)(2)(a)
  Testing of parts, 120.2
  Uninsulated overhead lines, 130.8(C)
Definitions, Art. 100
  Batteries and battery rooms, 320.2
  Electrolytic cells, 310.2
  Lasers, 330.2
  Lockout/tagout practices and devices, 120.2(F)(2)(k)
  Power electronic equipment, 340.2
  Research and development laboratories, 350.2
Device (definition), Art. 100
Direct-current ground-fault detection, batteries, 320.3(C)(1)
Disconnecting means, 120.1, 120.2(F)(2)(c), 130.2
  Definition, Art. 100
  Lockout/tagout devices, use of, 120.2(E)(3), 120.2(E)(4)(d), 120.2(E)(6)
  Routine opening and closing of circuits, 130.6(L)
Disconnecting (or isolating) switches (disconnector, isolator); *see also* Disconnecting means
  Definition, Art. 100
  Safety-related maintenance requirements, Art. 210
Documentation
  Arc flash risk assessment, 130.5(A)
  Electrical safety program, 110.1(A), 110.1(I)(3), 110.3(C)
  Equipment labeling, 130.5(D)
  Of maintenance, 205.3
  Training, employees, 110.2(C), 110.2(E), 120.2(B)(4)
Doors, secured, 130.6(G)
Drilling, 130.10
Dust, combustible, 130.6(J)
Dwelling unit (definition), Art. 100

E
Electrical hazard, Art. 130; *see also* Arc flash hazard; Risk; Shock hazard

Categories of
   General, Annex K
   Personal protective equipment required for; *see* Personal
     protective equipment (PPE)
   Definition, Art. 100
   Evaluation procedure, F.4
   Identification procedure, 110.1(G)
Electrically safe work condition, Art. 120
   Definition, Art. 100
   Lockout/tagout practices and devices, 120.1, 120.2
   Temporary protective grounding equipment, 120.3
   Verification of, 120.1
Electrical safety (definition), Art. 100
Electrical safety program, 110.1, Annex E
   Auditing, 110.1(I)
   Awareness and self-discipline, 110.1(C)
   Contractors, relationship with, 110.3
   Controls, 110.1(E), E.2
   Documentation of, 110.1(A), 110.1(I)(3), 110.3(C)
   General, 110.1(A)
   Hazard identification and risk assessment procedure, 110.1(G)
   Job briefing, 110.1(H)
   Maintenance, 110.1(B)
   Principles, 110.1(D), E.1
   Procedures, 110.1(F), E.3
   Risk assessment procedure, Annex F
Electrolyte (definition), 320.2
Electrolyte hazards, storage battery, 320.3(B)(1)
Electrolytic cell, Art. 310
   Auxiliary nonelectric connections, 310.6(B)
   Employee training, 310.3, 310.4, 310.5(D)(6)
Electrolytic cell line working zone
   Attachments and auxiliary equipment, 310.5(D)(10)
   Cranes and hoists, 310.5(D)(9)
   Employee training, 310.3, 310.4(A), 310(4)(B)(2), 310.5(D)(6)
   Portable equipment and tools, use of, 310.6
   Safeguards, employee, 310.5, Annex L
Elevated equipment, 130.8(F)(1)
Emergency response, training in, 110.2(C)
Employees
   Electrical safety program, 110.1
   Lockout/tagout procedure, 120.2(B)(1), 120.2(B)(2)
   Responsibilities, 105.3, 130.8(D)
     Lasers, 330.5
     Power electronic equipment, 340.7(B)
     Special equipment, 300.2
   Safeguarding; *see* Safeguarding
   Training; *see* Training, employees
Employers
   Electrical safety program, 110.1
   Responsibilities, 105.3
     Host and contract employers, 110.3
     Lockout/tagout procedure, 120.2(C)(1)
     Power electronic equipment, 340.7(A)
     Safety related design requirements, Annex O
     Special equipment, 300.2
     Uninsulated overhead lines, work on or near, 130.8(D)
Enclosed (definition), Art. 100
Enclosures
   Definition, Art. 100
   Maintenance of, 205.7, 210.1, 210.2
Energized
   Definition, Art. 100
   Electrical conductors or circuit; *see* Working on energized

electrical conductors or circuit parts
   Electrolytic cells; *see* Electrolytic cell; Electrolytic cell line
     working zone
Energized electrical work permit, 130.2(B), Annex J
Equipment; *see also* specific equipment
   Batteries, for work on, 320.3(C)(2)
   Definition, Art. 100
   Grounding
     Portable equipment within energized cell line working zone,
       310.6(A)
     Vehicle or mechanical equipment, 130.8(F)(3)
   Grounding-type, 110.4(B)(2)
   Overhead; *see* Overhead lines and equipment
   Spaces about, maintenance of, 205.5, 205.9
   Special; *see* Special equipment
   Use of, 110.4
Equipment grounding conductors, 110.4(B)(2)(a)
   Definition, Art. 100
Explanatory material, 90.5
**Exposed (as applied to energized electrical conductors or circuit
    parts),** 105.1, 130.2(2), 130.3
   Definition, Art. 100
   Safe work practices; *see* Working on energized electrical
     conductors or circuit parts; Work practices, safety-related
Exposed (as applied to wiring methods) (definition), Art. 100
Extension cords; *see* Flexible cord sets
Eye wash apparatus, 240.2

F

Fail safe (definition), 330.2
Fail safe safety interlock (definition), 330.2
Fiberglass-reinforced plastic rods, 130.7(D)(1)(d)
Fiber or flyings, combustible, 130.6(J)
Field evaluated (definition), 350.2
Fittings (definition), Art. 100
Flame arresters, battery cell, 320.3(D)
Flammable gases, 130.6(J)
Flammable liquids, 130.6(J)
Flexible cords and cables
   Grounding-type utilization equipment, 110.4(B)(2)(a)
   Handling, 110.4(B)(1)
   Maintenance of, 205.14
Flexible cord sets, 110.4(B)
   Connecting attachment plugs, 110.4(B)(5)
   Visual inspection, 110.4(B)(3)
Formal interpretation procedures, 90.6
Fuses, 130.6(L), 130.6(M)
   Current-limiting fuses, calculating arc-flash energies for use
     with, D.4.6
   Definition, Art. 100
   Fuse or fuse-holding handling equipment, 130.7(D)(1)(b)
   Safety-related maintenance requirements, 225.1

G

Gases, flammable, 130.6(J)
Ground (definition), Art. 100
Grounded, solidly (definition), Art. 100
Grounded conductors (definition), Art. 100
Grounded (grounding), 120.1
   Definition, Art. 100

Equipment
  Portable equipment within energized cell line working zone, 310.6(A)
  Vehicle or mechanical equipment, 130.8(F)(3)
Lockout/tagout procedures, 120.2(F)(2)(g)
Maintenance of, 205.6
Safety grounding equipment, maintenance of, 250.3
Ground fault (definition), Art. 100
Ground-fault circuit-interrupters, 110.4(C)
  Definition, Art. 100
  Testing, 110.4(D)
Ground-fault protection, battery, 320.3(C)(1)
Grounding conductors, equipment, 110.4(B)(2)(a)
  Definition, Art. 100
Grounding electrode (definition), Art. 100
Grounding electrode conductors (definition), Art. 100
Grounding-type equipment, 110.4(B)(2)
Guarded, 205.7; *see also* Barriers; Enclosures
  Definition, Art. 100
  Rotating equipment, 230.2
  Uninsulated overhead lines, 130.8(C), 130.8(D)

H
Handlines, 130.7(D)(1)(c)
Hazard (definition), Art. 100; *see also* Electrical hazard
Hazardous (classified) locations, maintenance requirements for, Art. 235
Hazardous (definition), Art. 100
Hinged panels, secured, 130.6(G)
Hoists, 310.5(D)(9)

I
Identified/identification
  Equipment, field marking of, 130.5(D)
  Lasers, 330.4(G)
  Maintenance of, 205.10
Illumination
  Battery rooms, 320.3(A)(2)(b)
  Working on energized electrical conductors or circuits, 130.6(C)
Implanted pacemakers and metallic medical devices, 310.5(D)(11)
Incident energy
  Calculation methods, Annex D
  Definition, Art. 100
  Equipment, labeling of, 130.5(D)
Incident energy analysis, 130.5(C)(1), 130.7(C)(15)
  Definition, Art. 100
Inspection, visual
  Cord- and plug-connected equipment, 110.4(B)(3)
  Safety and protective equipment, 250.2(A)
  Safety grounding equipment, 250.3(A)
  Test instruments and equipment, 110.4(A)(4)
Insulated (definition), Art. 100
Insulated conductors, maintenance of, 210.4
Insulating floor surface, L.1
Insulation, electrolytic cells, 310.5(D)(1)
Insulation rating, overhead lines, 130.8(B)
Interlocks, safety
  Fail safe safety interlock (definition), 330.2
  Maintenance of, 205.8
Interrupter switch (definition), Art. 100
Interrupting rating (definition), Art. 100

Isolated (as applied to location)
  Definition, Art. 100
  Electrolytic cells, 310.5(D)(5)
Isolating devices
  Control devices as, 120.2(E)(6)
  Lockout device, acceptance of, 120.2(E)(1)
Isolating switches
  Definition, Art. 100
  Safety-related maintenance requirements, Art. 210

J
Job briefing, 110.1(H), 130.2(B)(2)
  Checklist, Annex I

L
Labeled (definition), Art. 100
Laboratory
  Definition, 350.2
  Safety-related work requirements, Art. 350
Laser
  Definition, 330.2
  Energy source (definition), 330.2
  Product (definition), 330.2
  Radiation (definition), 330.2
  System (definition), 330.2
  Work practices, safety-related, Art. 330
    Employee responsibility, 330.5
    Safeguarding employees in operating area, 330.4
    Training, 330.3
Limited approach boundary; *see* Boundary, limited approach
Listed
  Definition, Art. 100
  Research and development laboratory equipment or systems, 350.5
Live parts
  Guarding of; *see* Guarded
  Safe work conditions; *see* Electrical safety program; Working on energized electrical conductors or circuit parts; Work practices, safety-related
Lockout/tagout practices and devices, 120.2
  Audit, 120.2(C)(3)
  Complex procedure, 120.2(C)(2), 120.2(D)(2), 120.2(F)(1)(e), Annex G
  Control, elements of, 120.2(F)(2)
  Coordination, 120.2(D)(3)
  De-energized conductors or circuit parts with, 120.2
  Definitions, 120.2(F)(2)(k)
  Equipment, 120.2(E)
  Grounding, 120.2(F)(2)(g)
  Hazardous electrical energy control procedures, 120.2(B)(6), 120.2(C)(2), 120.2(D)
  Identification of devices, 120.2(B)(7)
  Maintenance of devices, 205.8
  Plans for, 120.2(B)(5), 120.2(F)(1)
  Principles of execution, 120.2(B)
  Procedures, 120.2(F), Annex G
  Release
    For return to service, 120.2(F)(2)(m)
    Temporary, 120.2(F)(2)(n)
  Removal of devices, 120.2(E)(3)(e), 120.2(E)(4)(d), 120.2(F)(2)(l)
  Responsibility, 120.2(C)

Simple procedure, 120.2(C)(2), 120.2(D)(1), 120.2(F)(1)(d), Annex G
Testing, 120.2(F)(2)(f)
Voltage, 120.2(B)(8)
Working on/near conductors or circuit parts with, 120.1
Luminaires (definition), Art. 100

M
Maintenance requirements, Chap. 2, 110.1(B)
Batteries and battery rooms, Art. 240
Controller equipment, Art. 220
Fuses and circuit breakers, Art. 225
General, Art. 205
Hazardous (classified) locations, Art. 235
Introduction, Art. 200
Personal safety and protective equipment, Art. 250
Portable electric tools and equipment, Art. 245
Premises wiring, Art. 215
Rotating equipment, Art. 230
Substation, switchgear assemblies, switchboards, panelboards, motor control centers, and disconnect switches, Art. 210
Mandatory rules, 90.5
Marking; *see* Identified/identification
Mechanical equipment, working on or near uninsulated overhead lines, 130.8(F)
Motor control centers
Definition, Art. 100
Personal protective equipment required for tasks, 130.7(C)(15)
Safety-related maintenance requirements, Art. 210
Multi-employer relationship, 120.2(D)(2)

N
Nominal voltage (definition), 320.2
Nonelectric equipment connections, electrolytic cell line, 310.6(B)

O
Occupational health and safety management standards, alignment with, Annex P
Open wiring protection, 215.2
Outdoors, GFCI protection, 110.4(C)(3)
Outlets (definition), Art. 100
Overcurrent (definition), Art. 100
Overcurrent protection
Maintenance of devices, 205.4, 210.5
Modification, 110.4(E)
Overhead lines and equipment
Clearances, maintenance of, 205.15
Industrial procedure for working near overhead systems, example of, Annex N
Insulation rating, 130.8(B)
Working within limited approach boundary of uninsulated, 130.8
Overload (definition), Art. 100

P
Pacemakers, implanted, 310.5(D)(11)
Panelboards
Definition, Art. 100
Personal protective equipment required for tasks, 130.7(C)(15)
Safety-related maintenance requirements, Art. 210

Permissive rules, 90.5
Personal protective equipment (PPE), 110.1(G)(3), 130.2(B)(2), 130.9, 130.10(3), L.1
Arc flash protection, 130.5(C), 130.5(D), 130.7, C.1.1, C.1.2.1, C.1.2.3
Batteries and battery rooms, 320.3(A)(5), 320.3(B)
Body protection, 130.7(C)(6)
Care of, 130.7(B)
Clothing characteristics, 130.7(C)(11), Annex H
Eye protection, 130.7(C)(4), 330.4(A)
Flash protection; *see* Arc flash protective equipment
Foot and leg protection, 130.7(C)(8)
Hand and arm protection, 130.7(C)(7)
Head, face, neck, and chin protection, 130.7(C)(3)
Hearing protection, 130.7(C)(5)
Labeling of, 130.5(D)
Lasers, use of, 330.4(A), 330.4(H)
Maintenance, Art. 250
Required for various tasks, 130.7(C)(15), 130.7(C)(16)
Safeguarding of employees in electrolytic cell line working zone, 310.5(C)(1), 310.5(D)(2)
Selection of, 130.7(C)(15), Annex H
Shock protection, 130.7, C.1.1
Standards for, 130.7(C)(14)
Pilot cell (definition), 320.2
Planning checklist, Annex I
Portable electric equipment, 110.4(B)
Connecting attachment plugs, 110.4(B)(5)
Electrolytic cells, 310.6(A), L.2
Grounding-type, 110.4(B)(2)
Handling, 110.4(B)(1)
Safety-related maintenance requirements, Art. 245
Visual inspection, 110.4(B)(3)
Power electronic equipment, safety-related work practices, Art. 340
Definitions, 340.2
Hazards associated with, 340.6
Human body, effects of electricity on, 340.5
Reference standards, 340.4
Specific measures, 340.7
Power supply
Cell line working area, L.2
Portable electric equipment, circuits for, 310.6(A)
Premises wiring (system)
Definition, Art. 100
Maintenance of, Art. 215
Prohibited approach boundary; *see* Boundary
Prospective fault current (definition), 320.2
Protective clothing, 130.4(D), 130.7(C), L.1
Arc flash protection, 130.7(C)(9), 130.7(C)(10), 130.7(C)(13), 130.7(C)(16), C.1.1, Annex H
Care and maintenance, 130.7(C)(13)
Characteristics, 130.7(C)(11), Annex H
Layering of, M.1, M.2
Prohibited clothing, 130.7(C)(12)
Selection of, 130.7(C)(9), Annex H
Total system arc rating, M.3
Protective equipment, 130.7, 130.8(D), 130.8(F)(2)
Alerting techniques, 130.7(E)
Arc flash protection; *see* Arc flash protective equipment
Barricades, 130.7(E)(2)
Barriers; *see* Barriers
Batteries, maintenance of, 320.3(A)(5)
Care of equipment, 130.7(B)

Insulated tools, 130.7(D)(1)
Maintenance, Art. 250
Nonconductive ladders, 130.7(D)(1)(e)
Personal; *see* Personal protective equipment (PPE)
Rubber insulating equipment, 130.7(D)(1)(g)
Safety signs and tags, 130.7(E)(1)
Shock protection, 130.7, C.1.1
Standards for, 130.7(F)
Temporary protective grounding equipment, 120.3
Voltage-rated plastic guard equipment, 130.7(D)(1)(h)
Purpose of standard, 90.1

Q

Qualified persons, 130.2(B)(3), 130.3(A)
Approach distances, 130.4(C), 130.4(D), C.1.1, C.1.2
Definition, Art. 100
Electrolytic cells, training for, 310.4(A)
Employee training, 110.2(D)(1)
Lockout/tagout procedures, 120.2(C)(2), 120.2(C)(3), 120.2(D)(1)
Maintenance, performance of, 205.1
Overhead lines, determining insulation rating of, 130.8(B)

R

Raceways
Definition, Art. 100
Maintenance, 215.3
Radiation worker (definition), 340.2
Readily accessible
Battery enclosures, 320.3(A)(2)(a)
Definition, Art. 100
Receptacles
Definition, Art. 100
Electrolytic cell lines, L.2
Maintenance, 245.1
Portable electric equipment, 110.4(B)(2)(b), 110.4(B)(3)(c), 110.4(B)(5)(b)
References, Annex A, Annex B
Research and development
Definition, 350.2
Safety-related work requirements for research and development laboratories, Art. 350
Restricted approach boundary; *see* Boundary, restricted approach
Risk
Control, 110.1(G)(3)
Definition, Art. 100
Reduction, F.3, F.5
Risk assessment, Annex O
Arc flash, 130.2(B)(2), 130.3(A)
Definition, Art. 100
Procedure, 110.1(G), Annex F
Shock, 130.2(B)(2), 130.3(A), 130.4(A)
Ropes, 130.7(D)(1)(c)
Rules, mandatory and permissive, 90.5

S

Safeguarding
In cell line working zone, 310.5, Annex L
Definition, 310.2
In laser operating area, 330.4
Safety grounding equipment, maintenance of, 250.3

Safety interlocks
Fail-safe safety interlock (definition), 330.2
Maintenance of, 205.8
Safety-related design requirements, Annex O
Safety-related maintenance requirements; *see* Maintenance requirements
Safety-related work practices; *see* Work practices, safety-related
Scope of standard, 90.2
Service drop (definition), Art. 100
Service lateral (definition), Art. 100
Service point (definition), Art. 100
Shock hazard, 320.3(5), K.2
Definition, Art. 100
Protection from, 130.7, C.1.1
Shock protection boundaries, 130.4
Shock risk assessment, 130.2(B)(2), 130.3(A), 130.4(A)
Short circuit, 320.3(C)(1), O.2.3(1)
Short circuit current
Arc-flash energies, effect on, 130.5(3)
Arc flash PPE, maximum available current for, 130.7(C)(15), H.2
Calculations, Annex D
Prospective, 320.2, 320.3(A)(5)
Unintended ground, caused by, 320.3(C)(1)
Short-circuit current rating (definition), Art. 100
Short circuit interruption devices, F.2.4.2
Signs, electrolytic cell areas, 310.5(B)
Single-line diagram
Definition, Art. 100
Maintenance of, 205.2
Special equipment, Chap. 3; *see also* Batteries; Electrolytic cell; Laser; Power electronic equipment
Battery rooms
Definition, 320.2
Requirements, 320.3(A)(2)
Organization, 300.3
Responsibility, 300.2
Special permission (definition), Art. 100
Standard arrangement and organization, 90.3, 90.4
Step potential (definition), Art. 100
Stored energy, 120.1, 120.2(F)(2)(b)
Structure (definition), Art. 100
Substations, safety-related maintenance requirements, Art. 210
Switchboards
Definition, Art. 100
Safety-related maintenance requirements, Art. 210
Switches; *see also* Switching devices
Disconnecting (or isolating) switches (disconnector, isolator); *see also* Disconnecting means
Definition, Art. 100
Safety-related maintenance requirements, Art. 210
Load-rated, 130.6(L)
Portable cutout type, 310.5(D)(8)
Switchgear
Arc-resistant, 130.7(C)(15)
Definition, Art. 100
Metal-clad, 130.7(C)(15)
Definition, Art. 100
Metal-enclosed, 130.7(C)(15)
Definition, Art. 100
Personal protective equipment required for tasks, 130.7(C)(15)
Safety-related maintenance requirements, Art. 210
Switching devices (definition), Art. 100; *see also* Circuit breakers; Disconnecting means; Switches

T

Tagout; *see* Lockout/tagout practices and devices
Temporary protective grounding equipment, 120.3
Terminals, maintenance of, 230.1
Testing
    De-energized parts, 120.1
    Energized equipment, 130.2(B)(3)
    Equipment safeguards, 310.5(D)(12)
    Ground-fault circuit-interrupters, 110.4(D)
    Lockout/tagout procedure, 120.2(F)(2)(f)
    Personal protective equipment, 310.5(D)(2)
    Safety and protective equipment, insulation of, 250.2(B)
    Safety grounding equipment, 250.3(B), 250.3(C)
Test instruments and equipment, 110.4(A)
    Cell line working zone, 310.6(D)
    Maintenance of, 250.4
    Visual inspection, 110.4(A)(4)
Thermal hazard, 320.3(5)
Tools
    Batteries, for work on, 320.3(C)(2)
    Electrolytic cells, safe work practices, 310.5(D)(7), 310.6, L.2
Touch potential (definition), Art. 100
Training, employees, 105.3
    Documentation, 110.2(C), 110.2(E), 120.2(B)(4)
    Emergency responses, 110.2(C)
    Lockout/tagout practices, 120.2(B), Annex G
    Qualified persons, 110.2(D)(1)
    Retraining, 110.2(D)(3)
    Unqualified persons, 110.2(D)(2)
    Work practices, safety-related, 105.3, 110.2, 110.3(B), 310.3, 310.4, 310.5(D)(6), 330.3

U

Underground electrical lines and equipment, 130.9
Ungrounded (definition), Art. 100
Uninsulated overhead lines, working within limited approach boundary of, 130.8(A)
Unqualified persons, 130.2(B)(2)
    Approach distances, 130.4(C), 130.8(E), C.1.1
    Definition, Art. 100
    Electrolytic cells, training for, 310.4(B)
    Employee training, 110.2(D)(2)
Utilization equipment
    Definition, Art. 100
    Grounding-type, 110.4(B)(2)(a)

V

Valve-regulated lead acid cell (definition), 320.2
Vehicular equipment, working on or near uninsulated overhead lines, 130.8(F)
Ventilation, batteries, 240.1, 320.3(D)
Voltage
    (Of a circuit) (definition), Art. 100
    Electrolytic cells, voltage equalization, 310.5(D)(4)
    Lockout/tagout procedures, 120.2(B)(8)
    Nominal (definition), Art. 100
VRLA (valve-regulated lead acid cell) (definition), 320.2

W

Warning signs
    Battery rooms and enclosures, 320.3(A)(5)
    Lasers, 330.4(B)
    Maintenance of, 205.11
Welding machines, 310.6(C)
Wiring, premises
    Definition, Art. 100
    Maintenance of, Art. 215
Working on energized electrical conductors or circuit parts, Art. 130; *see also* Work practices, safety-related
    Alertness of personnel, 130.6(A)
    Approach boundaries; *see* Boundary
    Blind reaching by employees, 130.6(B)
    Conductive articles being worn, 130.6(D)
    Conductive materials, tools, and equipment being handled, 130.6(E)
    Confined or enclosed work spaces, 130.6(F)
    Definition, Art. 100
    Electrically safe working conditions, 130.2
    Energized electrical work permit, 130.2(B)
    Failure, anticipation of, 130.6(K)
    Flash risk assessment, 130.5
    Housekeeping duties, 130.6(I)
    Illumination, 130.6(C)
    Insulated tools and equipment, 130.7(D)(1)
    Occasional use of flammable materials, 130.6(J)
    Opening and closing of circuits, routine, 130.6(L)
    Overhead lines, working within limited approach boundary of, 130.8
    Portable ladders, 130.7(D)(1)(e)
    Protective shields, 130.6(F), 130.7(D)(1)(f)
    Reclosing circuits after protective device operation, 130.6(M)
    Safe work conditions, 130.3
Working spaces
    Clear spaces, 130.6(H), 205.9
    Maintenance of, 205.5, 205.9
Work permit, energized electrical, 130.2(B), Annex J
Work practices, safety-related, Chap. 1; *see also* Electrically safe work condition; Working on energized electrical conductors or circuit parts
    Approach distances; *see* Approach distances
    Batteries and battery rooms, Art. 320
    Contractors, relationship with, 110.3
    De-energized equipment; *see* Electrically safe work condition
    Electrical conductors or circuit parts that are or might become energized, work on or near, 130.3
    Electrical safety program, 110.1
    Electrolytic cells, Art. 310, L.1
    Lasers, Art. 330
    Power electronic equipment, Art. 340
    Purpose, 105.2
    Research and development laboratories, Art. 350
    Responsibility for, 105.3
    Scope, 105.1
    Special equipment; *see* Special equipment
    Training requirements, 105.3, 110.2, 110.3(B)
    Use of equipment, 110.4

IMPORTANT NOTICES AND DISCLAIMERS CONCERNING NFPA® STANDARDS

NOTICE AND DISCLAIMER OF LIABILITY CONCERNING THE USE OF NFPA STANDARDS

NFPA® codes, standards, recommended practices, and guides ("NFPA Standards"), of which the document contained herein is one, are developed through a consensus standards development process approved by the American National Standards Institute. This process brings together volunteers representing varied viewpoints and interests to achieve consensus on fire and other safety issues. While the NFPA administers the process and establishes rules to promote fairness in the development of consensus, it does not independently test, evaluate, or verify the accuracy of any information or the soundness of any judgments contained in NFPA Standards.

The NFPA disclaims liability for any personal injury, property or other damages of any nature whatsoever, whether special, indirect, consequential or compensatory, directly or indirectly resulting from the publication, use of, or reliance on NFPA Standards. The NFPA also makes no guaranty or warranty as to the accuracy or completeness of any information published herein.

In issuing and making NFPA Standards available, the NFPA is not undertaking to render professional or other services for or on behalf of any person or entity. Nor is the NFPA undertaking to perform any duty owed by any person or entity to someone else. Anyone using this document should rely on his or her own independent judgment or, as appropriate, seek the advice of a competent professional in determining the exercise of reasonable care in any given circumstances.

The NFPA has no power, nor does it undertake, to police or enforce compliance with the contents of NFPA Standards. Nor does the NFPA list, certify, test, or inspect products, designs, or installations for compliance with this document. Any certification or other statement of compliance with the requirements of this document shall not be attributable to the NFPA and is solely the responsibility of the certifier or maker of the statement.

ADDITIONAL NOTICES AND DISCLAIMERS

Updating of NFPA Standards

Users of NFPA codes, standards, recommended practices, and guides ("NFPA Standards") should be aware that these documents may be superseded at any time by the issuance of new editions or may be amended from time to time through the issuance of Tentative Interim Amendments or corrected by Errata. An official NFPA Standard at any point in time consists of the current edition of the document together with any Tentative Interim Amendments and any Errata then in effect. In order to determine whether a given document is the current edition and whether it has been amended through the issuance of Tentative Interim Amendments or corrected through the issuance of Errata, consult appropriate NFPA publications such as the National Fire Codes® Subscription Service, visit the NFPA website at www.nfpa.org, or contact the NFPA at the address listed below.

Interpretations of NFPA Standards

A statement, written or oral, that is not processed in accordance with Section 6 of the Regulations Governing the Development of NFPA Standards shall not be considered the official position of NFPA or any of its Committees and shall not be considered to be, nor be relied upon as, a Formal Interpretation.

Patents

The NFPA does not take any position with respect to the validity of any patent rights referenced in, related to, or asserted in connection with an NFPA Standard. The users of NFPA Standards bear the sole responsibility for determining the validity of any such patent rights, as well as the risk of infringement of such rights, and the NFPA disclaims liability for the infringement of any patent resulting from the use of or reliance on NFPA Standards.

NFPA adheres to the policy of the American National Standards Institute (ANSI) regarding the inclusion of patents in American National Standards ("the ANSI Patent Policy"), and hereby gives the following notice pursuant to that policy:

NOTICE: The user's attention is called to the possibility that compliance with an NFPA Standard may require use of an invention covered by patent rights. NFPA takes no position as to the validity of any such patent rights or as to whether such patent rights constitute or include essential patent claims under the ANSI Patent Policy. If, in connection with the ANSI Patent Policy, a patent holder has filed a statement of willingness to grant licenses under these rights on reasonable and nondiscriminatory terms and conditions to applicants desiring to obtain such a license, copies of such filed statements can be obtained, on request, from NFPA. For further information, contact the NFPA at the address listed below.

Law and Regulations

Users of NFPA Standards should consult applicable federal, state, and local laws and regulations. NFPA does not, by the publication of its codes, standards, recommended practices, and guides, intend to urge action that is not in compliance with applicable laws, and these documents may not be construed as doing so.

Copyrights

NFPA Standards are copyrighted. They are made available for a wide variety of both public and private uses. These include both use, by reference, in laws and regulations, and use in private self-regulation, standardization, and the promotion of safe practices and methods. By making these documents available for use and adoption by public authorities and private users, the NFPA does not waive any rights in copyright to these documents.

Use of NFPA Standards for regulatory purposes should be accomplished through adoption by reference. The term "adoption by reference" means the citing of title, edition, and publishing information only. Any deletions, additions, and changes desired by the adopting authority should be noted separately in the adopting instrument. In order to assist NFPA in following the uses made of its documents, adopting authorities are requested to notify the NFPA (Attention: Secretary, Standards Council) in writing of such use. For technical assistance and questions concerning adoption of NFPA Standards, contact NFPA at the address below.

For Further Information

All questions or other communications relating to NFPA Standards and all requests for information on NFPA procedures governing its codes and standards development process, including information on the procedures for requesting Formal Interpretations, for proposing Tentative Interim Amendments, and for proposing revisions to NFPA standards during regular revision cycles, should be sent to NFPA headquarters, addressed to the attention of the Secretary, Standards Council, NFPA, 1 Batterymarch Park, P.O. Box 9101, Quincy, MA 02269-9101; email: stds_admin@nfpa.org

For more information about NFPA, visit the NFPA website at www.nfpa.org.